FOREST PRODUCTS MARKETING

McGraw-Hill Series in Forest Resources

Avery and Burkhart: Forest Measurements
Brown and Davis: Forest Fire: Control and Use
Dana and Fairfax: Forest and Range Policy
Daniel, Helms, and Baker: Principles of Silviculture
Davis and Johnson: Forest Management
Dykstra: Mathematical Programming for Natural Resource Management
Ellefson: Forest Resources Policy
Harlow, Harrar, Hardin, and White: Textbook of Dendrology: Covering the Important
  Forest Trees of the United States and Canada
Knight and Heikkenen: Principles of Forest Entomology
Laarman and Sedjo: Global Forestry: Issues for Six Billion People
Panshin and De Zeeuw: Textbook of Wood Technology
Sharpe, Hendee, and Sharpe: Introduction to Forestry
Sinclair: Forest Products Marketing
Stoddart, Smith, and Box: Range Management

Walter Mulford was Consulting Editor of this series from its inception in 1931 until
January 1, 1952.

Henry J. Vaux was Consulting Editor of this series from January 1, 1952 until July 1,
1976.

Paul V. Ellefson, University of Minnesota, is currently our Consulting Editor.

FOREST PRODUCTS MARKETING

Steven A. Sinclair
Professor of Forest Products Marketing
Virginia Polytechnic Institute and State University

McGRAW-HILL, INC.

New York St. Louis San Francisco Auckland Bogotá
Caracas Lisbon London Madrid Mexico Milan Montreal
New Delhi Paris San Juan Singapore Sydney Tokyo Toronto

This book was set in Times Roman by General Graphic Services, Inc.
The editors were Anne C. Duffy and John M. Morriss;
the production supervisor was Leroy A. Young.
The cover was designed by Fern Logan.
Project supervision was done by The Total Book.
R. R. Donnelley & Sons Company was printer and binder.

FOREST PRODUCTS MARKETING

1 2 3 4 5 6 7 8 9 0   DOC   DOC   9 0 9 8 7 6 5 4 3 2 1

ISBN 0-07-057546-0

Library of Congress Cataloging-in-Publication Data

Sinclair, Steven A. (Steven Allen), (date).
   Forest products marketing / Steven A. Sinclair
      p.   cm. — (McGraw-Hill series in forest resources)
   Includes bibliographical references and index.
   ISBN 0-07-057546-0 — ISBN 0-07-057547-9 (IM)
   1. Forest products—Marketing.   2. Forest products industry.
I. Title.   II. Series.
HD9750.5.S56   1992
634.9'8'0688—dc20   91-21927

ABOUT
THE AUTHOR

STEVEN A. SINCLAIR is Professor of Forest Products Marketing at Virginia Polytechnic Institute and State University. He received his B.S. in Wood Science and Technology from North Carolina State University, and his Ph.D. in Forest Products and MBA from Virginia Polytechnic Institute and State University. He has held faculty appointments at the University of Minnesota and has worked in the southern pine plywood and furniture industry. Dr. Sinclair has published over 100 articles focusing on different aspects of forest products marketing. He has had consulting or research support from Georgia-Pacific, Louisiana-Pacific, Weyerhaeuser, Citibank, Alberta Research Council, Noranda Forest Sales, P & M Cedar Products, and more than fifty other forest products companies. Dr. Sinclair is Project Leader for the Forest Products Marketing Research Project Group of the International Union of Forestry Research Organizations, Contributing Editor to the Forest Products Research Society's *Forest Products Marketing and Economics Newsletter,* and on the Board of Directors of the Society of Wood Science and Technology.

To Patrick and Brian

CONTENTS

PREFACE

C. W. Bingham, Senior Executive Vice President, Weyerhaeuser, in an address to the Forest Products Research Society noted that forest products firms must differentiate themselves in the marketplace and lessen their emphasis on commodities. He urged the industry to seek out more discrete market segments, enhance its bond to the customer, and develop innovative specialty products. It is clear that marketing is becoming more crucial to the forest products industry each passing year. Marketing is the income-producing side of the business and is critical to business success. This is especially true for the forest products industry as it moves from its traditional mode of being a production-oriented manufacturer of industrial commodity products to a more market-oriented producer of specialty products.

This textbook represents the only comprehensive text written in the last 20 years for forest products marketing education. Unique among forestry books, a complete instructor's guide with test banks, overhead masters, case analyses, course outlines, and course organization suggestions is available to assist first-time and experienced instructors alike. This book is designed to allow forestry/forest products seniors to gain a basic understanding of forest products marketing in a single course without previous marketing course work. However, it could be used equally well as a supplemental text to a forest economics course or by industry or government professionals who are changing jobs.

Marketing in this text is treated as a critical activity necessary for business success. It is not shown as a collection of academic theories but rather is presented through examples of real companies and real data on major markets. A vice president of marketing at James River Corporation has called it required reading even for experienced industry hands.

The textbook begins with an explanation of the concept of marketing and proceeds to explain the historical development of marketing in the forest products industry. This provides a background perspective and appreciation of marketing in the dynamic forest products context.

For those unfamiliar with marketing, several chapters are devoted to the fundamentals of marketing and marketing/business strategy development. Each of the major forest products markets is then examined in depth, with numerous real-world examples of how specific companies utilize marketing in their corporate strategy. Growing international markets are also discussed along with an overview of export marketing basics. The last chapter explains how marketing is organized in forest products firms.

Appendix A provides case studies of real forest products firms struggling to implement marketing strategies and solve business problems. These are included to give students the opportunity to apply their new marketing skills to real problems in the industry.

In Appendix B is a series of short biographical sketches of major North American forest products producers along with selected retailers and specialty manufacturers. These sketches provide students with the basic facts on major players in the industry and can assist them in making job decisions. Each sketch can also be combined with a copy of the company's current annual report (provided free for the asking by companies) to provide students with a unique understanding of many firms over the course of a school term. Somehow, I never could understand how to teach forest products marketing without discussing the companies themselves. In my own classes, we discuss two or three companies each week using the short sketches and annual reports as background material.

Some have said marketing is less an art or science than a philosophy or practice. There are few absolutely right or wrong answers in a philosophy or practice, just hard thinking and work. This will frustrate a few but help others to flourish. I hope you will flourish.

ACKNOWLEDGMENTS

A book is rarely ever solely the thinking of a single person. This one is no different and I am indebted to many individuals whose writings have influenced the development of this text. The books of McCarthy and Perreault, Day, Shaw and Semenik, Ries and Trout, Porter, Buell, Stanton, Bonoma, and Rich were all influential. A special acknowledgment is due to Orie Beucler and Jim Bowyer, whose fine seminar book provided a framework for Chapters 3, 4, and 6. Several scholars were kind enough to allow me to use their work in this book. Stuart Rich, David Cohen, and Ed Pepke contributed case studies and Bob Bush coauthored Chapter 13.

My many friends and colleagues who were honest enough to point out my short-comings and suggest many improvements deserve more than thanks. To Jack Muench, Bill Luppold, Phil Araman, Bob Bush, Robert Anderson, Alberto Goetzl, John Griffith, Fred Kurpiel, Harlan Petersen, Dave Cohen, Ramsey Smith, Tim Powers, and others, thank you! Many of the ideas and examples come from my former graduate students. I count myself fortunate to have been associated with the likes of Bob Govett, Ed Stalling, Kevin Seward, Chris Meyer, Ed Cesa, Charlie Blinn, Dave Cohen, Cindi West, Paul Smith, Curt Hassler, Celeste Barnes, Bruce Hansen, Paul Chambers, Mark Trinka, Frick Stureson, Tom Garrahan, and many others.

The support I received from my department and its leader, Geza Ifju, made my efforts on this book possible. Jo Buckner deserves special acknowledge for typing the many drafts of this text and Judd Michael for his development and refinement of many of the figures. My appreciation also goes to McGraw-Hill for its interest in this book and the efforts of its fine editors, as well as the following reviewers: Robert C. Abt, North Carolina State University; Keith A. Blatner, Washington State University; Hugh O.

Canham, State University of New York, College of Environmental Science and Forestry; Paul V. Ellefson, University of Minnesota; Robert W. Rice, University of Maine; and John H. Syme, Clemson University. The excellent work of Annette Bodzin and Anne Duffy made my task much more pleasant.

And, my very special thanks go to Karen and Brian for putting up with my long hours and mental preoccupations during the writing of this text.

Steven A. Sinclair

FOREST
PRODUCTS
MARKETING

PART ONE

OVERVIEW AND INTRODUCTION

Marketing is the whole business when viewed through the eyes of the customer.

Peter Drucker

CHAPTER 1

WHAT IS MARKETING ANYWAY?

Somewhere around the time all garbage collectors became sanitation engineers and all janitors became maintenance supervisors, a funny thing happened to a lot of sales people: they became "marketing" people instead. They were transformed into marketing managers, marketing engineers and marketing associates, and some even became marketing representatives.

Blake, 1983

As can be seen from the quotation that opens this chapter, marketing ranks right up there with engineering as one of the English language's most misused words. Most attempts to define it struggle to list all its various functions such as sales, distribution, pricing, promotion, products, and many others. Here is an example of such a definition: "marketing is the discovery or identification of needs and the execution of those activities necessary to plan and provide need-satisfying products and services and to price, promote, distribute, and effect exchange of these products at an acceptable cost and in a socially responsible manner" (Shaw and Semenik, 1985).

Such attempts to list everything that marketing may involve usually fall short. It is a bit like attempting to define football as blocking and tackling. But for a moment let's put marketing on a more personal level. It has been said that anytime you try to persuade some person to do something you are in essence engaged in marketing (Stanton, 1978). Think for a moment about preparing a resume for a job interview. Don't you try to determine what the interviewer and the hiring organization need in a new employee, and then try to emphasize your qualifications which match those needs in your resume? Aren't you trying to persuade the hiring organization to give you the job?

Some, when asked to define marketing, call it selling, but marketing is clearly more than that. Theodore Levitt has said that the difference between marketing and selling is more than semantic. Selling focuses on the needs of the seller, marketing on the needs of the buyer (Levitt, 1986). When you prepare your resume for an interview do you emphasize your (the seller's) needs (i.e., company car, big salary, executive office, etc.) or do you emphasize the hiring organization's (the buyer's) needs? Which strategy is likely to be most successful?

The entire sales role is important but only one function of marketing. Marketing involves not just the sales staff and its activities, but rather the entire organization. Marketing has been called a philosophy—"a disciplined way of thinking about organizational goals and their relationship to overall business goals" (Ghosh, 1988).

Marketing can be thought of as a total system of business activities which is designed to determine customers' needs/desires, then to plan and develop products and services to meet those needs/desires, and then to determine the best way to price, promote, and distribute the products and services to the customers (Stanton, 1978). When examined this way as a system, marketing is sometimes referred to as the "marketing concept." This concept means that an organization aims all its efforts at satisfying its customers at a profit (McCarthy and Perreault, 1987). The marketing concept is a relatively simple one in theory, but very important to the success of the firm or organization and difficult at times to implement, as we will see in upcoming chapters. A major portion of the forest products industry for years operated under the notion that customers exist to buy products. Contrast this with the marketing concept of a firm existing to satisfy customer needs.

A firm soon ceases to exist without customers. And any firm in today's competitive environment that ignores its customers' needs will not flourish. You must have a customer-based orientation for long-term success.

Implementing the marketing concept is difficult and requires an effort by the entire organization, not just the marketing managers. For firms where the marketing concept has been successfully implemented, the chief executive officer has played an active role. The chief executive officer and other members of top management must set values and examples for other employees within the firm to adopt. A study of TJ International's annual report most years will leave no doubt about the positive example its management team is setting. The marketing concept must permeate the entire organization and not stop at the marketing manager.

Profit is a clear objective of a marketing-oriented firm. A marketing-oriented firm designs its product and service offerings to meet customer needs with a profit. It doesn't allow profit to be just that part of revenues which remains after all costs are covered.

Table 1-1 shows some key differences in attitude between those managers which have fully adopted a marketing orientation to business and those managers which still believe customers exist to buy their products, i.e., the production/sales-oriented managers.

HOW WILL I KNOW MARKETING WHEN I SEE IT?

How many times have you been shopping with someone and asked politely what they were looking for? And they replied, "I'm not sure, but I'll know it when I see it." If we reject the notion of defining marketing by listing all its parts or subfunctions and call it a philosophy, a way of thinking, even a concept, then how will we know it when we see it?

Don't feel alone in grappling with marketing. For years, academics and practitioners have disagreed among themselves and with each other on whether marketing is an art or a science and the answer is still not clear. There is some agreement that marketing is the function by which a firm (or organization) can encourage the exchange of goods or services for money in a way that is profitable and at the same time satisfying to its customers. Two key points are raised about marketing here. First, marketing is the income-generating activity of the firm/organization and this should never be overlooked. The second key item to note is that marketing is the function by which the firm reaches out to customers and by which customers reach in to the firm. If viewed in this manner, marketing is a revenue-producing function which is served by other activities within the firm/organization which can be considered cost centers (Bonoma, 1985).

TABLE 1-1
DIFFERENCES BETWEEN MANAGEMENT'S ATTITUDE WHEN OPERATING UNDER A
MARKETING ORIENTATION VERSUS A PRODUCTION/SALES ORIENTATION

Topic	Marketing orientation	Production/sales orientation
Attitudes toward customers	Customer needs are the prime determinant of company plans	They should be glad we exist; we are working hard to cut costs and bring out more efficient products
Product offering	We try to make what our customers need with a profit for our firm	We sell what our plants can produce
Role of marketing research	To determine customer needs and how well our company is satisfying them	To determine customer reaction, if used at all
Interest in innovation	Focus on locating new opportunities	Focus is on technology and cost cutting
Importance of profit	A critical objective	A residual, what's left after all costs are covered
Role of customer credit	Seen as a customer service	Seen as a necessary evil
Role of packaging	Designed for customer convenience and as a selling tool	Seen merely as protection for the product
Inventory levels	Set with customer requirements and costs in mind	Set with production requirements in mind
Transportation arrangements	Seen as a customer service	Seen as an extension of production and storage activities, with emphasis on cost minimization
Focus of advertising	Need-satisfying benefits of products and services	Product features and quality, maybe how products are made
Role of sales force	Help the customer to buy if the product fits his/her needs and is profitable to firm while coordinating with rest of firm—including production, inventory control, advertising, etc.	Sell, sell, sell; don't worry about coordination with other promotion efforts or rest of firm or even if profitable

Source: Adapted from McCarthy and Perreault, 1987.

OK, what do we know about marketing? It is a managerial system of business activity which should be customer-oriented. For success, customer needs/wants must be recognized and satisfied effectively. Marketing is not any one activity, nor is it a simple sum of many activities; rather it is the integration of these activities into a dynamic system.

DON'T PUT MARKETING IN A BOX

What do I mean, don't put marketing in a box? Think broadly! Marketing is used by nearly every individual, firm, and organization to persuade people to take an action or to develop a point of view. In this book, we will center our discussion on the marketing of physical products manufactured from trees. This is due in part to space and time limitations, but do not let this limit your thinking and your understanding of marketing. When you write to a college or school you are interested in, do you receive a smudged, poorly photocopied handout about what they have to offer?

No, not at all! In most instances, you receive an attractive, professionally done color brochure describing the school and its facilities in the most favorable way. Guess what— they are trying to persuade you to pay your tuition and attend their school.

Marketing can and is applied successfully to service organizations. Banks, hospitals, insurance firms, churches, YMCAs, campgrounds, and consulting foresters all use marketing as a system of managerial activity. Service activities marketing in the forestry field include recreation, land management, timber management, wildlife management, and others. Think broadly—the principles of marketing apply equally to these activities.

In the chapters to come, we will examine the main functional areas within marketing such as distribution, promotion, pricing, and product policy. Perhaps through a deeper understanding of these functional areas you will develop your own philosophy of marketing. But first, let's develop an understanding of the origins of the forest products industry and the beginnings of forest products marketing.

BIBLIOGRAPHY

Blake, L. C. 1983. " 'Marketing,' What's in a Name?" *Industrial Marketing* (March):110.
Bonoma, T. V. 1985. *The Marketing Edge.* The Free Press. New York.
Ghosh, A. 1988. Presentation at the 1988 AMA Educators Meeting. San Francisco, CA.
Levitt, T. 1986. *The Marketing Imagination.* The Free Press. New York.
McCarthy, E. J., and W. D. Perreault, Jr. 1987. *Basic Marketing.* Irwin. Homewood, IL.
Shaw, R. T., and R. J. Semenik. 1985. *Marketing.* South-Western Publishing. Cincinnati, OH.
Stanton, W. J. 1978. *Fundamentals of Marketing.* McGraw-Hill. New York.

CHAPTER 2

HISTORICAL DEVELOPMENT OF FOREST PRODUCTS MARKETING

Then Hiram sent Solomon the following message: "I have received your message and I am ready to do what you ask. I will provide the cedars and pine trees. My men will bring the logs down from Lebanon to the sea and will tie them together in rafts to float them down the coast to the place you choose. There my men will untie them, and your men will take charge of them. On your part, I would like you to supply the food for my men." So Hiram supplied Solomon with all the cedar and pine logs that he wanted and Solomon provided Hiram with 100,000 bushels of wheat and 110,000 gallons of olive oil every year to feed his men.

1 Kings, 5 8–11

Wood and wood products have been part of the history and culture of the human race for thousands of years. Forest resources have been used to develop countries economically, to provide shelter and worth, to provide foreign exchange or, in the case of the quotation above, to ensure the friendship of a powerful king.

Many of the major products of the forest products industry such as paper, lumber, veneer, and furniture date from antiquity. The Chinese are first credited with making paper from the pounded bark of the mulberry tree. Around 4000 B.C. the Egyptians were known to have had 8-foot-long metal saws which they used to cut both stone and wood. In approximately 1500 B.C., the Egyptians were manufacturing veneered furniture which was later found by archaeologists in the tombs of the Pharaohs.

LUMBER

Early lumber manufacturing was accomplished using pit sawing. A two-person team of sawyers were reported to average about 200 board feet per day. The first sawmills were wind-, water-, or treadmill-powered frame saws in use around 1200 A.D. in Europe. This new automation technology of its day was slow to be adopted and caused significant management problems. In 1663, pit sawyers in London prevented the construction of a sawmill and later in 1803 pit sawyers in New Orleans destroyed a steam-powered sawmill. These earlier frame mills could produce about 4000 board feet per day or the equivalent of 40 pit sawyers.

The first sawmills in the United States are believed to have been established at Jamestown, Virginia, around 1625 and at Berwick, Maine, around 1631. The initial lumber industry of the early colonies was centered in the white pine forests of the New England states. As the nation grew and expanded it increasingly needed more wood, so the lumber industry moved to the white pine forests of Michigan, Wisconsin, and Minnesota and later in the early 1900s to the southern pine forests of the southeastern United States. The great timber stands of the Pacific northwest were also being opened up during this time and by the 1920s they were supplying a large portion of the nation's lumber needs.

The early lumber industry in the United States was blessed with a low-cost plentiful supply of raw material and relatively strong markets because the nation needed considerable quantities of lumber and other wood products to support its rapid growth. The introduction of steam power in the 1800s, coupled with new saws, made high-production sawmills possible. At the turn of the century, the largest mills could cut 1 million board feet per day. By 1909, lumber production in the United States reached its all-time peak of 45 billion board feet.

PULP AND PAPER

Although paper had been manufactured dating from the Chinese mulberry bark paper and the Egyptian papyrus, the early manufacture of paper was not from wood. It was not until approximately 1840 that the manufacture of wood pulp was accomplished. This early manufacture of wood pulp was done by forcing a short bolt or log against a rotating stone in a water slurry. The rotating stone ground off wood fibers which were carried away in the water slurry. This new process for making pulp was called the "groundwood process" and was introduced into the United States in the early 1860s.

However, the first U.S. paper mill had been built much earlier in 1690 to produce paper from the fibers of old rags and not from wood pulp. These early paper mills produced paper sheet by sheet in a slow time-consuming process. In 1827, the first paper machine was installed in the United States, allowing the continuous production of a long, wide sheet of paper. This new technology was a significant boost to the paper industry. By the 1880s, chemical methods of pulping wood had appeared. In 1880, the sulfite process was available and by 1884 the Kraft or sulfate process was available to produce wood pulp. Both of these methods relied on using chemicals to dissolve the part of the wood (lignin) that holds its fibers together. In 1909, the first Kraft mill using southern pine was built in Roanoke Rapids, North Carolina, beginning the growth of the strong southern pulp and paper industry.

INTEGRATED FIRMS

Between the two world wars, early leaders in the forest products industry began to utilize their raw material resources more completely. This was accomplished by several large predominantly western lumber companies such as Weyerhaeuser moving into the pulp and paper field. Prior to this point, the industry had largely been segmented into lumber companies and paper companies.

The addition of pulp and paper capacity to a traditional lumber manufacturing firm allowed it to make optimum use of its resources. This was accomplished by using the

waste material from lumber production and nonsawlog material from harvesting operations as raw material for the pulp and paper mills. Increasing timber prices further
encouraged better utilization practices, and other western lumber companies followed the
lead of Weyerhaeuser and purchased or built pulp and paper facilities in the 1940s and
1950s.

At the same time, companies which had traditionally been pulp and paper companies
such as St. Regis Paper Company (now part of Champion International) and International
Paper Company began to move into the lumber business. The paper companies moved
into the lumber business to enable them to process their high-quality sawtimber into more
valuable lumber products rather than into paper.

Diversifying into both lumber and paper products also helped to diversify company
earnings by producing both paper and lumber sales, which have tended to be somewhat
countercyclical. In addition, paper prices have exhibited more stability than lumber and
plywood prices. By the later 1950s and early 1960s, the forest products industry was
taking on the appearance that it has today: an industry increasingly dominated in most
product and market segments by large, fully integrated companies which produce a wide
range of lumber, panel, and paper products from their timberlands.

COMPETITIVE BEGINNINGS

The forest products industry typically produces large, bulky products which are relatively
expensive to transport. As a result, the early stages of development in the U.S. forest
products industry centered on local sawmills and paper mills supplying products for local
markets.

The early years of competition in the forest products industry focused on production.
Some have termed this "reciprocal competition," where industries with similar strategic
positions rely on operating differences to separate the successful firms (South, 1981). The
early forest products industry had many factors which led to a production approach to
management, including low-cost timber, plentiful timber and power, strong consumer
demand, and limited competition. This resulted in the dominant goal of management
being to decrease costs through more efficient production.

However, with the advent of rail and steamship transportation strong competition was
possible between firms some distance apart. This competition has intensified and has
continued to this day, resulting in global competition. Other factors which have spurred
competition in the United States have been:

- Growth of large industrial centers
- Decline of local timber base
- New wood processing technologies, resulting in smaller/drier timbers for
construction

The growth of the large industrial centers on the northeast created heavy concentrated
demand for wood products which could not be met by the local timber base. This resulted
in the decline of local timber bases around industrial centers and the need to ship in wood
products for local needs. Now mills some distance apart were competing for customers in
the same marketplace. New sawing and processing technologies aided this change by

allowing smaller and drier timbers to be used for construction. These new smaller and lighter timbers were, of course, easier to transport and, by the mid-1920s, Weyerhaeuser had four steamships carrying brand-name lumber to east coast ports. However, competition (and company management) was still largely based on a production orientation with decreasing costs as the basic management objective.

World War II and the strong domestic economy following the war brought with it a strong demand for forest products. However, by the late 1950s and 1960s the task of growing and converting timber into useful products at the lowest cost was no longer the total strategy of management. The industry made two significant changes: (1) it diversified into unrelated businesses and (2) it attempted to adopt the marketing concept.

Diversification Outside the Forest Products Industry

In the 1960s and early 1970s, the forest products industry was swept up into an undisciplined diversification binge, as were many other large industries in the United States (*Forbes,* 1984). Georgia-Pacific, International Paper, and Boise Cascade, to name a few companies, moved into businesses they knew little about. These acquisitions included oil and gas, land development operations, chemicals, and, in the case of Boise Cascade, even such exotic adventures as Latin American utilities (*Business Week,* 1979). The forest products industry has always been highly dependent upon housing starts and general economic trends; however, the diversification schemes of most companies during this time frame failed to produce the expected result of protecting the companies against the swings in their forest products business. Beginning in the mid-1970s and continuing until the present time, many companies have been selling assets not related to their core forest products business.

The failure of this diversification strategy has been linked to management's slighting of the core forest products business during this period (Rich, 1979a; Hinton, 1983). Additionally, few forest products companies had the critical skills for managing the wide variety of diversified assets that were required. The problems of managing diversified assets were frankly stated by a Union Camp vice chairman (*Dun's Review,* 1975, p. 43):

> You must have someone in the parent organization who really understands the (new) business before it gets heavy. Other paper companies have had troubles along these lines. They had these MBAs who, I am sure, were intelligent. But once they failed, there wasn't a damn thing anyone could do back at headquarters. They didn't know the business, and they were completely out on a limb.

In the 1980s, among the major integrated forest products companies, a desire to concentrate on the core business of growing timber and converting it to useful forest products appears to have been the path for success for many companies. Boise Cascade, Georgia-Pacific, Weyerhaeuser, International Paper, and others in recent years have all shed subsidiaries which were not related to their core forest products business to enable them to generate cash for reinvestment and modernization of their forest products production facilities.

Adoption of the Marketing Concept

In the 1960s, many major industries adopted a marketing approach to management and corporate strategy. That is, all decisions were made with the perceived needs of the consumers as the final guide. Even the government jumped on this band wagon with many university forestry schools adding marketing-utilization faculty and the U.S. Forest Service establishing a Forest Products Marketing Laboratory in Princeton, West Virginia. Large forest products corporations were also enticed by this idea and tried to change from a historic preoccupation with production and cost reduction to a marketing approach and serving the customers' needs.

The marketplace for forest products firms was changing as a result of the growing threat of new product competition, the emergence of new consumer groups, and changing distribution patterns required to meet these needs (Rich, 1970). The change in management orientation was a partial reaction to these marketplace changes. No longer could management largely ignore marketing. *But,* neither could it ignore production and timberlands in making decisions. As is the case in many instances, changing an entire corporation's orientation to doing business is tough at best and disastrous at worst. Coupled with the ill-fated diversification schemes, this poorly implemented shift in management style proved nearly fatal to some firms.

The degree to which some firms tried to change is indicated in the quotations below from this period:

Weyerhaeuser: "Shift from commodity selling of lumber to end user marketing of wood products."

International Paper: "All parts of the company are part of the marketing function."

U.S. Plywood: "Every major policy decision at our company is a marketing decision."

Boise Cascade: "Our objective is to be out in the marketplace as close as possible to where the building materials are used."

Marketing and market planners were given free rein in many forest products companies and received minimal guidance from top management. Product lines were expanded with little regard to or coordination with manufacturing facilities. As might be expected, the failure to integrate marketing with the other functional areas such as manufacturing and timberlands produced near disastrous results for some firms (Rich, 1980, 1979b).

Most firms realized that their positions were extreme by the mid-1970s and began again to give more attention to their major fixed assets, i.e., timberlands, production facilities, and distribution networks. Attention was also given to their product mix to achieve a more realistic balance between specialty and commodity products.

By the early 1980s, most of the industry was suffering through a period of extremely weak demand and fluctuating prices. The late 1970s had seen strong housing starts, high timber prices, and strong product prices. But the early 1980s saw housing starts drop below 1 million units and product prices retreat, leaving many firms stuck with high-cost timber contracts. The need for better marketing to cope with these problems was again apparent. Companies began to more strongly consider export markets, to enhance their specialty product lines, to reexamine their need for timberlands, to increase product quality, and so on.

Business Week in a 1982 article said, ". . . the ability of companies to create specialty products and develop export markets will become almost as important as their ability to maintain timber supplies and operate at a low cost." Clearly the management pendulum has moved back toward the marketing function, but not to the exclusion of manufacturing and timberlands. The lessons of the 1960s and 1970s were remembered.

C. W. Bingham, Executive Vice President, Weyerhaeuser, in a major address to the Forest Products Research Society in 1986 recounted the following turmoil in the industry between 1979 and the end of 1985 (Bingham, 1986):

1 Both record lumber demand and excess supply;

2 Deflated lumber prices lower than in 1970;

3 Eleven million acres of land and timber for sale in the United States—and remaining unsold;

4 Large integrated companies having the lowest profit margins in the industry;

5 A tariff war between Canada and its most important customer and supplier, the United States;

6 Interest rates which have, within a short period, hit both the highest and lowest levels in 20 years;

7 Runaway inflation followed by inflation rates similar to those of the 1960s;

8 The surge of the value of the dollar, peaking in 1985, then sliding back below its level of 1978;

9 The rush to the Sun Belt, followed by chaos in the oil patch, and a rejuvenation of the Northeast and North Central markets of the United States?

Bingham went on to say that beyond these factors new and growing markets emerged in the 1980s: industrial, repair and remodel, and the do-it-yourself segments were clearly making their presence felt. Wood products firms will have to differentiate themselves in the marketplace and lessen their emphasis on commodities. The industry must seek out more discrete market segments, enhance its bond to the customer, and develop innovative specialty products.

Another industry executive, Roger Wiewel, Senior Vice President, Marketing Group, MacMillan Bloedel, provided the following insights into what successful marketing and management of forest products firms will look like in the future (Wiewel, 1986):

During the post-war period, North Americans profited immensely in a global market where there was no competition from the other half of the industrial world which had been mostly destroyed, and from a third-world that hadn't started to compete. They had a level playing field all right—the EEC (Europe) and Japan were leveled.

Now we are all in the pro circuit and we have to play scratch. For some, the competition has proven to be fun and exhilarating . . . and very profitable.

Others just haven't got the stomach for the championship fight.

Managements with vision will run this industry in the near future because managements without it will be out.

Fortunately, North American forest industry managements are changing—winners are taking over, executives who are gearing to compete globally and do better by the North American consumer, our major customers.

Now what I have said so far is that there are two types of citizens here in North America—those who say "stop the world, I want to get off," and those who want to get in and tangle in the world marketplace.

Only one will survive, and it will be the latter because I don't believe anyone can escape the global marketplace.

So Monday morning when we all go back to work, what are we going to do?

1 Let's upgrade our product line. The greatest waste of our natural resources is not getting the full potential out of these resources.

Analyze your mills and your product lines. Those products that get a very low profit return on fiber, including interest on total capital employed, shouldn't consume the fiber. Upgrade or downsize that facility. Leave that low return product to the third world. If we don't buy their products, they can't pay their debts.

2 Stop thinking commodity. There are no commodities, only commodity-thinking. Foster a culture that encourages corporate commitment to operational excellence. My company converts 9,000,000 cubic meters of wood in the B.C. base, and I can assure you that I don't expect even one cubic meter to be treated as a commodity product. Make what the customer wants—every problem is an opportunity.

3 Let's promote wood and wood-based products and expand our markets. Wood is good.

Japan is buying American 2 × 4 platform construction, as are other nations. The modern office equipment—the office of the future—cries for new paper grades.

This forest industry in North America probably spent $200 million in legal fees in the last decade fighting trade issues. If we had used this $200 million to promote our forest products, we wouldn't have had the trade issues in the first place.

4 We have to become more innovative in the North American industrialized world. Put some right-brained people in top management. Even good market research doesn't create markets for products that don't exist.

5 Provide service to your customers. Manufacturing is not all that sick in North America. Twenty million Americans work in manufacturing. Manufacturing jobs are expected to remain high during the 1980's. Maybe service jobs are growing because they give a service their customer wants—"Smokestack America" could thrive if we shipped on time every time to the customer's specifications. Develop performance standards (such as the Crosby Quality Management Program) for shipping against customer orders and a measurement program for proper feedback on nonperformance.

SUMMARY

Marketing is a key aspect of success for U.S. forest products firms. Marketing as a management strategy was emphasized in the 1960s and early 1970s, then abandoned in the late 1970s and early 1980s; but by the mid-1980s marketing had become a strong fixture of company strategy.

We now need to turn our attention to marketing fundamentals in the next several chapters. After these fundamentals are mastered, we will apply them to the forest products industry.

BIBLIOGRAPHY

Bernsohn, K. 1943. *Cutting up the North*. Hancock House Publishers, Inc. Vancouver, B.C.

Bingham, C. W. 1986. "Market Trends from a Producer's Perspective." *Forest Products Journal* 36(10):11–13.

Birkner, E. 1963, "The Distribution Dilemma." *House & Home* (September):119–127.

Business Week. 1982. "The Battered Fortunes of the Forest Products Industry. *Business Week* (September 13):70–74.

Business Week. 1979. "Boise Cascade: Expansionism That Now Sticks Close to Home." *Business Week* (February 19):54–55.

Cox, T. R. 1974. *Mills and Markets*. University of Washington Press. Seattle, WA.

Dun's Review. 1975. "New Growth at Union Camp." *Dun's Review* (March) 105(3):40–43.

Forbes. 1984. "Paper Heart." *Forbes* (May 7):19–201.

Hidy, R. W., F. E. Hill, and A. Nevins. 1963. *Timber and Men, The Weyerhaeuser Story*. Macmillan Co. New York.

Hinton, T. L. 1983. "Capital Investment: International Paper Fights Back." *Management Accounting* LXV(1):41–44.

Maxwell, R. S., and R. D. Baker. 1983. *Sawdust Empire*. Texas A & M University Press. College Station, TX.

Morgan, M. 1982. *The Mill on the Boot*. University of Washington Press. Seattle, WA.

Rich, S. U. 1980. "Retreat from the Marketing Concept." *Forest Products Journal* 30(2):9.

Rich, S. U. 1979a. "Implementing the Investment Portfolio Concept." *Forest Products Journal* 29(8):11.

Rich, S. U. 1979b. "Integration of Marketing Planning with Corporate Strategic Planning." *Forest Products Journal* 29(12):10.

Rich, S. U. 1970. *Marketing of Forest Products: Text and Cases*. McGraw-Hill Book Company. New York.

Sinclair, S. A. 1990. "A Note on the Forest Products Industry." In *Marketing Management: A Choice Approach*. Dryden Press. Orlando, FL. pages 849–866.

South, S. 1981. "Competitive Advantage: The Cornerstone of Strategic Thinking." *Journal of Business Strategy* 1(48, Spring):15–25.

Twining, C. E. 1975. *Downriver*. The State Historical Society of Wisconsin. Stevens Point, WI.

Wiegner, K. K. 1982. "A Giant Shapes Up." *Forbes* (July 19):43–44.

Wiewel, R. N. 1986. "What Are You Going to Do on Monday Morning?" In *Proceedings, North American Wood/Fiber Supplies and Markets. Strategies for Managing Change*. Proceedings No. 47351. Forest Products Research Society. pages 12–14.

Wynn, G. 1946. *Timber Colony*. University of Toronto Press. Toronto, Canada.

Youngquist, W. G., and H. O. Fleischer. 1977. *Wood in American Life: 1776–2076*. Forest Products Research Society. Madison, WI.

PART TWO

TWO

MARKETING FUNDAMENTALS

This section is designed to provide a basic background in marketing fundamentals. It will be very useful for those individuals who have not had a marketing course, and for those who have had a course it will serve as a refresher. When we consider marketing from a strategic context, it can be viewed as a collection of decisions which must be made in various areas. These decisions, taken collectively, can be termed the "marketing mix." This section is organized around the various areas which require fundamental decisions: markets, products, distribution, promotion, and pricing. First let's consider markets.

Marketing is loaded with subtleties and nuances. Just as one thinks he really understands how to market effectively, he realizes that he has just scratched the surface. To understand marketing is to understand human nature and how to influence it. Becoming skilled in marketing is a lifelong process. It involves observation and listening, coupled with analytical ability and information sensitivity. All of this, then, is carefully blended with creativity, empathy and persistence.

D. F. Healy

CHAPTER 3

MARKETS

Every organization—in fact every individual—engages in some form of marketing . . . What eludes people is the translation of these ideas into practical, customer-oriented programs and offerings.

D. F. Healy

A market is typically defined as the set of all current and potential buyers of a particular product or service. Inherent in this definition is a key assumption that buyers are able financially to make a purchase.

Marketing strategy is increasingly organized around target markets. A target market can be defined as that segment of the market toward which a firm or organization targets its products or services. Many large markets can be redefined into several smaller market segments. Market segmentation permits the definition of one or more target markets and the ability to tailor marketing programs to each market segment or target (Beucler and Bowyer, 1987).

Markets can be segmented according to a wide array of characteristics such as consumer demographics (i.e., age, sex, income, etc.), product application, geography, or customer preferences for certain product features; more recently markets have been segmented based upon the benefits which customers wish to derive from the product or service they purchase. Product application has been used by many wood building products producers to segment the market into segments such as new home construction, repair and remodeling, do-it-yourself, manufactured housing, and commercial construction. Markets are many times broadly segmented into consumer and industrial/organizational markets. In consumer markets, individuals purchase goods and services for their personal or family needs. In industrial/organizational markets, goods and services are marketed from one business to another business.

DEFINING TARGET MARKETS

Let's next consider three alternative methods in defining target markets: mass marketing, market segmentation, and multiple segmentation.

Mass Marketing

Perhaps we are most familiar with mass marketing. It has been used for a broad range of consumer products such as soft drinks, Kleenex tissues, and fast food. Mass marketing represents a single marketing program aimed at as broad a range of customers as possible.

17

Mass marketing assumes that customers have very similar product needs and wants and seeks to maximize sales by having the same product for everyone. Due to the necessity of appealing to a broad range of customers, there can be difficulty in achieving focus in mass marketing programs.

Mass marketing requires resources beyond the means of most small firms for mass production, mass distribution, and, many times, mass advertising. However, fewer numbers of products can lead to lower production costs because product uniformity is emphasized and diversification is not. Mass-marketed products are typically distributed to a wide range of outlets. Try going into a drugstore or a supermarket and not finding Kleenex tissues for sale.

The classic example of a mass-marketed product is the Model T Ford. You could purchase the Model T Ford in any color you wanted as long as it was black. There were very few, if any, options and prices were low. Within the wood products industry, sheathing grade ⅜ inch softwood plywood has evolved in the North American market into a mass-produced product which is mass-marketed along with SPF No. 2 and better studs. Mass marketing is not generally applicable to smaller and secondary wood products producers, but sometimes can be used by smaller specialized producers of wood product commodities such as high-speed stud mills.

Following the 1950s and 1960s, there was increasing recognition among North American firms that the consumer market was not a simple homogeneous market. Improved marketing research facilitated the pinpointing of specific customer segments and, to a certain extent, decreased the applicability of mass marketing. In fact, some writers have concluded that those businesses that offer mass market functional products and services for sale will find it increasingly more difficult to survive (Sheth, 1983).

Market Segmentation or Niche Marketing

Mass marketing is a shotgun approach to marketing while market segmentation or niche marketing focuses on a single, well-defined customer group (i.e., more like a rifle approach). Niche marketing permits a strong focus on a given market segment, making it particularly applicable for smaller firms with limited financial and time resources. It emphasizes specialization for the niche market and not diversification to multiple markets.

Greater care is needed in differentiating company products from other competitors' products in niche marketing. When a company targets a market segment as a potential niche market, the strength of the competition and the total potential volume of sales in the market are key considerations. There can be a niche which is just too small and/or too competitive to be profitable.

Within a given market niche it is possible for a small firm to establish brand name recognition. Focusing on a narrow market niche can enhance the ability of a small company to meet the needs of its customers in that particular market segment. This can allow a small firm to compete very effectively with large firms in niche markets. As a result, there is an increasing emphasis being placed on the careful definition and pursuit of market niches by both large and small firms.

Some examples of past successes in niche markets include Redi-to-Use Hardwoods

A customer selects a hardwood
board from a Weyerhaeuser
ChoiceWood display rack.

and Frank Purcell Walnut Flooring. Redi-to-Use Hardwoods was established as a venture
of Hammermill Paper. They developed a line of clear S4S hardwood boards[1] for sale at a
premium price to the home center marketplace for do-it-yourself customers. Through a
high-quality product and exceptional service, the Redi-to-Use Hardwood Division of
Hammermill Paper grew very rapidly and was acquired by Weyerhaeuser around the time
Hammermill was purchased by International Paper. Weyerhaeuser continued the niche
market strategy and renamed the product line ChoiceWood.

Frank Purcell Walnut Flooring grew out of the need for Frank Purcell Walnut Lumber
Company to market its lower-grade walnut lumber. After some significant market
research, it was concluded that walnut flooring was a niche market in which the company
might be able to become a substantial player even though Frank Purcell Walnut was
relatively small. The walnut flooring market proved to be a good strategic move for Frank
Purcell Walnut even though the niche is quite small.

A market niche which cannot be defended or which may grow very large can turn into
a mass market. The market for radius edge decking is an example of this phenomenon.

---

[1]S4S is an abbreviation for "surfaced on all four sides." "Clear" refers to the fact that the boards have a
clear surface, i.e., no knots, splits, checks, bark pockets, etc.

Radius edge decking is a high-quality board with the edges rounded to improve appearance and stop splintering. This product is typically pressure-treated with CCA (chromated copper arsenate) preservative and is used for residential and commercial decks. When initially introduced to the marketplace, it was largely sold in a niche market for upscale decks; however, radius edge decking became a commodity product sold in a mass market situation.

When a firm uses a market niche strategy, caution must be exercised about blind dependence on a single niche. If a company operates in only one market niche and that niche suddenly becomes unprofitable or indefensible due to circumstances beyond the control of the company, it can become devastating to the firm. One-niche firms must always be aware of these dangers and plan for new niches should theirs become crowded and unprofitable.

Multiple Segmentation

Multiple segmentation represents a blend of a niche strategy and mass marketing. Marketing plans are focused on two or more well-defined customer groups. This allows a firm to avoid having all its eggs in one basket.

Multiple segmentation can fit independent, primary, and secondary forest product producers rather well. Different marketing programs are developed to appeal to different segments within the overall marketplace. This strategy tends to require the availability of more products because of marketing in different segments. However, it permits diversification and the lessening of risks associated with a decline in any single market segment. Multiple segmentation requires greater resources than niche marketing and is most feasible when the segments are truly unique (Beucler and Bowyer, 1987).

To better understand multiple-segmentation marketing, let's consider a typical hardwood lumber producer. A variety of different grades of lumber can be produced from a given hardwood log. These different grades (i.e., products) can require different marketing strategies. A hardwood lumber producer may market its lumber products to distribution yards, furniture case goods producers, flooring/millwork producers, export markets, and pallet producers. These different market segments may be needed to effectively market the range of species, grades, lengths, and sizes that a hardwood lumber mill may produce.

EFFECTIVE SEGMENTATION

Successful market segmentation requires sufficient differences between groups of customers to identify the segment. But within each group of customers (target market segments), there must be enough similarity to develop a targeted marketing program. Additionally, it must be possible to clearly define the product needs of the customer group. Many times segments can be devised, but there is no way to reach the customer group in terms of defining the product needs or advertising. We were once involved in an extensive study of ready-to-assemble furniture customers. Using data from nearly 1500 households, we developed a statistically sophisticated segmentation scheme. The market segments were real. But after all our work we analyzed the segments and while their attitudes and buying patterns were different, they were nearly alike in almost all demographic variables. We had no way to single out any specific segment and reach it.

A separate market segment must also be large enough to be economically viable. And the customers in the market segment must be able to be reached in an efficient manner. As the marketplace becomes increasingly splintered with various groups and different customer needs, market segmentation will become increasingly important. Most major wood products firms are using various forms of market segmentation to address their customers. But multiple market segments can create confusion among potential customers. And segments can be defined which are undefendable and therefore open to strong competition.

Ways to Segment Wood Products Markets

Various approaches exist for segmenting markets for wood products. One method of segmenting wood products markets is by customer class. For major wood building products producers, customer classes can be defined as do-it-yourselfers, stick builders, remodeling contractors, residing contractors, tract builders, manufactured housing, and others. Each of these customer classes has somewhat different needs and desires in the marketplace. For example, in the residential siding market there are differences in the preference for various siding materials by builder class (Stalling and Sinclair, 1989).

Perhaps the most common method for segmenting wood products markets is geography. Markets can be segmented using geography in a number of ways including international markets, single-country or national markets, regional markets, state or provincial markets, and local markets. Many times a geographic segmentation scheme works well for wood products. Within the U.S. marketplace for wood products, numerous differences have been shown to exist based upon region of the country (Stalling and Sinclair, 1989; Smith and Sinclair, 1990). In general, the northeast and the west have a stronger demand for solid wood items and higher-quality wood products.

A major method for segmenting markets for forest products is based on whether the product is sold in an industrial or consumer market. In industrial markets, products are marketed from one firm to another firm. For example, hardwood lumber is marketed by sawmills, wholesalers, or distribution yards to furniture firms which then use the lumber to manufacture furniture. As a result, the volume of hardwood lumber purchased by furniture producers is derived from the demand they have for their furniture. This is known as "derived demand" and it is a characteristic of industrial markets. Industrial customers generally buy more frequently and in larger quantities than consumers. Credit needs, promotion, delivery, and most other aspects of marketing are different to meet industrial needs. Each industrial customer group which buys forest products for re-manufacturing, resale, or internal use has different needs and desires which can form the basis of an effective segmentation scheme.

Consumer markets for forest products are wide and varied, ranging from diapers, facial tissue, and writing papers to prefab lawn storage buildings, shelving boards, and many do-it-yourself wood products at the local home center/lumberyard. Individual consumers buy in smaller amounts and have different transportation, credit, promotional, and technical requirements. Again, these markets require a different approach for success from industrial markets.

Benefit segmentation is another segmentation method which has been successfully applied to wood products markets. Smith (1988) used a benefit segmentation approach to segment the markets for CCA-treated wood products. He found five distinct market segments for retail customers of CCA-treated products. These five segments were:

1 *Price-insensitive, kiln-dried after-treatment, brand name customers.* A segment that is willing to pay a premium for kiln-dried after-treatment services and brand name lumber materials and that does not feel appearance is more important than grade/species.

2 *Kiln-dried after-treatment/appearance customers.* A quality-oriented segment that perceives kiln-dried after-treatment, general appearance, and moisture content, but not quality marks as representative of a quality treated product. This segment will not pay more for branded treated lumber.

3 *Price-insensitive brand name/appearance customers.* A market segment willing to pay a premium for high-appearance, branded products. These customers incorporate surface cleanliness, promotional materials, and dealer selection into the purchase decision and feel that additional cost for kiln-dried after-treatment materials is unwarranted.

4 *Price-sensitive, low-tech bargain hunters.* A segment that looks for the best appearance, brand name treated lumber at the lowest price. The product's technical attributes are seemingly unimportant to this customer.

5 *Price-sensitive, appearance/grade customers.* A customer who makes treated lumber purchases based on both appearance and grade. Branded products are considered to be superior but not worth more money. Price is important; however, this customer will pay for quality/appearance.

When focusing on benefits, marketers can examine the product and service attributes that should be offered as opposed to what is currently available. Both new and old products should be tailored to specific customer segments seeking a specific benefit pattern (Haley, 1968).

FINDING NEW MARKETS

When analyzing new markets, the concept of a "product-market" is useful. Beucler and Bowyer (1987) define a product-market as a product or product line that satisfies particular customer needs or wants in a market. Day et al. (1979) go further and define a product-market as the set of products judged to be substitutes, within specific situations where similar patterns of benefits are sought by a group of customers.

Ultimately, all product-market boundaries are arbitrary. Market and product class definitions appropriate for tactical marketing decisions tend to be narrow, reflecting the short-run concerns of sales and product managers. For example, hardboard manufacturers might tend to limit their view of competitors to other hardboard manufacturers. While appropriate for short-run tactical decisions, manufacturers' perceptions of product-marketing boundaries should be stretched far enough so that significant threats and opportunities are not missed. A lack of understanding of product-market boundaries can result in an inadequate and delayed understanding of emerging threats in a competitive environment.

The product-market opportunity matrix is a method of investigating alternative ways of approaching new markets and maintaining and/or increasing sales for an individual firm (Beucler and Bowyer, 1987). Figure 3-1 shows the four-cell product-market matrix. These cells represent different starting points for strategy development which are based on the position of the product in the matrix.

Market Penetration (Current Market, Current Product)

When you have a current product and a current market, the typical strategy is to expand the sales of the product in its present market. This can be done through expanded distribution, expanded promotional efforts, and/or more intensive price competition. Sales can be increased among existing customers. You can attract your competitors' customers or you can attract new nonusers. This strategy works best in a growing market.

Market Development (New Market, Current Product)

In this cell of the matrix, we are interested in increasing sales of existing products in new markets or in new product applications. This can be done through enhanced or enlarged distribution systems and expanded promotional efforts. The expansion can be to new geographical markets or to new markets for existing products. One of the more unique examples of market development was accomplished by Blandin Wood Products in Grand Rapids, Minnesota. Blandin was one of the very first producers of waferboard/oriented strandboard in the United States. However, it was a relatively small firm as compared to

FIGURE 3-1
The product-market matrix. (*Adapted from Evans and Berman, 1985.*)

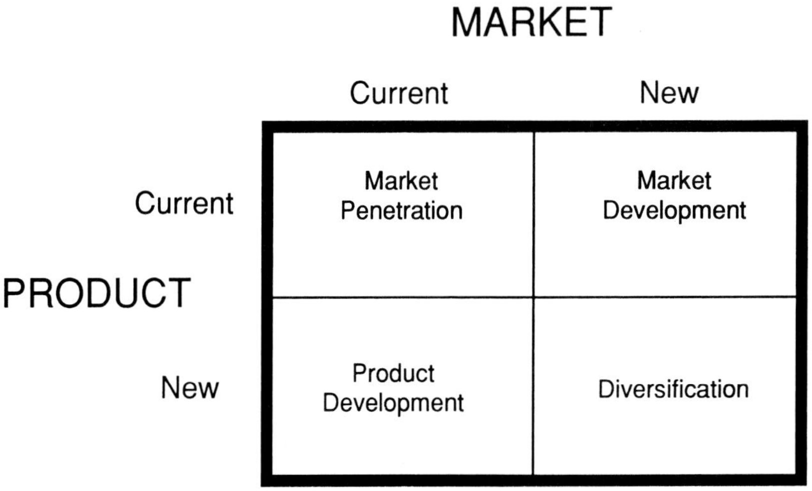

MARKET

	Current	New
Current	Market Penetration	Market Development
New	Product Development	Diversification

PRODUCT

major structural panel producers such as Weyerhaeuser, Georgia-Pacific, or Louisiana Pacific. Not wishing to compete with the major producers in the commodity markets, Blandin set out to find new markets for its products. One market it developed is the one-piece floor system for recreational vehicles. This allowed the company to use the advantage of its large opening press to create a single piece of floor with no seams for recreational vehicle manufacturers.

Product Development (Current Market, New Product)

In this cell of the matrix new or modified products are developed for sale in current markets. This can be done by adding quality improvements and small product changes which are then marketed to existing customers using the existing channels of distribution. You have seen the television ads, "New and improved." This strategy works well for a company which has a strong product line and loyal customers. One recent example of this strategy has been the effort by home centers and wood treaters to develop treated lumber deck kits for their existing customer base. These kits are typically CCA-treated southern pine which comes precut with all the necessary hardware. It's essentially an existing product which has been enhanced and is being sold to a similar customer base.

Diversification (New Market, New Product)

This is the most risky of the four cells, involving new products and aimed at new markets. A common way to implement this strategy is through acquisition, which allows the firm to buy not only the new product but also, to a degree, the new customers. This is what happened in the previous example of Weyerhaeuser purchasing Redi-to-Use Hardwoods.

Typically this strategy's goal is to broaden the product base, spreading sales over more product-markets and thereby minimizing risk. Distribution and promotional efforts are in many cases new to the firm, and there is some risk when getting into a new product or market which is not familiar to the firm. The forest products industry has a history of getting into product-markets it knew little about. These ventures were more often than not failures.

EVALUATING MARKET ATTRACTIVENESS

A method that enables a company to evaluate the relative strength of existing and potential products is the growth/share matrix developed by the Boston Consulting Group. A hypothetical growth/share matrix for a wood products firm's product lines might appear as shown in Fig. 3-2.

Each product is positioned on the matrix by its market growth and relative market share. The "market growth" is the present or forecasted annual growth of the market in which the product competes. The "relative market share" is the company's market share for that product expressed as a ratio of the largest competitor's share.

The theory behind this matrix is that products with large market shares are more profitable and that products with high growth rates require more cash in order to sustain growth. The matrix is organized into four cells. A product is examined for market growth

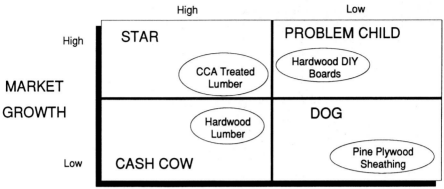

FIGURE 3-2
Growth/share matrix for a hypothetical wood products firm.

and market share, then placed in the appropriate cell. Each cell has a generic strategy associated with it and strong labels which are now part of common business jargon.

1 Cash cows have strong market shares in a slow or no-growth market. They are typically profitable and require little cash to maintain market share—hence the name "cash cow." Companies usually try to preserve their cash cows and use the extra cash to support growth in other parts of the firm.

2 Stars are also profitable with their large market share, but usually require significant cash to maintain their market share in growing markets. As stars move into slower or stable growth stages in the market, they can become cash cows.

3 Problem children have a relatively weak share of a growing market. This typically results in large cash needs to keep the operation going. Strategies can involve large amounts of investment to build share or divestment.

4 Dogs have small shares in stable markets. They are clearly a mixed bag in terms of strategy. Conventional wisdom has been to divest these low-profit products/businesses, and hence the negative name "dogs." However, more recent studies have shown some so-called dogs to be profitable in markets where share is not a critical determinant of profitability.

The inherent simplifications of the growth/share matrix have lead to serious questions concerning the assumptions and empirical support which have held the basic model together (Day, 1986). There have been numerous discussions in the literature concerning whether the growth/share matrix should be used as a diagnostic aid, a conceptual framework, or a prescriptive guide. As a diagnostic aid, it can be used to examine prior strategic judgments about the present or future position of a business or product. As a conceptual framework, it can guide the generation of various strategic options. And as a prescriptive guide, it can be used to develop the proper choice of strategic options and resource allocations within a firm (Day, 1986).

In spite of its simplification and some disenchantment with its basic premises, the

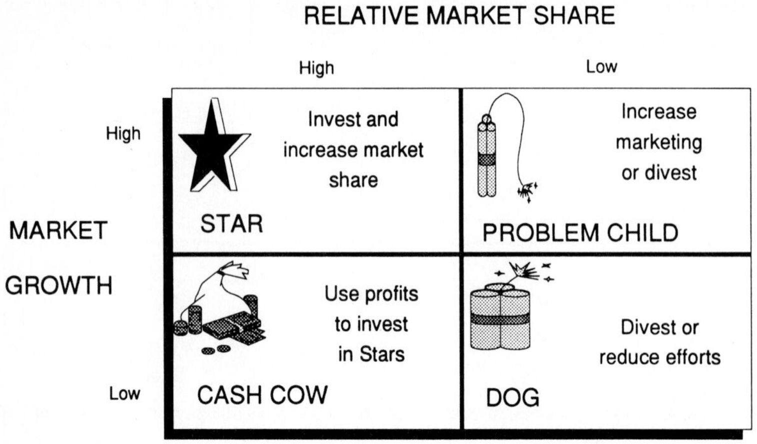

RELATIVE MARKET SHARE

	High	Low
High	Invest and increase market share **STAR**	Increase marketing or divest **PROBLEM CHILD**
Low	Use profits to invest in Stars **CASH COW**	Divest or reduce efforts **DOG**

MARKET GROWTH

FIGURE 3-3
Growth/share matrix strategy summary.

growth/share matrix remains a useful basic tool for analyzing a firm's strategies. The growth/share matrix can be used to identify products, product lines, or mixes of products for more investment or deletion. This can serve a useful purpose for many firms. The growth/share matrix can be a structured way to develop a market-oriented first pass at strategy development. For a summary of generic strategy options, see Fig. 3-3.

MARKET ATTRACTIVENESS/COMPETITIVE POSITION

Another approach to investigating the strength of a firm's portfolio of products considers the market attractiveness for a product or product line and the strength of a company's competitive position. This type of analysis was developed by McKinsey Company and General Electric in the 1970s (Day, 1986), and is sometimes known as the ''GE business screen.'' It has been used by several major forest products firms to analyze their portfolio of products and make strategic decisions. The basic premise of the analysis is that a more attractive market should, in the long run, be more profitable than less attractive markets and that firms with stronger competitive positions should be more profitable than firms in weaker competitive positions.

To apply the analysis, the relevant factors which determine market attractiveness and competitive position must be listed for a given product-market to be examined. Developing the list of factors may require some time. Abell and Hammond (1979) provide the chart on p. 27 as a place to start.

In the early formulations of the analysis all the relevant factors are weighted according to their impact on either market attractiveness or competitive position and a mathematical score is calculated for each axis to allow the product-market to be positioned on the matrix shown in Fig. 3-4. After a product-market is positioned on the grid, various strategy decisions are possible. Many of the generic decision rules are given in Figure 3-4.

Attractiveness of your business

A. Market factors
- Size of product-market (dollars, units)
- Market growth rate
- Stage in life cycle
- Diversity of market (potential for differentiation and segmentation)
- Price elasticity
- Bargaining power of customers
- Cyclicality/seasonality of demand

B. Economic and technological factors
- Investment intensity
- Nature of investment (facilities, working capital, leases)
- Ability to pass through effects of inflation
- Industry capacity
- Level and stability of technology utilization
- Barriers to entry/exit
- Access to raw materials

C. Competitive factors
- Types of competitors
- Structure of competition
- Substitution threats
- Perceived differentiation among competitors

D. Environmental factors
- Regulatory climate
- Degree of social acceptance
- Human factors such as unionization
- International factors

Strength of your competitive position

A. Market position
- Relative share of market
- Rate of change of share
- Variability of share across segments
- Perceived differentiation of quality/price/service
- Breadth of product line
- Company image

B. Economic and technological position
- Relative cost position
- Capacity utilization
- Technological position
- Patented technology, product, or process

C. Capabilities
- Management strength and depth
- Marketing strength
- Distribution system
- Labor relations
- Relationships with regulators
- Quality of raw material/timber supply

In analyzing both the product-market attractiveness and a firm's competitive position, understanding the nature of the competition is essential. Present and potential competitors directly affect the attractiveness of a product-market. A company's position can be evaluated only relative to its competitors'. Some users of this tool analyze their current position on the grid and then, looking at the market and their competitors, project future positions on the grid.

SOURCES OF WOOD PRODUCTS MARKET DATA

No single best source exists for market data on wood products. However, a wide variety of agencies and groups do publish an amazing assortment of market information on various

MARKET
ATTRACTIVENESS

	PROTECT POSITION	INVEST TO BUILD	BUILD SELECTIVELY
high	**PROTECT POSITION** * invest to grow at maximum manageable rate * concentrate effort on maintaining strength	**INVEST TO BUILD** * challenge for leadership * build selectively on strengths * reinforce vulnerable areas	**BUILD SELECTIVELY** * specialize around limited strengths * seek ways to overcome weaknesses * withdraw if indications of sustainable growth are lacking
medium	**BUILD SELECTIVELY** * invest heavily in most attractive segments * build up ability to counter competition * emphasize profitability by raising productivity	**SELECTIVITY/MANAGE FOR EARNINGS** * protect existing program * concentrate investments in segments where profitability is good and risk is relatively low	**LIMITED EXPANSION OR HARVEST** * look for ways to expand without high risk, otherwise, minimize investment and rationalize operations
low	**PROTECT AND REFOCUS** * manage for current earnings * concentrate on attractive segments * defend strengths	**MANAGE FOR EARNINGS** * manage position in most profitable segments * upgrade product line * minimize investment	**DIVEST** * sell at time that will maximize cash value * cut fixed costs and avoid investment meanwhile
	strong	medium	weak

COMPETITIVE POSITION

FIGURE 3-4
Generic strategy options in a competitive framework. (*Adapted from Day, 1986.*)

aspects of the forest products industry. A reference book, *Business Data and Market Information Source Book for the Forest Products Industry,* was published by the Forest Products Research Society in 1979. This book is basically a listing of many sources of market data for the forest products industry. The Forest Products Research Society in Madison, Wisconsin, can be contacted for further information.

Government

One source of marketing and business data is the U.S. government, which publishes a wide variety of information, much of which is collected under the auspices of the Department of Commerce. Much of the forest products information produced by the Department of Commerce is synthesized by various researchers in the U.S. Forest Service to provide a more useful basis of information for the forest products industry. A few examples of the publications provided through the U.S. Forest Service and the Department of Commerce are listed below:

• U.S. Department of Agriculture, Forest Service: *U.S. Timber Production, Trade Consumption, and Price Statistics 1950–199x.* Published annually through the Washington office.
  • U.S. Department of Commerce:
    Census of Manufactures
    Current Industrial Report
    Census of Business
    Timber Trends in the United States
    C40, *Housing Authorized in Individual Permit-Issuing Places*
    C42, *Construction Report*
    C25, *Sales of New One-Family Homes*
    C30, *Value of New Construction Put in Place*
    C50, *Expenditures for Residential Upkeep and Improvement*

Trade Associations

Many trade associations have economists and market analysts on their staffs who collect and analyze data on the industry segment they represent. For example, the American Plywood Association (Box 11700, Tacoma, WA 98411) publishes an economics report series which represents probably the best available data on wood structural panel production, distribution, and markets. The National Forest Products Association (1250 Connecticut Ave., Suite 200, Washington, DC 20036) publishes a statistical series on the wood products industry covering a wide array of basic wood products. The Western Wood Products Association (Yeon Building, 522 SW 5th Ave., Portland, OR 97204), the Southern Forest Products Association (Box 52468, New Orleans, LA 70152), and the American Paper Institute (260 Madison Ave., New York, NY 10016) all collect and periodically report market and production data for their members. Many other associations exist for almost every major wood or fiber product. A list of such associations is contained in the *Directory of the Forest Products Industry* (see below).

Trade Publications

Several for-profit organizations also collect and publish production and market information on the forest products industry. The many price-reporting organizations such as Crow's, Madison's, Random Lengths, *Weekly Hardwood Review,* and the *Hardwood Market Report* publish newsletters and sometimes annual synopses. Some organizations which publish trade magazines also collect and sell reports, directories, and fact books on the industry or industry segments. Some examples are:

Directory of the Forest Products Industry, published annually by Miller Freeman, 500 Howard Street, San Francisco, CA 94105

Fine Paper Directory, Grade Finders, Inc., 662 Exton Commons, Exton, PA 19341

Forest Industries North American Factbook, published annually by Miller Freeman, 500 Howard Street, San Francisco, CA 94105

Crow's Buyers and Sellers Guide, published annually by C. C. Crow's Publications, Box 25749, Portland, OR 97225

Green Book's Softwood Lumber Marketing Directory, Box 34908, Memphis, TN 38184

Lockwood-Post's Directory of the Pulp, Paper and Allied Trades, published annually by Miller Freeman, 500 Howard Street, San Francisco, CA 94105

Random Length's Buyers and Sellers Guide, published annually by Random Lengths Publications, Box 867, Eugene, OR 97401

Secondary Wood Products Manufacturers Directory, published annually by Miller Freeman, 500 Howard Street, San Francisco, CA 94105

The Lumbermen's Red Book, published twice a year by Lumbermen's Credit Association, Inc., 111 West Jackson Blvd., Chicago, IL 60604

Hardwood Review (an annual yearbook of hardwood prices and market statistics) published by *Weekly Hardwood Review,* Box 471307, Charlotte, NC 28247

Random Lengths Yearbook (an annual yearbook of softwood prices and market statistics) published by Random Lengths Publications, Box 867, Eugene, OR 97401

Other Industry Sources

There are many other sources for information concerning the forest products industry. Most major brokerage firms have research departments which collect and publish a wide variety of market information concerning various segments of the forest products industry. Many of these firms will, for a fee, provide a subscription to their periodic reports on the forest products industry. Additionally, most major firms in the forest products industry have market research departments which collect information and develop reports on their segments in the industry.

Another source of information is credit rating agencies or firms like Lumbermen's Credit or Dun and Bradstreet. For specific wood products companies, these credit rating firms can provide a great deal of information. Many also can supply lists of wood products firms by Standard Industrial Classification (SIC) code, numbers of employees, or other criteria.

Major wholesale and retail firms also develop background information not only on wholesaling and retailing but also on the manufacturing side of the industry. Various industry consultants, university professors, forest service researchers, and other industry experts are available for consultation on specific items.

A rather large variety of private research organizations exist which provide information on the forest products industry as well as other industries. The National Industrial Conference Board, the National Bureau of Economic Research, Predicasts (part of Standard & Poors), Dodge Construction Report, Resource Information Systems, Inc., Frost & Sullivan, and FIND/SVP are just some of the various organizations which provide information on the wood products industry.

On a worldwide basis, the Food and Agriculture Organization (FAO) of the United Nations collects and distributes a variety of data on the world's forest products industries. Other international agencies and organizations like the World Bank and the International Union of Forestry Research Organizations also collect and publish market data. Many countries have government agencies which collect data on their industries and markets. This data can sometimes be had for the asking.

FIGURE 3-5
The SIC system at the two-digit level.

Industries Classified	First two-digit SIC Numbers for Major Industry Groups
Agriculture, Forestry, and Fishing	01,02,07,08,09
Mining	10-14
Construction	15-17
Manufacturing	20-39
Transportation, Communications, Electric, Gas, and Sanitary Services	40-49
Wholesale Trade	50-51
Retail Trade	52-59
Finance, Insurance, and Real Estate	60-67
Services	70,72-73,75-76,78-86,88-89
Public Administration	91-97
Nonclassifiable Establishments	99

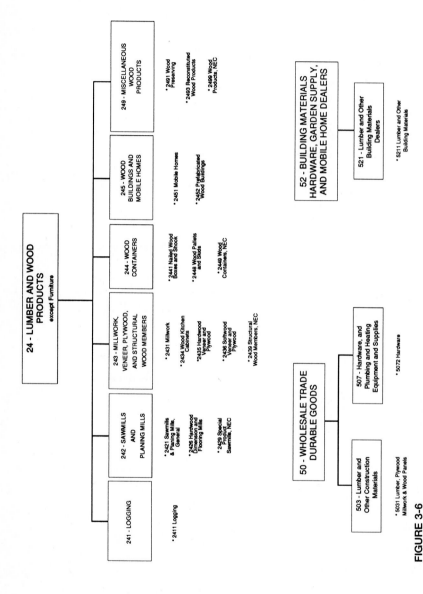

FIGURE 3-6
Wood products market as organized by SIC Codes 24, 50, and 52. (*Adapted from Beucler and Bowyer, 1987.*)

32

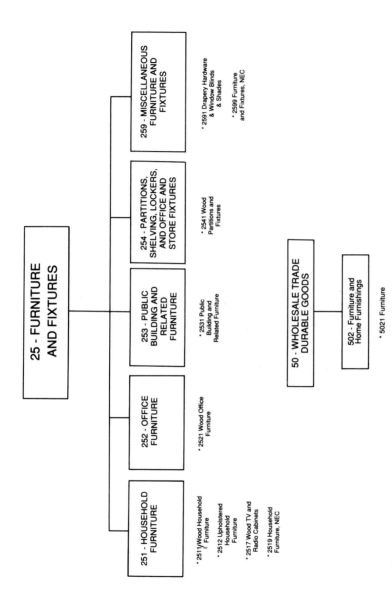

FIGURE 3-7
Furniture and fixtures market as organized by SIC Codes 25 and 50. *(Adapted from Beucler and Bowyer, 1987.)*

33

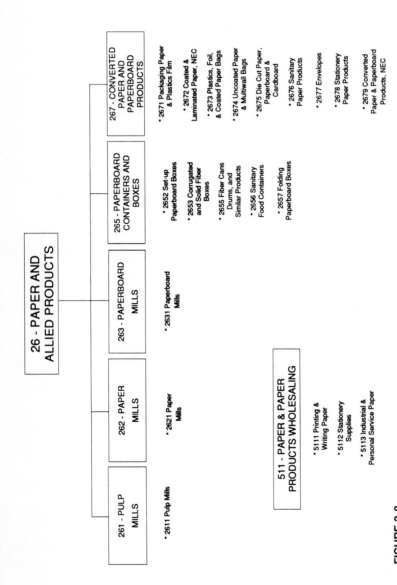

FIGURE 3-8
Paper and allied products as organized by SIC Codes 26 and 511.

34

Standard Industrial Classification

Most government data and much of the privately collected market data on the U.S. marketplace is organized by the Standard Industrial Classification System. SIC is a uniform numbering system for classifying enterprises according to their economic activity. SIC divides the U.S. economy into eleven divisions, as shown in Fig. 3-5. Within each division major industry groups are classifed by a two-digit number. For example, SIC 25 includes all manufacturers of furniture and fixtures. Within a major two-digit group, industry subgroups are further defined by three- and four-digit groups. For example, SIC 2511 includes wood household furniture and SIC 2448 wood pallets and skids. Further classifications are available down to a seven-digit code.

Figures 3-6, 3-7, and 3-8 provide a breakdown of SIC 24, 25, and 26, the major two-digit codes for forest products.

SUMMARY

Market segmentation is a method to examine markets from the demand perspective to seek out logical groups of customers which require similar products and levels of service. By identifying these groups better matches can be made between the products and services offered and the customers' needs. Various tools exist to help analysts identify and evaluate new market segments. Many sources of data are available to the wood products market analyst to assist in this job of evaluating markets. In the next chapter, we'll see how to tie our knowledge of markets to specific products.

BIBLIOGRAPHY

Abell, D. F., and J. S. Hammond. 1979. *Strategic Market Planning: Problems and Analytical Approaches.* Prentice-Hall. Englewood Cliffs, NJ.

Beucler, O., and J. L. Bowyer. 1987. *Marketing of Manufactured Wood Products.* Seminar Notebook #1. University of Minnesota, Department of Forest Products. St. Paul, MN.

Bingham, F. G., and B. T. Raffield. 1990. *Business to Business Marketing Management.* Irwin. Homewood, IL.

Day, G. S. 1986. *Analysis for Strategic Market Decisions.* West Publishing. St. Paul, MN.

Day, G. S., A. D. Shocker, and R. K. Srivastava. 1979. "Customer-Oriented Approaches to Identifying Product-Marketing." *Journal of Marketing* 43(4):8–19.

Evans, J. R., and B. Berman. 1985. *Marketing.* Macmillan Publishers, New York.

Haley, R. I. 1968. "Benefit Segmentation." *Journal of Marketing* 35(July):30–35.

Healy, D. F. 1987. *New Strategies for Growth: Paper Distribution Plans for the Future.* Paper and Plastics Education Research Foundation, Great Neck, NY.

Hutt, M. D., and T. W. Speh. 1981. *Industrial Marketing Management.* Dryden Press. Chicago, IL.

Rumelt, R. 1981. "In Search of the Market Share Effect." *Proceedings of the Academy of Management.* pages 2–6.

Sheth, J. N. 1983. "Emerging Trends for the Retailing Industry." *Journal of Retailing* 59(3):6–18.

Smith, P. M. 1988. *An Analysis of the Retail Customer of CCA Pressure Treated Lumber, Timbers and Plywood.* Unpublished Ph.D. dissertation. Department of Wood Science and Forest Products, Virginia Polytechnic Institute and State University. Blacksburg, VA.

Smith, P. M., and S. A. Sinclair. 1990. "The Do-It-Yourself Customer for CCA Treated Lumber Products." *Forest Products Journal* 39(7/8):35–41.

Stalling, E. C., and S. A. Sinclair. 1989. "The Competitive Position of Wood as a Residential Siding Material." *Forest Products Journal* 39(4):8–14.

Stern, L. W., and A. I. El-Ansary. 1977. *Marketing Channels*. Prentice-Hall. Englewood Cliffs, NJ.

CHAPTER 4

PRODUCTS

The ability of the firm to put together a mix of products and services that responds to customer needs and competitive pressures lies at the heart of successful marketing.

Adapted from Hutt and Speh, 1981

Developing this mix of products and services is clearly a critical decision for the firm and requires a broad view of the concept of a product. This broad concept is sometimes termed the ''augmented product,'' ''extended product,'' or ''total product,'' which not only includes the actual physical product but also the variety of service features and reputation that impacts on satisfying customer needs and meeting the competition (Fig. 4-1).

To meet customer needs, a wood products firm must offer more than just the physical product. The relationship encompasses the entire set of benefits for the customer (Beucler and Bowyer, 1987). In fact, with relatively standardized grading rules in North America for many wood commodity products, it is often easiest to set a firm apart from its competitors on the nonproduct dimensions in the total product concept.

A product can be defined as a physical object, a service, a place, an organization, an idea, or a personality that satisfies a customer want or need (Busch and Houston, 1985). In other words, the friendly salesperson, the helpful delivery truck driver, and the convenient location of the lumberyard are all part of the product.

When considering just the physical product, most forest products firms have to make product decisions at three levels. A firm producing softwood lumber must decide which specific items it will produce: precision end-trimmed (PET) 2×4 studs, 2×4s random length (R/L), or maybe larger sizes like 2×6s and 2×10s. And what about 4/4 (1-inch-thick) boards? The second-level decision is how these various items fit into a product line. The firm may decide to have only two product lines, softwood-dimension lumber and appearance-grade boards. The third-level decision is what mix of products (items and product line) best meets the customers' needs. The firm, at this junction, may go back and develop a third product line of items to be marketed to the wood-treating market if it has treating plants as customers or potential customers.

THREE LEVELS OF PRODUCT DECISIONS

1 Item—a specific single product:
I.e., a 2 × 4 PET stud
2 Product line—a group of closely related products:
I.e., softwood-dimension lumber
3 Product mix—a collection of items and product lines marketed by the company:
I.e., 2 × 4 PET stud
   Softwood-dimension lumber
   Market pulp
   Structural panels

PRODUCT DIFFERENTIATION

Product differentiation and market segmentation are closely related, especially in industrial markets for forest products. Product differentiation seeks to develop and promote differences between a given product and competing products. This process occurs many times in response to market segmentation as a separate differentiated product is developed and promoted to each market segment.

The most successful differentiation strategies can be used to help firms avoid direct price competition. Such factors as service level, quality of product, quality of service, distribution methods, credit, and others can be employed to develop a differentiated product.

A strong differential advantage can lead to real or perceived barriers for competitors entering another company's market niches. Without a differential advantage, a company offers its customer no reasons to select its products over competitors with the exception of price (Beucler and Bowyer, 1987). The essence of competition is differentiation—providing something different and providing it better than your competitor (Levitt, 1986). A low-price strategy rarely generates the best return unless you are the high-volume, lost-cost producer in the market.

TYPES OF PRODUCTS

There are three broad categories of products which we will consider: commodity products, specialty products, and differentiated products.

Commodity Products

A commodity product is one which is manufactured to a more or less standard set of specifications. There is very little difference in commodity products based upon their manufacturer. In other words, a commodity product produced by one manufacturer is essentially the same as one produced by any other manufacturer. For example, a 2 × 4 stud produced by MacMillan Bloedel is essentially the same as a stud produced by Weyerhaeuser or Sam's Pretty-Good Studs, Inc.

The commodity product is typically distributed in a mass market approach to all

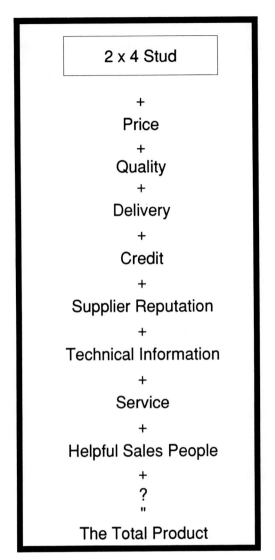

2 x 4 Stud

+

Price

+

Quality

+

Delivery

+

Credit

+

Supplier Reputation

+

Technical Information

+

Service

+

Helpful Sales People

+

?

"

The Total Product

FIGURE 4-1
The total product concept.

customers, who are presumed to have similar needs. Competition within a grade category is essentially based on price and service. Most producers and marketers of wood products commodities have little control over pricing, which is set by the market in an auction-like manner. Higher profits for a firm producing commodities typically depend upon higher volumes resulting in increasing economies of scale. For long-term success in the wood products commodity market, a manufacturer needs to pay careful attention to its production costs.

Many marketing tools have limited usefulness when applied to commodity products because the physical products are so similar. However, some manufacturers, and to a greater degree wholesalers and distributors, have differentiated themselves in the marketplace on nonproduct factors such as service, credit, delivery lead time, and others.

Specialty Products

Specialty products are at the other end of the spectrum from commodity products. Specialty products are developed and offered to a small group of customers or small market segment. Since specialty products are offered to a more narrowly defined market niche, they can satisfy the needs of target market customers more effectively but typically at higher prices. Specialty product producers are largely niche marketers producing such products as custom kitchen cabinets, custom millwork, and others. Producers of specialty products are, most times, smaller and more flexible than large-scale commodity product producers. With specialty products, the competition is based more on the total product and much less on price, allowing for potentially higher margins.

Commodity grades of structural panels inventoried at a home center.

Differentiated Products

Differentiated products lie between specialty products and commodity products on our spectrum of product types. Differentiated products are produced with differences or variations in order to satisfy different market segments, but they are not marketed to as narrowly a defined niche as specialty products. Product differentiation may not take the form of actual product modifications but can relate to such factors as quality, credit offerings, service, and others (Beucler and Bowyer, 1987).

When differentiated products become very successful in the marketplace, they can become commodity products. CCA-treated radius edge decking and PET studs are examples of former differentiated products which are now considered commodity products.

Management Implications

Most forest products corporations are developing a strong emphasis on products in the differentiated product spectrum. They are clearly moving away from a commodity product orientation. These firms hope to capture the higher margins of the differentiated and specialty products and to avoid the debilitating price competition of the commodity markets.

Theodore Levitt (1986) had the following to say about commodity products:

> There is no such thing as a commodity. All goods and services can be differentiated and usually are. The only exception to this proposition is in the minds of the people who profess that exception.

PRODUCT BRANDING

A brand can be a name, term, symbol, design, or combination of these which is used to identify a product. A brand name is a name or group of letters used to identify a product. A

Differentiated structural panels cut to a convenient size for sales to do-it-yourself homeowners.

trademark is a symbol legally registered for the use of only one company. This symbol can be a graphic design or a word or group of letters written in a certain style.

Branding began in the Middle Ages as craftworkers put their marks on the goods they produced. This allowed consumers to trace faulty products back to the producer or to identify good products for purchase. Brands are now used mainly for identification to help consumers recognize a specific company's product.

In addition to being an aid to identification, brands also assist the marketing process in a number of ways. Brands serve as a focal point for advertising. They can help reduce price comparisons because of the difficulty in comparing one brand to another. As consumers become accustomed to the quality and consistency of a good brand, it saves them time and reduces their risk when shopping. This makes marketing easier and encourages repeat sales. Although more apparent on consumer goods, brands are heavily used on industrial products also.

Selecting a good brand name is very difficult. Stanton (1978) gives the following characteristics of a good brand:

1 It should suggest something about the product's characteristics, benefits, use or action.
2 It should be easy to pronounce, spell or remember.
3 It should be distinctive.
4 It should be versatile enough to be used on new products added to the product line.
5 It should be capable of being registered and legally protected.

In the forest products industry, paper producers have made better use of brands than have wood product producers. Many basic wood products are not branded, although some brands such as Potlatch's Oxboard brand of OSB have been successful. As you examine the Oxboard advertisement in Fig. 4-2, try to pick out how many of the characteristics of a good brand it possesses.

Product Packaging

Coupled with branding, product packaging or, as it is sometimes called in commodity wood products, presentation, is growing in importance to forest products marketing. Packaging serves two main purposes: (1) to protect the product from dirt, moisture, and sometimes pilferage, and (2) to improve the product's appearance while keeping the firm's brand prominent. For consumer products such as interior strip paneling and others, packaging has long been important. However, as export markets have grown, careful packaging/presentation has become more important in commodity products such as lumber and plywood.

PRODUCT LIFE CYCLE

The product life cycle represents the sales history of a product, product class, or even an entire industry. Figure 4-3 shows it as a series of four stages: introduction, growth, maturity, and decline. The biological analogy of the product life-cycle concept makes it easy to grasp. The concept provides a useful basis for classification of products and

FIGURE 4-2
Advertisement for Potlatch's Oxboard.

strategy development. Some authors have gone so far as to state that a company's marketing success often depends on its ability to understand and manage the life cycles of its products (Stanton, 1978). According to others, it suffers from a lack of supporting research to validate the concept.

Most commentators believe the product life cycle to be a useful concept which allows us to better structure our understanding of the interplay of market, technological, and competitive forces as a basis for forecasting and strategy development (Day, 1986).

The life cycle can take many shapes in the real world, but Fig. 4-3 gives the classic shape and four stages. This shape is defined by the cash flow or sales on the y axis and time on the x axis. Profits generally grow with sales, but as a result of increased competition in the mature stage profits tend to drop before sales and profit margins narrow considerably for mature or declining products.

Uses of the Product Life Cycle

Product life cycles can be of some interest for product planning for a number of reasons. In recent years, product lives, particularly those of consumer products, have tended to shorten. This has placed an increasing emphasis on and need for better planning. Also, new products are tending to require increased amounts of capital. Further, the recognition of the approximate position of a product along the product life cycle permits better strategic planning concerning advertising, continued product development, distributor support, and increased capacity additions.

The concept of a product life cycle suggests that changes in the mix of products produced by a firm can be made to achieve a balance of new, growing, and mature products. If the product life cycle for a product or a product line can be conceptualized,

FIGURE 4-3
The product life cycle.

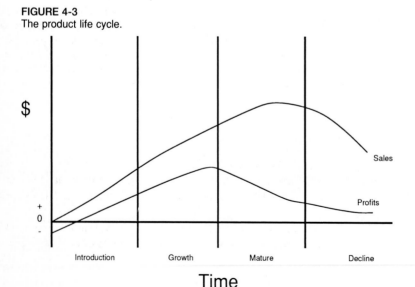

perhaps steps can be taken to facilitate acceleration through the development or introductory stage and to prolong the growth and maturity stages. The important thing is to recognize the current stage, then anticipate the next stages and prepare to react to them in a timely fashion.

Many forest products firms let other firms develop new products, and then place similar competing products into the market during the growth stage. Other firms introduce products during the growth phase with significant innovations and they may benefit from higher pricing and better margins in this stage. Some examples of this phenomenon can be found in the waferboard/OSB market. Blandin Wood Products of Grand Rapids, Minnesota, and Elmendorf of Clairemont, New Hampshire, were the first firms to successfully produce waferboard and oriented strandboard, respectively, in the United States. Neither of these firms was considered a big player in the U.S. forest products industry. However, after the market was established Louisiana-Pacific clearly identified this as a market and product to bet the company on. It constructed many plants to produce oriented waferboard early in the growth stage of this product category. Later, it switched to an improved oriented strandboard utilizing an isocyanate resin and at the time of this writing was clearly the dominant producer in North America.

Types of Life Cycles

The life-cycle analogy can be developed for a wide variety of planning needs. For example, life cycles can be developed based upon the sales growth for an entire industry. If one were to plot the growth of the forest products industry in terms of the number of units of products produced, for instance, it would likely be in a mature stage. On the other hand, the microcomputer industry is still clearly in a growth stage.

Life cycles can also be developed for a segment of an industry or a broad product line. The growth of do-it-yourself building products clearly puts it in the growth stage of its life cycle while softwood constructional lumber is in the mature stage. Life cycles can also be developed for a single product or product line. This is the most common textbook application of a product life cycle.

Length of the Product Life Cycle

The lengths of the various stages of a life cycle can be difficult to predict. Fortunately, it's most important just to recognize the stage a product is in and to maximize the advantages and minimize the disadvantages during each stage (Beucler and Bowyer, 1987).

The rate of technical change, the rate of market acceptance, the ease of competitive entry, and other factors influence the overall length of a product life cycle. Some evidence supports the notion that product life cycles are becoming shorter.

The progression through a product life cycle may be altered and even reversed due to changes in marketing strategy, expansion into new markets, new technologies (especially those that drastically lower costs and/or increase quality), changing consumer values, and other factors.

Product life-cycle patterns for wood products can be greatly affected by the following:

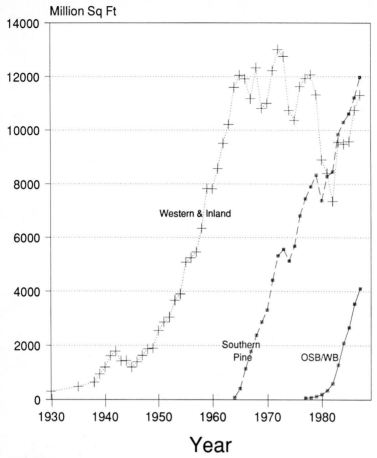

FIGURE 4-4
Product life cycles of wood-based structural panels.

- Periods of strong recession or economic growth
- Changing customer tastes and preferences
- Currency exchange rates
- Housing starts and existing home sales
- Technological innovations
- Interest rates
- Disposable income levels

Product Life Cycle for Structural Panels

Product life cycles can be illustrated with an example. Figure 4-4 shows a family of life cycles for wood-based structural panels. In a broad sense, each of these panel types is in the same broad product/market and can be viewed as a substitute. The introduction stage for the wood-based structural panel was dominated by western plywood and ran for many years ending around the time of World War II. Western plywood then enjoyed tremendous growth until the mid-1960s. An innovation in adhesives triggered by more costly timber caused western plywood to enter the maturity stage. New technologies in adhesives allowed southern pine plywood to be produced. Southern pine plywood skipped the introduction stage, already pioneered by western plywood, and grew very rapidly until the early 1980s. Then another technological innovation became economically feasible due to high peeler log prices and oriented strandboard/waferboard emerged. It took several years of introduction, but this product moved into the growth stage in the early 1980s.

Obviously these trends are much clearer to us now in hindsight than they were to folks trying to predict the future 30 years ago. However, imagine the immense benefit to a firm if it understands the life cycles of its products and the factors that can potentially influence those cycles, and can adjust its strategy to take advantage of this knowledge.

PRODUCT POSITIONING

Positioning a product appropriately is a major element of marketing success. Product positioning is the process by which the manufacturer/wholesaler/retailer creates an image of a product in relationship to competitive products. For example, Volvo cars have the image of safety and durability. Figure 4-5 shows a Louisiana-Pacific advertisement

FIGURE 4-5
Louisiana-Pacific product positioning advertisement for its Inner-Seal brand OSB.

which was designed to position its Inner-Seal panel as a less expensive alternative to plywood. The image to be created here is one of "It's just as good as plywood, but more economical." Many natural and recycled products are positioned based on their environmental appeal. Some forest products firms provide a guarantee of sorts that their products come from sustained-yield forests. Several paper producers clearly position their recycled paper product lines for environmental appeal.

In order to evaluate the position of a given product, it is necessary to understand how customers perceive the product vis-à-vis its competitors. Too frequently, assessments of strengths and weaknesses of competing products are limited to tangible characteristics such as price and physical attributes, disregarding the intangibles such as customer perceptions and attitudes (Dickson, 1974). Customer perceptions do not always correspond to what manufacturers believe about their own products, yet it is precisely these perceptions which determine success in the marketplace. Perceptions result in beliefs about products which combine as the basis for an attitude, which directs buying behavior.

Perceptions also influence public opinions and the legislative process. For example, don't forget that the public's perception of the profession of forestry, especially its perception of the profession as a good steward of our natural resources, can influence the role forestry will be allowed to have in future natural resource management.

FIGURE 4-6*a* and *b*
Professional builder/contractor perceptions of siding products portrayed on four dimensions.

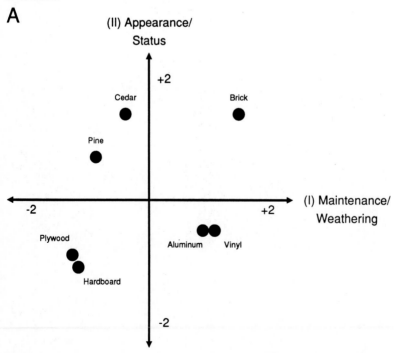

Perceptual Mapping

Analyzing these images or perceptions in an objective manner to facilitate informed decision making can be difficult. A technique is available to solve this problem—perceptual mapping. Perceptual mapping results in a geometric representation of how competing products are perceived. It has been used to assist in the development of product positioning strategies, as a tool to generate new product ideas and to predict which products customers will regard as substitutes, and in promotional strategy decisions.

Residential Siding: An Example   The following perceptual maps are for residential siding products (Fig. 4-6*a* and *b*). They were developed through a nationwide survey of residential contractors. Let's examine these perceptual maps of siding products to better understand their potential usefulness to marketing managers.

The distance between two products may be loosely interpreted as a measure of perceived substitutability of each product for any other. Vinyl and aluminum in Fig. 4-6*a* are clearly substitutes. But in Fig. 4-6*b* aluminum is perceived as performing poorly in dent resistance, giving vinyl an advantage. Similarly, Fig. 4-6*a* shows plywood and hardboard to be substitutes while Fig. 4-6*b* shows plywood in an advantageous position compared to hardboard.

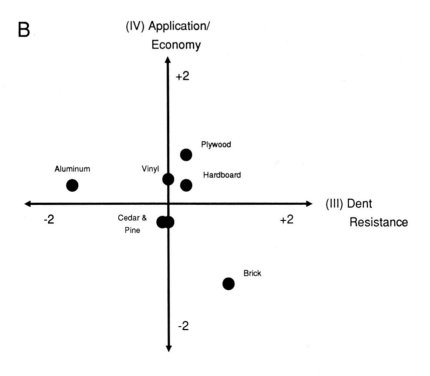

Wood Composites   Hardboard, plywood, vinyl, and aluminum occupy the negative half of the appearance/status dimension (Fig. 4-6*a*) and the positive half of the application economy dimension (Fig. 4-6*b*). Builders feel these products are economical to apply but lack a status image, and thus are most competitive in the lower home-price ranges.

What is the competitive advantage of the wood composite products? The competitive strengths of the wood composites appear to be application economy and dent resistance (Fig. 4-6*b*). Hardboard's only perceived advantage over aluminum is dent resistance. Plywood shares hardboard's dent resistance advantage and is additionally perceived as being faster and easier to apply. Customers placing a higher value on application economy than the other dimensions should prefer plywood.

Solid Wood   The solid wood sidings fell together into their own niche (Fig. 4-6*a*). Their competitive strength (and only positive score) rests on the most intangible dimension of the four: beautiful appearance and a high-status image. Cedar sidings rated the highest of all seven products in this dimension. Consumers also perceived the cedar sidings to be easier to maintain and more weather-resistant than pine.

Brick is the lone occupant of the most desirable quadrant (Fig. 4-6*a*). Brick is perceived as having a beautiful appearance/high-status image, yet is easy to maintain and weather-resistant. Brick's disadvantage is its low position on the application economy dimension (Fig. 4-6*b*).

Solid wood sidings, particularly cedar, appear to hold a niche for those seeking a high-status, quality image over easy maintenance, and remain competitive against brick because of brick's weak rating on application economy.

While vinyl's position did not indicate much of a threat to solid wood in these maps, the competitive situation is far from stable. As products are repositioned through product changes or promotional campaigns, the threat of substitution changes. Products repositioned toward the upper right quadrant of Fig. 4-6*a* should enjoy increased competitive advantages and increased market shares. If vinyl manufacturers successfully repositioned vinyl in the appearance/status dimension, vinyl's threat to solid wood, particularly pine, would increase.

PRODUCT DELETION

For most companies, product deletion is a difficult task. It can appear to be admitting failure. But for most firms there are products that are losers. A periodic product review should critically evaluate whether or not to retain such products. Deleting a product need not be viewed as a failure of the product or the strategy behind it but rather as a natural evolution in the business cycle.

When considering products for deletion, caution is advised. Even though a product may not be profitable by itself, it may be desperately needed to round out a product line. Many times a complete product line is required for a successful sales program. Deletion of a particular product may result in a competitor having the only comparable product. This may put the firm at a competitive disadvantage. It should also be determined whether or not the poor sales performance of the product is due to the product itself or poor marketing efforts on behalf of the product (Beucler and Bowyer, 1987).

NEW PRODUCT MANAGEMENT

The development of new products and the process by which it is done has been popularized in business literature. Several interesting accounts were published by Peters and Waterman (1982) in *In Search of Excellence*. Additionally, academic textbooks have been written to cover this field with one of the better ones being *Essentials of New Product Management* by Urban, Hauser, and Dholakia (1987). This section will only touch on the highlights of developing and managing new products. Readers with more interest in the subject should consult the books previously mentioned.

Many alternative organizational forms exist to facilitate new product development and management. These range from the famous skunk works at Lockheed Aircraft producing secret military jets to new product departments, product managers, new product committees, and new product task forces. One thing that seems to be important organizationally is that one individual be responsible for the new product, i.e., a product champion. However, it is also important that the broad variety of individuals involved in the product be concerned about it and take ownership in the new product development.

What is a New Product?

New products can be divided into three categories (Stanton, 1978):

1 Products which are truly innovative and unique. A good example here is plywood. When the first plywood was manufactured after the turn of the century, there was really nothing quite like it.

2 Replacements for an existing product which involve a significant differentiation. OSB has replaced plywood in many applications. It is used like plywood, but it is different.

3 Imitative products which are new to a company, but are not new to the market. This is a me-too category. After OSB enjoyed a degree of market acceptance, a number of firms added OSB to their product mix.

As with most marketing, the key is customer perception. If customers perceive a product as new, then it is new as far as they are concerned. Another thought: most people resist change and the truly innovative products can be tough to sell. Look at the long introduction stage for plywood shown in Fig. 4-4. Conversely, imitative or evolutionary products which retain a degree of familiarity may be more easily accepted in the market.

The Importance of New Products

New products are necessary and important for a wide variety of reasons, only some of which we will discuss here. New products are needed to develop a competitive advantage over other firms in the marketplace, to protect a market niche, and to contribute to growth in sales and profitability. And growth may be necessary to attract and retain good employees. Excess plant capacity may require the development of new products. A competitor's product may require the development of a new product in order to remain competitive. When a given product becomes obsolete or deficient, a new product may be required to replace it. In seasonal businesses, new products may be required to even out production schedules and sales during the year. Other firms may need to develop new products in order to differentiate their product lines and to supply total packages to their customers.

In the forest products industry, perhaps the dominant reason behind new product development has been changes in the timber resource base. As certain timber species have become too costly or had declines in available quality, the industry has innovated to solve the problem. As western peeler logs became too expensive, southern pine logs were used for plywood. As softwood pulpwood became scarce and costly, softwood sawmill waste and hardwood pulpwood was utilized.

New Product Planning

The new product planning process is illustrated in Fig. 4-7. It can be useful to divide the process into a series of stages and hurdles for the new product to pass through. Let's examine these stages.

Development of New Product Ideas Various estimates are available that indicate how many new product ideas must be generated for every successful new product. Estimates range all the way up to 100 new product ideas generated per successful new product. There are many sources of new product ideas (Beucler and Bowyer, 1987):

   Internal marketing and sales force
   New product committee
   Customers directly, or through information gained by sales reps or wholesalers
   Competitor product development
   Trade shows
   Equipment suppliers
   Brainstorming sessions
   Internal research and development laboratory or engineering development group

It is sometimes worthwhile to remember that evolutionary changes in existing products can create new products. An example of this evolutionary process includes high-quality hardwood boards for do-it-yourselfers in home centers. Hardwood boards had been available for a number of years; however, the "new product" was clear on both faces, kiln-dried, surfaced, edged, and end-trimmed, then displayed in attractive racks. This evolutionary but new product created strong interest and a much higher demand for hardwood boards.

Other times, factors which are not part of the physical product can be changed to create product innovations and new products. Nonphysical factors which can be changed include credit, place of sale, delivery, technical advice, and service. For example, one of the more innovative ideas to come along in years involved the Weyerhaeuser computer design center in home improvement centers. This computer-aided design package allowed the individual consumer to design his or her own deck or other home improvements on the computer. This gave flexibility in designing a project to meet a

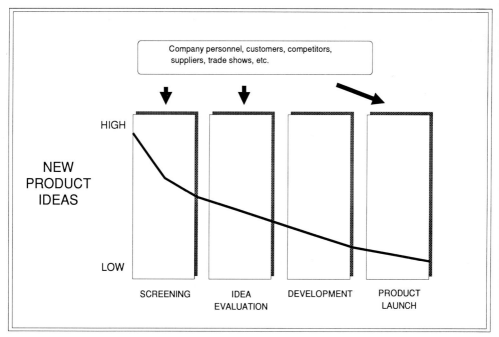

Company personnel, customers, competitors, suppliers, trade shows, etc.

HIGH

NEW PRODUCT IDEAS

LOW

SCREENING   IDEA EVALUATION   DEVELOPMENT   PRODUCT LAUNCH

FIGURE 4-7
Stages of new product development. (*Adapted from McCarthy and Perreault, 1987.*)

certain budget, allowed a visual display of the design on the screen, and provided a blueprint and list of required materials.

We have already discussed the importance of generating new product ideas. But it's good to remember that this is an absolutely critical first step in new product development. Industrial marketers should pay close attention to their customers, because research has shown that they are the source of many new ideas in industrial products (McCarthy and Perreault, 1987).

Screening   New product ideas should be screened for their consistency with company objectives. They must be compatible with the company's technical, financial, and managerial resources. The idea should match company skills and resources. A good idea should provide the company with a competitive advantage.

Safety and potential product liability must also be considered at this stage. A firm practicing the real marketing concept will have no problem with safety. It will simply be of utmost concern. An unsafe product will not long stand the glare of today's marketplace and will hurt the firm in the long run. Unsafe products or products which fail to perform can create tremendous liability problems for companies and careful attention needs to be given to this concern early in the new product process.

Of course, the product must be screened financially also. New products are really investments of significant company resources and they should stand up to an evaluation of their expected returns on investment.

Idea Evaluation   After an idea gets by the screening process it needs to be more carefully evaluated. This evaluation can take many forms. For many firms the customer is playing a larger and larger role. Firms are integrating their customers into the product design and evaluation process through the use of focus groups, advisory councils, and other mechanisms. Some companies use their competitors to evaluate products by allowing them to be first with new products. They use a me-too strategy which lets their competitors take the first steps, then they themselves enter the market with basically the same product or one with enhancements. An alternative strategy is to offer a product that is complementary to the competitor's product.

Of course, in the evaluation process it is important to get the proper feedback from customers and the firm's own engineering/design staff, as well as continuing financial analysis, to determine if the idea should proceed to development.

Development   Development involves taking what is already known about the idea, gathering more information if needed, then turning the idea into a prototype product. Then it's usually back to the customer with the prototype product to get more feedback and reactions. There is still time to make changes to better meet customer needs.

If all signs are still positive, market testing may be done. Batches of new products can be selectively distributed in specific areas to gauge customer reactions and pinpoint any remaining problems. For industrial products, a market test might consist of providing sample products to key customers for their evaluation.

The decision to test-market the product is not an easy one. Test marketing does cause time delays and allows your competition to better understand what you are doing. Further, you incur the cost of the test. However, market testing is becoming increasingly important, especially when the product is priced higher than comparable competitor products, if it is a significant departure from an existing product, or if startup costs are high (Beucler and Bowyer, 1987).

Product Launch   If the product idea survives this far, it may be put on the market. Final decisions must be made about the exact features of the product and all parts of the marketing mix must be considered to develop a marketing plan for the product. And not least, the financial returns must be carefully analyzed.

In consumer product companies, a product launch frequently involves much advertising and marketing fanfare. Often, companies will ''roll out'' the product in successive markets. This allows production to build gradually and allows some success in early markets.

For an industrial product, the procedure can be similar but is usually much quieter with limited fanfare. Initial product launch will typically be with the key customers that had input into the product's development.

Cannibalization

In the forest products industry, where many new products are evolutionary, another factor should be considered when developing estimates of new product sales and their impact on company performance. New product sales are usually thought to be from new customers or expanded markets, or gained at the expense of a competitor. But many times, especially for evolutionary products, the new product takes sales away from the company's own existing products. For example, OSB took sales from plywood. This is called "cannibalization." Cannibalization must be considered in the planning process and its potential to occur factored into the product decision.

Why New Products Fail

It has been estimated that eight out of ten new products fail (Swasy, 1990). Products fail for a broad variety of reasons and, in one study of 125 firms, the following reasons were given in order of importance (Hopkins and Bailey, 1971):

1 Inadequate market analysis
2 Product deficiencies
3 Lack of effective marketing effort
4 Higher costs than anticipated
5 Competitive strength or competitors' reaction
6 Poor timing of introduction
7 Technical or production problems

Interestingly, most factors which led to product failure were controllable by the companies and the majority were marketing-related rather than technical.

To avoid these problems in new product development, several things need to be done. Better market research can help reduce the uncertainties of the marketplace and enhance the opportunities for a new product's success. Better and more realistic estimates of potential customers and the likely actions of competitors can help also. Finally, more realistic estimates of costs and pricing can help ensure new product success.

One of the more publicized product failures for a forest products company occurred for Weyerhaeuser's new diaper brand, UltraSofts. Test marketing and market research showed consumer acceptance was good if not great (Swasy, 1990). But a host of problems killed the product. The diaper was positioned at the high end of the market against Procter & Gamble and Kimberly-Clark products. It was introduced just before the two main competitors launched aggressive national advertising campaigns as they rolled out newly improved diapers with discount coupons. If this weren't bad enough, big problems surfaced in the manufacturing of UltraSofts and when Weyerhaeuser couldn't give longer-term contracts to suppliers their supplies of raw materials were jeopardized (Swasy, 1990). This, of course, resulted in higher-than-planned-for costs and the price of the new diaper was raised 22 percent, killing retailer interest. In this case, underestimating the competition, poor supply planning, and manufacturing bugs combined to doom a good product to failure.

New Product Acceptance

Consumers accept new products for a variety of reasons, some of which are understood well and some of which are not. However, it is generally believed that consumers accept products better if they perceive them as better than other products in terms of price, quality, or some other critical factor. Evolutionary new products can be accepted quickly because they are perceived as being compatible with current needs and past product experiences. If there is greater ease in understanding and using a product, it is more likely to be accepted. Greater customer acceptance can be developed for a new product if free samples are given out, money back guarantees are provided, or initial purchases are reasonably priced.

SUMMARY

Developing the right mix of products and services is absolutely critical to a firm—so critical that most decisions about products and services are made in corporate offices by executives and top managers. Most firms try to differentiate their products from other firms'. This can be done in many ways other than changing the physical product. Products have a life cycle which, if conceptualized properly, can have management implications and strategic benefits. Product positioning can be a powerful tool when used to position a product correctly in the marketplace. It is a tool which has been successfully used by forest products firms. New product development and management is the lifeblood of many firms. Most forest products firms could do a better job in new product development.

Once you have the right products, with the correct positioning strategy, you still must get the goods to the customer. Chapter 5 helps explain that process.

BIBLIOGRAPHY

Beucler, O., and J. L. Bowyer. 1987. *Marketing of Manufactured Wood Products.* Seminar Notebook #1. University of Minnesota, Department of Forest Products. St. Paul, MN.

Busch, P. S., and M. J. Houston. 1985. *Marketing/Strategic Foundations.* Irwin. Homewood, IL.

Copper, R. G. 1975. "Why New Products Fail." *Industrial Marketing Management* 4(6):315–326.

Day, G. S. 1986. *Analysis for Strategic Market Decisions.* West Publishing. St. Paul, MN.

Day, G. S., A. D. Shocker, and R. K. Srivastava. 1979. "Customer-Oriented Approaches to Identifying Products-Marketing." *Journal of Marketing* 43(4):8–19.

Dickson, J. 1974. "Use of Semantic Differentials in Developing Products Marketing Strategies." *Forest Products Journal* 24(6):12–16.

Gavish, B., D. Horsky, and K. Srikanth. 1983. "An Approach to the Optimal Positioning of a New Product." *Management Science* 29(11):1277–1297.

Hauser, J. R., and F. S. Koppelman. 1979. "Alternative Perceptual Mapping Techniques: Relative Accuracy and Usefulness." *Journal of Marketing Research* 16(November):494–506.

Hooley, G. J. 1979. "Perceptual Mapping for Product Positioning: A Comparison of Two Approaches." *European Research* (January):17–23.

Hopkins, D. S., and E. L. Bailey. 1971. "New Product Pressures." *Conference Board Record* (June):16–24.

Hutt, M. D., and T. W. Speh. 1981. *Industrial Marketing Management.* Dryden Press. Chicago, IL.

Jacobson, R., and D. A. Aaker. 1987. "The Strategic Role of Product Quality." *Journal of Marketing* 51(October):31–44.

Levitt, T. 1986. *The Marketing Imagination.* The Free Press. New York.

Markell, S. J., T. H. Strickland, and S. E. Neeley. 1988. "Explaining Profitability: Dispelling the Market Share Fog." *Journal of Business Research* 16:189–196.

McCarthy, E. J., and W. D. Perreault, Jr. 1987. *Basic Marketing.* 9th edition. Irwin. Homewood, IL.

Peters, T. J., and R. H. Waterman. 1982. *In Search of Excellence.* Harper & Row Publishers. New York.

Ray, M. L. 1982. *Advertising and Communication Management.* Prentice-Hall, Inc. Englewood Cliffs, NJ.

Rumelt, R. 1981. "In Search of the Market Share Effect." *Proceedings of the Academy of Management.* pages 2–6.

Shocker, A. D., and V. Srinivasan. 1974. "A Consumer-Based Methodology for the Identification of New Product Ideas." *Management Science* 20(6):921–937.

Stalling, E. C., and S. A. Sinclair. 1989. "The Competitive Position of Wood as a Residential Siding Material." *Forest Products Journal* 39(4):8–14.

Stanton, W. J. 1978. *Fundamentals of Marketing.* McGraw-Hill Book Company. New York.

Swasy, A. 1990. "Diaper's Failure Shows How Poor Plans, Unexpected Woes Can Kill New Products." *Wall Street Journal* (October 9):B-1, B-3.

Urban, G. L., J. R. Hauser, and N. Dholakia. 1987. *Essentials of New Product Management.* Prentice-Hall, Inc. Englewood Cliffs, NJ.

CHAPTER 5

DISTRIBUTION

If farms and factories are the heart of industrial America, distribution networks are its circulatory system.

E. Raymond Corey et al. (1989)

For many folks who watch television and read printed advertisements, the "middleman" has been pictured as virtually an enemy of the consumer. These folks are often unfairly portrayed as ripoff artists who occupy dark back rooms filled with cigar smoke. The mental image which is normally conjured up is not a pleasant or accurate one. These people and their institutions perform a very valuable service to the American consumer by improving the efficiencies of moving goods and services from producers to consumers.

These middlemen or if you will, channel intermediaries, accomplish this improved efficiency by specializing in their jobs. Established intermediaries may have years of experience and personal contacts which enable them to do their jobs with great expertise. Intermediaries also reduce the number of contacts required for a manufacturer in the marketing process. Consider, for example, a producer of softwood lumber. If every person who needed a ¾-inch-thick board or 2×4 came directly to the sawmill to purchase it, confusion and chaos would be the ultimate result. Most sawmills and their employees are just not equipped either mentally or physically to serve the dual functions of sawmill and retail yard in a very efficient manner. The intermediaries in this case can handle the retail task much more efficiently, which ultimately results in lower prices for consumers.

But just what do these folks do for us? They serve as the vital link between the manufacturer and the ultimate buyers of the products. They work in the channels of distribution, servicing the manufacturer and the ultimate buyers. They provide numerous services including maintaining contact with the final buyers, negotiating on delivery and price terms, establishing contractual agreements and arrangements, transferring titles, providing credit or financial arrangements, servicing products, providing local inventory, taking care of transportation needs, and many times arranging for storage of the goods. All of these are functions which must be performed. It's simply a matter of how best to do them.

We previously said intermediaries work in the channels of distribution. Now, just what is a channel of distribution or marketing channel? Some have defined a marketing channel as an interorganizational system made up of a set of interdependent institutions and agencies involved with the task of moving things of value (ideas, products, services) from points of conception, extraction, or production to points of consumption (Stern and El-Ansary, 1977). Now that's a mouthful! A simpler way is to imagine a pipeline full of

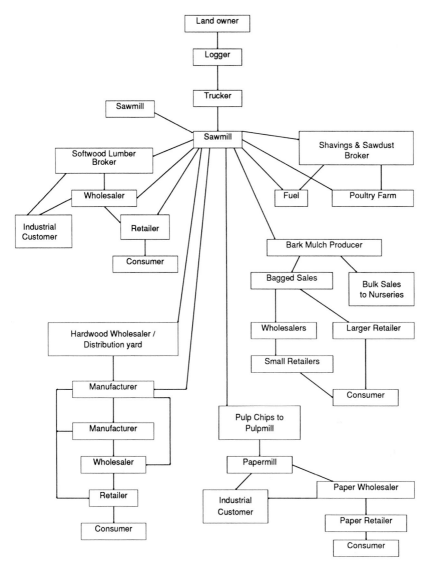

FIGURE 5-1
Channels of distribution for sawmill products.

goods and services moving to the final consumer. This pipeline is not a straight line but has lots of twists and storage areas along the way.

Perhaps at this point a picture can explain more easily what we are talking about. Figure 5-1 shows the channels of distribution that a tree may take as it is harvested and moves through the various processing channels during its conversion at a sawmill to lumber and other useful products. You can see that there are many channels through which lumber and the various sawmill by-products move on their way to the final consumer. The sawmiller would have a difficult time indeed maintaining the necessary expertise to efficiently serve all of these various consumers.

JUST WHO ARE THESE INTERMEDIARIES?

Agents

The various channel intermediaries can be subdivided into two rather broad categories: agents and merchants (Fig. 5-2). Agents typically take no title to the goods they are handling but rather concentrate on negotiating purchases or sales for another party, typically for a commission. Two common types of agents in the forest products industry are brokers and manufacturer's representatives. A broker takes no title to the actual goods but does negotiate purchases or sales between the producer and seller for a commission. Brokers are typically used when handling large quantities of bulk goods.

Another type of agent is the manufacturer's representative, sometimes known in the trade as a "manufacturer's rep" or just a "rep." This person is an independent salesperson usually knowledgeable about the products he or she sells and his or her customers. Salespeople typically handle complementary goods from several manufacturers and operate under contractual arrangements in a specific geographic area. Many furniture producers use manufacturer's reps.

FIGURE 5-2
Types of channel intermediaries.

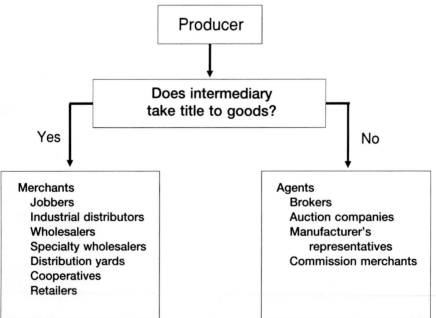

Merchants

Merchants, on the other hand, typically do take title to the products they are selling. One type of merchant, the wholesaler, buys, inventories, and resells goods to other wholesalers, manufacturers, or retailers. Many wood products wholesalers may do some dry kilning or remanufacturing of products also. These wholesalers do a very valuable service in sometimes buying products by the railcar load and then breaking down the materials purchased into smaller amounts for lumber dealers and home improvement centers. For example, the latest lumber-carrying railroad cars hold enough lumber to build eight houses, which is considerably more than a small retailer wishes to stock at any one time.

Another type of merchant is the industrial distributor. Distributors typically sell equipment, parts, and electrical, plumbing, and other supplies. This type of merchant is not prevalent in the distribution of wood products; however, distributors play a strong role in the distribution of maintenance items for many wood products manufacturers.

A jobber is another type of merchant. Jobbers work differently than traditional wholesalers. Jobbers normally obtain orders from customers and then purchase goods from manufacturers to meet those orders. The goods are then shipped directly from the manufacturers to the customers. Jobbers do not actually handle or inventory any goods. Usually they sell bulk loads of lumber, coal, chemicals, and other products.

The cooperative or buying service is yet another form of merchant. In this case a number of companies cooperatively buy goods and services to get a better price or sometimes cooperatively sell goods for a better price through larger quantities. There are numerous cooperatives in the wood products area.

A final form of merchant is the retailer. We are all familiar with the function of retailers. They typically inventory and take title to goods for ultimate resale to the final customers. More detail on channel intermediaries for specific forest products and markets is given later in Chapters 11, 12, 13, and 14.

CHANNEL STRATEGIES

Channel strategy must ultimately be consistent with the firm's marketing objectives and be used to reach the firm's target markets. The channel strategy selected by a firm can play a large role in the firm's success. In fact, in today's economy individual producers do not compete against each other as much as the entire channel of distribution for each producer competes. The entire channel has become a single competitive unit.

For example, Georgia-Pacific is the dominant firm in many areas of the United States in structural panels (especially plywood) not so much for its production capacity, but because it dominates the distribution channels as the largest building products distributor. On a smaller scale, Weyerhaeuser with its ChoiceWood product line maintains a captive company warehouse with enough inventory to meet its promised 10-day delivery anywhere in the United States. For retailers stocking the kiln-dried hardwood boards of the ChoiceWood product line, 10-day delivery is a real advantage in managing their inventory and, of course, this gives Weyerhaeuser an advantage over the competition in this market segment.

Importance of on-time distribution as a competitive strategy.

Channel Decisions

There are various channel strategy decisions which must be made. Perhaps the most obvious one is whether to sell the product directly to the consumer or to choose to use channel intermediaries. Clearly the direct-sale route offers enhanced control and the opportunity for potential profits. However, it clearly requires a different expertise than that used in manufacturing the product and may also require substantial investment of management time and financial resources by the firm. The alternative strategy, of course, is to use channel intermediaries. This strategy clearly offers fewer problems, requires less capital investment, and generates fewer management headaches. But it does offer much less control and the opportunity for enhanced profit is gone.

Another decision concerns the intensity of the distribution of the product. This can basically be described by noting the two extremes. One extreme is an exclusive distribution of the product, thereby limiting its distribution to a very few select distributors or retailers. Two examples of this are Curtis Mathis Electronics and Ethan Allen Furniture. For products of both companies you must go to an exclusive retail outlet which sells only the products of that firm and no others; in other words, you can buy Ethan Allen Furniture or Curtis Mathis products only at one of their retail stores. The other extreme is the intensive distribution of products. Intensive distribution involves distributing the product to as wide an array of retailers as possible. Kleenex tissue and Coca-Cola are two examples of products with intensive distribution.

A third decision revolves around the concept of channel ownership. A firm must decide whether it wishes to use independent and autonomous channels of distribution or whether it wishes to be vertically integrated. The conventional independent and autonomous channel serves members which have collectively agreed to cooperate on the basis of mutual need; however, there is typically no formal coordination or contract. On the other hand, a vertically integrated marketing system achieves coordination through the ownership or franchising of channel members. In regard to the previous example, both Ethan Allen and Curtis Mathis have control over their vertically integrated channel through a form of ownership or franchise. Many major wood products producers have become very vertically integrated from timberland ownership through log processing and up to the point of maintaining their own wholesale distribution centers.

Choosing the Right Channel

Choosing the right channel strategy has never been easy and many firms have experimented with various channel strategies over time. For example, Boise Cascade at one point owned numerous wholesale distribution centers and building products retailers in the western states, providing them with a nearly totally owned channel of distribution from the forests to the final consumer. However, in the late 1980s the company disbanded this system and centered its corporate efforts on more efficient manufacturing of wood products.

Figure 5-3 provides thumbnail decision rules for channel strategies. As can be seen, most producers of commodity wood products are advised to use a longer channel (i.e., intermediaries). Commodity products are ultimately sold to many customers over a wide geographic area and the product is simple and requires minimal product quality maintenance after shipment. Specialty products, on the other hand, might be better marketed through shorter channels, even going directly to the retailer.

DISTRIBUTION TRENDS

The distribution side of the forest products industry is changing rapidly. Many industrial buyers and retailers are moving toward less and less inventory. The expectation is that the distributor/wholesaler will carry the necessary inventory. Just-in-time systems have been slow in coming to secondary wood products manufacturers, but they are coming. Two-week shipment horizons will no longer be sufficient; rather, deliveries may be timed to the hour.

Many retailers and industrial buyers are reducing the number of their suppliers, but are developing and maintaining a much closer relationship with the remaining suppliers. These suppliers' performance is closely monitored and customer expectations are high. Real-time, direct computer links to suppliers' inventory and ordering systems are becoming common in these relationships to speed order processing and handling.

Bar coding, even of commodity wood building products, has begun. Its benefit to retailers in check out scanning and inventory control is enormous. Manufacturers are also realizing the benefits of bar coding in their own inventory management systems.

Number of potential customers

  *   Small number = short channel
  *   Large number = longer channel

Geographic concentration

  *   Highly concentrated = short channel
  *   Widely dispersed = longer channel

Order size per transaction

  *   Large $ value = short channel
  *   Small $ value = longer channel

Complexity of product

  *   Complex, requiring technical selling and service = short channel
  *   Simple = longer channel

Product quality maintenance

  *   Much control required = short channel
  *   Little or no control required = longer channel

Company resources

  *   If inadequate = longer channel

Intermediaries' availability and capability

  *   If inadequate = short channel

FIGURE 5-3
A summary of factors which influence channel choice. (*Shaw and Semenik, 1985.*)

PHYSICAL DISTRIBUTION

''Physical distribution''[1] is the term which encompasses activities needed to get the right quantity of goods to the right spot at the right time and at the right cost. This aspect of marketing has been estimated to represent 20 to 30 percent of the total cost of goods and employs about 14 percent of the U.S. labor force (Shaw and Semenik, 1985).

Each component of the channels of distribution is part of the physical distribution system. As noted earlier, these systems are now used as competitive units in the battle for the marketplace. Two concepts can form the basis for viewing an effective distribution system (Shaw and Semenik, 1985):

1 Total package of values concept
2 Total cost concept

Customers buy more than basic products. They also buy an expectation of a level of

---

[1]Physical distribution is synonymous with logistics or marketing logistics.

customer service as part of a total package of values. Customer service level is typically defined as the percentage of orders shipped in a specified time. Some firms even advertise and guarantee a specified service level, such as guaranteed receipt of orders in 10 days. However, the package of values concept is more than shipping. It is the value customers place on other things such as:

- Action on complaints
- Returns policy
- Accuracy of filling orders
- Ordering methods, 800 phone numbers, etc.
- Dealer lead time
- Frequency of stock outs
- Minimum size orders
- Mixed product orders

In order to manage the distribution system, a marketing manager needs to know how customers value the items above. For example, Caterpillar Company promises spare part needs will be met in hours, not weeks, even if it means flying a part into a remote site. Downtime on an expensive machine is costly and CAT customers place a high value on speedy delivery of spare parts.

The total cost concept focuses on the cost of the entire distribution system and serves to maximize the effectiveness of the whole rather than the individual pieces. Absolutely no stock outs maximizes the package of values but typically at an unacceptable cost to most companies. The systems approach serves the need to make the necessary tradeoffs between service and cost to achieve the correct balance.

Transportation Methods

A wide variety of transportation methods is available in the United States. However, because of the bulky nature of most wood products only three main methods are used for their transport: railroads, trucks, and waterways.

Railroads are extensively used for long-distance transportation of wood products. They are especially useful for transporting products from the west coast or British Columbia production facilities into the midwest or eastern parts of the North American continent. The recent development of new lumber-carrying railcars has made rail transportation even more efficient. Typically, railroads are considerably less expensive than truck transport on the longer routes.

Truck transportation can be cheaper than railroads on shorter hauls and many times provides faster delivery than railroads. Packaging is less expensive and simpler with trucks and truck transportation can provide more flexible scheduling. In addition, truck transportation allows products to be shipped to many places which do not have rail access.

Water borne transportation on ships and barges represents the lowest-cost method of moving a ton of products. However, water borne transportation is not as rapid as rail or truck transportation and in some instances may be impacted adversely by weather conditions. The recent use of containerized shipments has made water transportation a more viable option for many wood products firms. For overseas marketing of wood

Softwood lumber shipped by truck.

products, water borne transportation is typically the only option available. International shipping rates can vary widely and domestic producers are well advised to seek the assistance of an expert when shipping materials overseas.

SUMMARY

Moving goods to the market is a critical function of marketing, especially for bulky forest products where transportation costs can represent a significant portion of the final cost to the consumer. A good freight office with capable employees can be a source of competitive advantage.

Transportation costs, as we have said, can be a significant portion of the final costs, but what role should those costs play in setting the final price? Chapter 6 will address the issue of pricing.

BIBLIOGRAPHY

Bingham, F. G., and B. T. Raffield. 1990. *Business to Business Marketing Management.* Irwin. Homewood, IL.

Boyd, H. W., and O. C. Walker. 1990. *Marketing Management: A Strategic Approach.* Irwin. Homewood, IL.

Corey, E. R., F. V. Cespedes, and V. K. Rangan. 1989. *Going to Market: Distribution Systems for Industrial Products.* Harvard Business School Press. Boston, MA.

Hutt, M. D., and T. W. Speh. 1981. *Industrial Marketing Management.* Dryden Press. Chicago, IL.

McCarthy, E. J., and W. D. Perreault, Jr. 1987. *Basic Marketing.* 9th edition. Irwin. Homewood, IL.

Shaw, R. J., and R. J. Semenik. 1985. *Marketing.* 5th edition. Southwestern Publishing Co. Cincinnati, OH.

Stern, L. W., and A. I. El-Ansary. 1977. *Marketing Channels.* Prentice-Hall, Inc. Englewood Cliffs, NJ.

CHAPTER 6

PRICING

There is hardly anything in the world that someone cannot make a little worse and sell a little cheaper—and the people who consider price alone are this man's lawful prey.

John Ruskin

Price is one of the major marketing decision variables. Setting a proper price is critical because it influences how many units are sold and how much money a firm earns.

In simplistic terms, price can be viewed as the amount of money required to purchase a given good or service. But in the real world, price is not so easy to define because it has many aspects. Have you ever ordered something by mail and when the bill came the price was higher than expected because shipping and handling charges were added? What's the price? The amount you thought you were being charged or the total amount which includes shipping and handling? When purchasing 40,000 board feet of lumber for a home center, the effective price should include all the expenses required to get the lumber to the home center, including the lumber price, transportation, insurance, and any taxes. The point is, pricing is not a simple job.

Many factors can affect pricing decisions, including customer reaction to pricing, government actions, wholesaler and retailer needs, the competitive environment, and the costs of developing, manufacturing, distributing, and marketing products. It is important to note that costs are not the sole determinants of prices.

The degree to which customers are sensitive to a change in price is called "price elasticity." If a small change in price will result in a large change in demand, then the price elasticity of demand is said to be elastic. If changes in price have little impact on changes in demand, then the price elasticity of demand is said to be inelastic. Understanding the elasticity of demand for a firm's products has obvious benefits when developing pricing strategies. (We'll discuss this in more detail in Chap. 8.)

The distribution channel for products or services also affects pricing decisions because manufacturers must price their products to allow the members of the distribution channel also to sell the product for a profit.

Companies can influence pricing to varying degrees depending upon their competitive environment. For commodity products sold in a mass-marketing approach, the market environment essentially controls the price. No one firm has much influence over the price of the commodity product. To price a product much above or below the market price in this environment is foolish. In the forest products industry, some examples of products in this category include plywood sheathing and copy paper. For most specialty or differentiated products, the company can exercise a larger degree of control over price. There is typically a moderate level of competition in these markets between companies; however, the products are differentiated from each other, giving the competing firms more of an opportunity to control price. Some products in this category might include high-quality wooden windows and doors and higher-quality publication papers.

PRICING OBJECTIVES

Pricing has a significant impact on most strategies of a firm and can directly impact on a firm's success. As a result of this, it is important to establish pricing objectives in order to clearly state pricing policies. Pricing objectives should flow from overall company strategy and be consistent with it.

As shown in Fig. 6-1, there can be various pricing objectives including profit-oriented, sales-oriented, and status-quo-oriented objectives. Profit-oriented objectives can have two subobjectives, which are maximum profits or target rate of return on investment or sales. When maximizing profits, care needs to be taken as to whether the profits are maximized in the long run or short run. This decision will call for different strategies. Some companies price according to the life-cycle stage in regard to what the traffic will bear.

As products reach the maturity stage of the product life cycle, competition intensifies and pricing reflects this with lower margins. In contrast, during the growth stage prices may be less a basis of competition between firms and prices and margins will be better. A maximum profit objective does not necessarily mean a high price. Economic theory tells us that when demand is elastic profit may be maximized at a lower price.

Target return pricing is a method generally used to realize satisfactory profits on large capital investments. Within the forest products industry it is used primarily in the paper industry to generate a target rate of return on the capital investment in the paper-making facility during periods of strong markets (Rich, 1983). This method of pricing is also used predominantly in regulated utilities such as electric power generation and phone service. One managerially useful aspect is that target return pricing does set a goal for management to measure performance against.

Sales-oriented pricing objectives can be used to increase sales growth or increase market share. A retailer may price in order to generate additional traffic through the store, which may help to boost sales revenue. It's important to remember that higher sales don't always mean higher profits.

Japanese manufacturers of electronics and automobiles have been accused for years of pricing their products to achieve long-term market share growth; U.S. firms, on the other hand, appear to have been more interested in maximizing their profits than in boosting their growth in market share. In an industry with strong economies of scale a market share objective may be very appropriate. It also forces a manager to consider what the competition is doing.

Firms that are satisfied can use status quo pricing. Competition can still occur but usually it's on nonprice issues. Various forms of status quo pricing involve price leadership, pricing to stabilize the market, pricing to stabilize margins, pricing to maintain a market share or position, and pricing to charge a fair price. Status quo pricing can also be used to maintain the loyalty of intermediaries, to avoid government

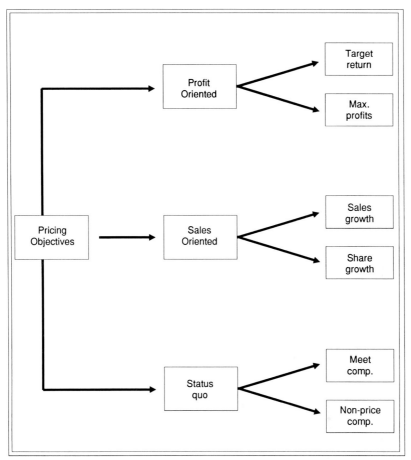

FIGURE 6-1
Various pricing objectives. (*McCarthy and Perreault, 1987.*)

interference, to discourage the development of products from alternative materials, and, lastly, to survive (Beucler and Bowyer, 1987).

PRICING STRATEGIES

Cost-Based Strategies

Markup Pricing Most retailers and wholesalers set prices using markup pricing. In markup pricing, a dollar amount (i.e., the markup) is added to the cost of the product to get the selling price. The markup is usually referred to as a ''percentage.'' For example, a $1.00 markup added to a $2.00 cost gives a $3.00 selling price and a markup of 33⅓ percent.

High markups don't always lead to high profits. The markups must cover the operating expenses and allow for a profit. Also, high markups make goods cost more and so typically result in fewer sales. Many merchants have realized that a lower markup, while providing less profit per unit sold, can result in more units being sold and a larger overall profit for products with an elastic demand (see Chap. 8 for more detail).

Manufacturers can also use markup pricing by adding a markup on the material and labor costs of a product. Manufacturers can also use an additional markup factor to cover overhead and/or general administrative costs and profits. This is especially common for small manufacturing firms.

Markup pricing can occur at each level of the distribution channel from manufacturer to consumer, compounding the effects of changes in the markup on the final retail price, as shown in Fig. 6-2.

Target Return Pricing   In target return pricing, a price is set so that the sale of the product results in a target level of return on investment or a specific dollar return given a predicted volume of product sales. A major weakness with this approach is what to do if demand falls short of the predicted volume (Fig. 6-3). As noted earlier, this is a common pricing technique in the capital-intensive paper industry. In some firms, especially less capital-intensive firms, the targeted return will be calculated as a return on sales.

Breakeven Pricing   Breakeven analysis evaluates the number of units required to be sold to break even (i.e., to cover all costs) given the price of an item. The breakeven point in units is expressed as follows:

$$\text{Breakeven point} = \frac{\text{total fixed costs}}{(\text{unit selling price}) - (\text{unit variable costs})}$$

The breakeven point is the number of units that must be sold at a given price to reach breakeven.

Traditional breakeven analysis focuses on the number of units to be sold to break even or, if you prefer, to have zero profit. But by adding a target amount of profit to the total fixed costs the formula can also be used to calculate the number of units necessary to be sold to reach a target profit.

Breakeven analysis is useful and can provide much information, especially when evaluating alternatives or how quickly a new product might become profitable. But its simplifying assumptions, such as a linear growth in profits with increasing unit sales, cause it to be a less-than-ideal tool for determining pricing solutions.

Learning Curve Effects   Many times as a firm gains experience in producing an item it finds ways to lower costs. This can happen as workers get better at their jobs, as the firm uses raw materials more efficiently, and for many other reasons. For firms that can predict the rate at which costs can be lowered based on prior experience or other factors, the dropping costs can be used to help set prices lower and build sales volume, counting ultimately on the lower costs to boost profits. This technique is frequently used to help set

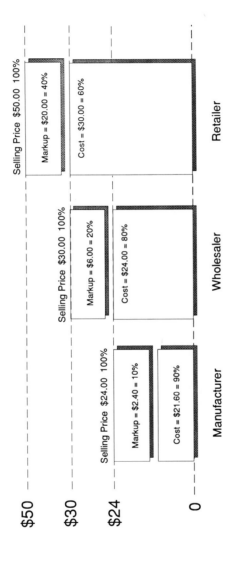

FIGURE 6-2
Example of a markup chain and channel pricing. (*McCarthy and Perreault, 1987.*)

71

prices for complex items such as airplanes, autos, and computers. It is a risky technique because the managers must not only estimate the number of units which can be sold at a given price but they must also estimate how fast and far costs will drop.

Marginal Cost Analysis   Marginal cost can be defined as the increase in total cost to produce one more unit of output and marginal revenue as the additional revenue gained by producing one more unit of output. As long as the marginal revenue exceeds the marginal cost the manager should produce the additional unit. This basic management rule can be used to assist in setting prices and output.

Marginal cost analysis can assist in profit maximization and may be very useful to a facility operating with excess capacity available, where a decision to produce more makes no addition to fixed costs. Marginal cost pricing is used in the paper industry during periods of weak markets in an attempt to keep operating levels up (Rich, 1983). Sales generated in this fashion are sometimes called "incremental business."

FIGURE 6-3
Cost-based pricing decisions. (*McCarthy and Perreault, 1987.*)

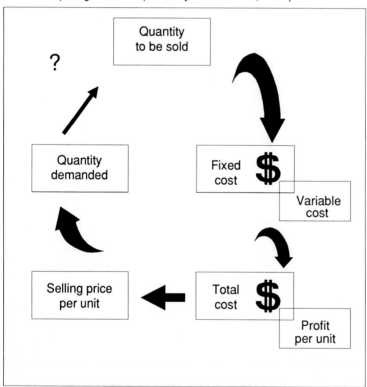

Demand-Based Strategies

Perceived-Value Pricing   Perceived-value pricing is setting prices consistent with customer perceptions and product positioning. The key factor in setting a price here is the customer's perception of the value of the product or service. Product positioning allows managers to view their products compared to competing products on customers' perceptions of various product characteristics. Considering product positioning assists managers in setting price relative to the competition. The amount customers are willing to pay for the value they perceive in the product considering quality, delivery time, service level, etc., is the perceived value price (Beucler and Bowyer, 1987).

Price/Quality Association   More and more, quality is becoming a critical issue for companies around the globe. Increased quality has been linked to lower overall costs, increased market share, and lower warranty and liability costs. For many products, however, quality is difficult to define and in the absence of other information research has shown that consumers associate higher price with higher quality and lower price with lower quality.

This is a useful concept when pricing certain luxury items where it is difficult to evaluate quality, such as exotic perfumes and furs. But manufacturers of custom cabinetry and furniture should also take note. For the more typical firm trying to produce a quality product but not a custom luxury product, the lesson might be to price your product at a fair price for the quality provided. Be careful not to price too low and run the risk of associating a poor quality image with your product. Studies have shown that raising quality while maintaining a fair price is a better way to increase long-term sales than lowering price with slipping quality.

Loss-Leader and Complementary Pricing   Sometimes a manager will set a price deliberately low, even below cost, to increase sales of other products. Loss-leader pricing sets low prices on commonly purchased items to attract customers to the store or to get calls to the lumber trading desk. The idea is to sell the customers other items also once they are in the store or on the phone.

Complementary pricing takes a different tack. One item is priced low (i.e., flashlights) to encourage the purchase of a complementary item (i.e., batteries). In home centers, lumber is typically priced very competitively with the goal of selling complementary items such as saws, nails, hammers, stain, brushes, etc.

Competition-Based Pricing (The Real World for Many Forest Products Producers)

In many commodity forest product markets, competition and the market are the main determinants of price. There can be small price differentials, to the extent that the product can be differentiated to have greater perceived value (Beucler and Bowyer, 1987). Most successful companies in these markets learn to compete on something other than price, such as service, delivery times, or accepting returns. While occasionally this may allow a firm a small price premium, it is rare.

In commodity markets which resemble the economist's model of pure competition (many buyers, many producers, similar products), the price is set by the market and most firms sell at that price. Pricing much higher will eliminate sales and pricing below will reduce profits.

In other commodity markets which resemble the economist's model of an oligopoly (many buyers, few producers), the few producers generally set the price. Many times this is done through a price leader. For most firms, there is little incentive to price differently than the leader. If you price higher, sales dry up, and if you price lower, competitors will match your lower price and little is gained.

A price leader firm typically has a strong market share and can literally lead the industry in terms of pricing. Competitors watch the price leader and set their prices to follow the leader's price. Price leadership is most common in strongly oligopolistic segments of the forest products industry, especially some paper markets (Dagenais, 1976; Rich, 1983).

Bidding results when competing firms bid on a job or an order. Some companies price a bid based on their notion of what their competitors are bidding. But competitive bidding need not be limited to price only. Other factors such as quality and delivery can also be very important. Much of the sales of wood products wholesalers results from competitive bidding. Unless the firm truly has a low-cost advantage, it is usually best to differentiate and develop a profitable business on the basis of quality, delivery, and other factors besides price.

Price Adjustments, Reductions, and Discounts

Discounts and Allowances   Various forms of discounts and allowances are common practice in the wood products industry. Some of the more common ones include trade discounts, cash discounts, promotional discounts, and discounts for quantity purchases.

Trade Discounts   These are reductions off the list price (usually expressed as a percentage) given to wholesalers and retailers to cover the cost of performing such functions as transportation, storage, promotion, and profit. Trade discounts are granted to retailers and wholesalers to varying degrees depending upon the amount of services that they are expected to provide. At first glance, trade discounts might seem to offer wood products manufacturers a certain degree of flexibility in their marketing strategy. Unfortunately, many trade discounts have become so well established by custom in the channels that many marketing managers have little flexibility and must accept them as essentially fixed pricing structures in the market.

Cash Discounts   These are reductions in the price offered in exchange for prompt payment of invoices (bills). A typical cash discount would be: 2/10, net 30. This means that the buyer may deduct 2 percent off the face value of the invoice if the invoice is paid within 10 days. If the buyer takes longer than 10 days to pay, the full price is due before the end of 30 days. There is some abuse of this with some buyers deducting 2 percent regardless of when they pay. Cash discounts can be beneficial for both buyers and sellers. They lessen the credit risk and improve the cash flow for sellers and benefit the buyers through offering them a discount.

Promotional Discounts   These are allowances given to wholesalers and/or retailers for use in promoting a manufacturer's product. The allowance may take the form of a price reduction, free display materials, shared advertising expenses, free samples, rebates, and others. Promotional discounts can be useful in offering wholesalers and/or retailers a special incentive to promote certain products. They are also used many times when introducing new products to the marketplace.

Quantity Discounts   These are offered to encourage customers to buy larger amounts. The degree of the discount or price reduction is many times based on the size of the purchase. Two types of quantity discounts generally exist. Cumulative quantity discounts apply to items purchased over a given period of time. The discount usually increases as the amount purchased increases. The purpose of the cumulative quantity discount is to encourage repeat buying by a given customer.

Noncumulative quantity discounts apply only to individual orders and have no cumulative effect. Their purpose is to encourage buyers to order in larger-size lots. Generally the paperwork costs for a large order are the same as those for a small order; therefore, it is more economical and efficient for manufacturers or wholesalers to process a few large orders versus many small orders.

Seasonal Discounts   Seasonal discounts are used to encourage wholesalers and retailers to purchase products earlier than their present demand requires. These discounts are offered to encourge preseason buying or advance orders. Seasonal discounts can be used to even out the production flows, thereby allowing better production planning.

Other Discounts   Many other forms of price reduction are available to the marketing manager. Orders can be shipped with invoices dated at some time in the future, effectively offering free financing of the inventory to the buyer. A seller can also offer a quantity discount and then agree to ship the material in smaller lots. A finance charge for invoices older than 30 days can be stated without any real intention to collect it. Special markdowns can be offered to meet the competition, to move inventory, or to temporarily increase sales volume. Rebates can be offered by the manufacturer to stimulate sales activity (Beucler and Bowyer, 1987).

Another area which can be particularly sensitive to wood products manufacturers is freight. A seller may quote a price with freight noted as FOB mill. This means that the freight is "free on board" (FOB) at the company's mill. In other words, the seller agrees to load the truck or railcar but it is up to the buyer to pay the freight charges from there to the buyer's place of business. However, if the seller wishes to offer a discount without lowering the price, it can do so by paying the freight. (For a more detailed discussion of freight and pricing, the reader is referred to Leckey, 1989.)

Pricing New Products

Skimming Price   A skimming price strategy uses a high price in the introduction and growth stages of a product's life cycle. The analogy is to skim the top off the market or the cream off the milk. The initial high price can allow the manufacturer to recoup investment and development costs of a new product early. It can be a successful strategy if some customers are willing to pay a premium for a new product. Additionally, if demand is high and production capacity low, a higher price can offer more flexibility for company management in keeping demand in line with productive capacity.

A skimming price strategy does encourage competitors to enter the market because of the high potential for profit. Therefore, many times as new competitors enter the market prices are lowered to preserve market share, but it is usually much easier to drop a price and increase the sales volume than to increase a price.

Penetration Pricing   Penetration pricing is introducing a new product at a low price in order to capture a large market share. The theory is that securing a large market share at a lower per unit profit will discourage other entries into the market. Sometimes this strategy is combined with learning curve effects and economies of scale.

A potential problem of using penetration pricing is the difficulty of raising the price at a later date if necessary to maintain margins. Sometimes temporary discounts are offered during the introductory stages to partially alleviate this problem.

LEGAL CONSIDERATIONS IN PRICING

Various state and federal laws impact on pricing. The goal of most of these laws is to help the marketplace operate more effectively in the consumers' interest (McCarthy and Perreault, 1987).

Many states have passed unfair trade practice legislation. These laws prohibit selling below cost in certain instances. Wholesalers and retailers are sometimes required to take a minimum markup over merchandise cost and shipping expense. These laws effectively stop some forms of low-price competition which could drive weaker firms out of business. On the other hand, high prices are typically okay as long as companies don't lie, collude to fix prices, or discriminate against certain buyers.

Companies making agreements with each other to sell a product at the same price (i.e., price fixing) is totally illegal and prohibited under the Sherman Act and other legislation. Many forest products companies have been prosecuted in federal courts for price fixing and large settlements have been paid.

In addition to price fixing between companies, it is illegal for one company to discriminate between buyers by charging different prices for the same product if it injures competition. This is prohibited under the Robinson-Patman Act, which does allow some price differences if they are the result of cost differences or the need to meet competition. This law makes the use of some special promotions and discounts tricky because they must be made available to all customers on proportionately equal terms.

SUMMARY

Effective pricing for many wood products firms should be the result of an ongoing strategy and not just the continuation of the status quo. While some firms do review their pricing policies on a frequent basis, others review discounts/allowances and other price adjustments infrequently if at all. With the changing and increasingly volatile nature of the wood products markets, all aspects of pricing strategy should be regularly evaluated.

This is particularly true for the wholesaling and distribution of commodity grades of wood and paper products. Wholesalers continue to face a squeeze on both the purchasing and selling sides of their business. Wood and paper products' prices and demand are based largely on economic conditions. As these conditions change, the price of the finished goods can also change. Further, many retailers and others who buy from wholesalers continue to use wholesalers as their bankers and warehouses. In such a competitive market, the temptation is to use pricing as the dominant form of differentiation; however, this can be a self-defeating strategy which progressively leads to a smaller and less profitable market share.

Real World Pricing

Many small non-commodity-producing wood products firms use a form of markup pricing. This has the advantage of being logical and simple to implement. The problems it involves are how to select the markup and whether or not it adequately covers costs. Most small firms are unable to provide costs by product, making it difficult to judge the degree to which the markup covers expenses.

For most small firms, pricing is clearly more art than science. Manufacturing or purchase costs, overhead, general administrative, and selling costs, as well as profit are considered in a general way, but more intuitively than as part of a formal procedure. Successful wholesalers achieve some differentiation from the competition through nonprice issues (Beucler and Bowyer, 1987).

A practical pricing approach would be to establish prices using a well-thought-out, frequently updated methodology, and then to use a lot of intuition in formalizing the prices that are quantitatively determined. Obviously, if you don't manage prices, they will manage you through low margins or lost sales (Beucler and Bowyer, 1987). Many wood commodity producers use the prices listed in the Random Lengths' and Crow's guides, the *Hardwood Market Report,* or other price reporting newsletters as a starting point and try not to price lower.

BIBLIOGRAPHY

Beucler, O., and J. Bowyer. 1987. *Marketing of Manufactured Wood Products.* Seminar Notebook #1. University of Minnesota, Department of Forest Products. St. Paul, MN.

Dagenais, M. G. 1976. ''The Determination of Newsprint Prices.'' *Canadian Journal of Economics* 59:442–461.

Day, G. S., A. D. Shocker, and R. K. Srivastava. 1979. ''Customer-Oriented Approaches to Identifying Products-Marketing.'' *Journal of Marketing* 43(4):8–19.

Jacobson, R., and D. A. Aaker. 1987. ''The Strategic Role of Product Quality.'' *Journal of Marketing* 51(October):31–44.

Leckey, D. 1989. *Trading Western Softwood Lumber.* Highland Press. Wilsonville, OR.

McCarthy, E. J., and W. D. Perreault, Jr. 1987. *Basic Marketing.* 9th edition. Irwin. Homewood, IL.

Oxenfelt, A. R. 1973. ''A Decision-Making Structure for Price Decisions.'' *Journal of Marketing* 37(1):48–53.

Rich, S. U. 1983. "Price Leadership in the Paper Industry." *Industrial Marketing Management* (12):101–104.

Shaw, R. J., and R. J. Semenik. 1985. *Marketing.* 5th edition. Southwestern Publishing Co. Cincinnati, OH.

Udell, J. G. 1973. "The Pricing Strategies of the United States Industry." *Combined Proceedings of the American Marketing Association.* Series No. 35, 151–155. Chicago, IL.

CHAPTER 7

PROMOTION

Good salespeople don't just try to sell the customer. Rather, they try to help the customer buy—by understanding the customer's needs. . . . Such helpfulness results in satisfied customers—and long-term relationships.

E. J. McCarthy and W. D. Perreault, Jr.

Promotion is the last of the major marketing functions which we will cover in this section on marketing fundamentals. Promotion has many misconceptions attributed to it courtesy of the popular press and, at times, unethical companies. However, promotion as a marketing function is any communication that creates a favorable disposition toward a good, service, or idea in the mind of the recipient of the communication (Shaw and Semenik, 1985).

A key point to note in this definition of promotion is that it only vaguely suggests that promotion is a selling tool. Communication is the appropriate role for the promotion function in marketing. A strategically designed and integrated marketing mix (i.e., markets, products, distribution, price, and promotion) results in a sale. Only when all the marketing functions are brought together in a synergistic fashion does a sale result. Promotion does not bear the entire responsibility for a sale.

THE MARKETING MIX

To better understand the thinking that goes into an effective promotional effort, one first must consider the other elements of the marketing mix: products, markets, price, and distribution.

Product

The product element of the marketing mix is perhaps the most important element in regard to promotion. The product must be thoroughly understood by the firm before successful communication can be achieved. In order to accomplish this, one must return to the total product concept. For example, when someone purchases a Mercedes he or she is not just buying transportation. Rather, the person is purchasing a whole bundle of benefits including, but not limited to, prestige, status, reliability, precise handling, engineering excellence, quiet ride, and many other factors.

Consider for a moment the Pope and Talbot advertisement shown in Fig. 7-1. This advertisement communicates the fact that many wood species are treated very similarly in the marketplace; therefore, the important benefits to many consumers do not revolve

79

FIGURE 7-1
Pope and Talbot advertisement.

around the actual product but rather around the ability of the manufacturer to service the customers' needs in terms of prompt shipments, accurate shipments, on-grade shipments, and many other factors.

Markets

Selecting the correct target market for a given product is critical. With improper market segmentation it can be next to impossible to have any form of promotion reach the right audience. Also, it may be difficult to effectively promote a product in the wrong market segment even if you do reach it. The bottom line is, effective promotion for most products depends to a large extent on effective market segmentation.

Price

Price is another key marketing function and it can present distinct opportunities in promotional programs. Either above average or below average pricing for a product can result in promotional opportunities. For example, consider the early advertisements for the Yugo automobile, which stressed its low cost. You might also remember that there is clearly a price/quality association in the minds of many consumers and later Yugo advertisements had to stress the durability and reliability of the Yugo to overcome some of the negative implications of a low-priced product. Oriented strandboard was at one time clearly positioned in the marketplace by many manufacturers as a lower-cost substitute for plywood. However, later advertising stressed the advantages of oriented strandboard over plywood rather than a price differential.

Generally speaking, promotional campaigns should stress competition between products based on nonprice characteristics. Strong competitors are many times willing to lower their prices to match yours if competition is strictly on a price basis. Furthermore, forest products firms which compete on strictly a price basis or a low-cost basis are typically less profitable than firms which compete in other avenues.

Distribution

Distribution is the final aspect of the marketing mix which must be carefully considered in developing a promotional program. As noted earlier, many firms not only compete with each other on a direct basis but use their entire distribution system as a competitive tool. This entire system must be considered in designing promotional activities. Perhaps the first aspect which must be considered is the intensity of the distribution system. An intensive distribution system which places the product in as many outlets as possible (i.e., Kleenex tissue) typically requires a large amount of advertising to support the retailers' acceptance of the product. On the other hand, more selective or exclusive distribution systems typically require a larger emphasis on personal selling and cooperative arrangements for advertising between manufacturer and retailer.

In addition, the geographic scope of a firm's distribution system should result in differences in the promotional program. Clearly the promotional program for a product distributed locally will be quite different from a product distributed on a national scale.

For retailers and to some extent manufacturers, there can be economies of scale derived from promotional programs as a result of the distribution system. For example, many building materials retailers will place several stores in a single media market. This allows them to advertise their offerings through the print and electronic media and split the cost of doing so among several stores rather than one store. As a result of this and other factors some building material chains almost "own" regional markets. Perhaps the best example of this can be seen in Florida with Scotty's Home Centers, which in the late 1980s owned over 150 stores with all but nine in the Florida marketplace. Imagine the efficiencies in promotion and advertising this can create.

THE PROMOTIONAL MIX

We have used the term "promotional program" earlier yet have failed to define all the components which could comprise this program. This idea is many times termed the "promotional mix." This promotional mix is a blend of four elements: advertising, personal selling, sales promotion, and publicity.

Advertising

First, let's consider advertising. We can see the advertisement for Pope and Talbot on the facing page and are very familiar with such advertisements in magazines and newspapers. Other media used in advertising include television, radio, billboards, and direct mail. For advertising to be effective, the correct customer audience needs to be determined and advertisements must be placed in the correct media that will reach this desired audience. One of the better examples of this was noted at a recent National Home Center Show in Chicago. On the freeways leading to the show at McCormick Place, Weyerhaeuser Company rented a large billboard to promote its commitment to 100 percent customer satisfaction. Each day thousands of home center buyers had to drive by the billboard in order to reach the Home Center Show. Talk about selecting the proper audience and then placing advertising media to reach it! Another factor that must be considered in advertising is developing a message which responds to customers in the ways in which they evaluate the product. Again, consider the Pope and Talbot advertisement. The species of the lumber was not important to consumers, but rather the level of service Pope and Talbot was able to offer them. And it was this high level of service which Pope and Talbot was advertising.

Knowing which message to send to customers and potential customers can be difficult. A four-cell matrix of promotional strategies can help with this task. This matrix, shown in Fig. 7-2, divides customers into groups based on their use of the product/service in question and their attitudes toward the product/service. For those who use the product and have a positive attitude about it, the recommended strategy is to reinforce the positive attitude. For those who use it but have a negative attitude, the strategy is to change their attitude. If a positive attitude is present but no product use, then incentives should be offered to promote product use. And last, if no product use is coupled with a negative attitude, the matrix suggests that these customers be left alone or that efforts be directed to make product use mandatory. Building-code-mandated product use is common in the wood building products industry with such items as preservative-treated sill plates and fire-retardant lumber and plywood in schools and churches.

FIGURE 7-2
Promotional strategy matrix. (*Adapted from Sheth, 1989.*)

PRODUCT USE

	USE	DO NOT USE
POSITIVE	Reinforce Positive Attitudes	Use Incentives to Promote Product's Use
NEGATIVE	Change Attitudes	Leave Alone or Make Use Mandatory

ATTITUDE TOWARD PRODUCT

A last item in advertising is the need to measure the effectiveness of the advertising effort. This can be done in many ways depending upon the criteria of measurement. For example, the famous Wendy's ''Where's the Beef '' advertising campaign produced a major increase in Wendy's sales. It was clearly an effective advertising effort. With forest products companies, however, the measurement of advertising effectiveness may require more sophistication. Many major wood products firms advertise in the mass media not simply to promote the products but to communicate to consumers their good citizenship and stewardship of the land they own. Measuring the effects of this advertising requires a more sophisticated approach.

Personal Selling

Personal selling has been defined as a face-to-face presentation of information relating to a firm's products or service. Personal selling is clearly more expensive on a per contact basis than advertising; however, it dominates the promotional mix of many major wood products corporations. The message (communication) delivered can be specialized to meet the specific customer's needs and the salesperson can get immediate feedback on the degree to which the product and its benefits are meeting those needs. When an individual sale can involve a large amount of money, either initially or over time, personal selling can be efficient and effective. For these reasons, it is used extensively in selling many industrial products including most basic forest products. The personal contact can help to build customer loyalty and encourge repeat purchases of forest products.

Today's salespersons must completely understand the business of their customers, because salespersons should be problem solvers for their customers. I teach forest products marketing short courses to industrial groups and almost without exception my audience will include someone from a railroad, a bank, a phone company, a consultant, or another company providing a service or product to the forest products industry. These people attend to further their understanding of the business of their customers.

Sales Promotion

Sales promotion, another component of the promotional mix, includes such tactics as free samples, coupons, cents-off approaches, point-of-purchase display materials, trade shows, and contests. Wood products firms use sales promotion techniques heavily. Trade shows are a constant theme in the industry, with major firms spending up to three-quarters of a million dollars and more to prepare their displays and rent space at major industry shows. Contests are widely used as sales promotion tools, with key customers being given free trips, etc., for a certain level of purchases. Many times sales personnel join the customers on the trip or at the big event to further build the relationship.

Publicity

Publicity is another component of the promotional mix. Publicity is usually thought of as information about a firm or its product which is disseminated to the public at basically no direct cost to the firm. Such publicity can include news releases regarding new product developments or new plants which create new jobs in given areas and the like. These news releases are typically sent to major news organizations, newspapers, magazines, and trade publications. One innovative use of publicity was done at the Weyerhaeuser Company. It introduced a brand name line of CCA-treated lumber products. Some consumers were unsure about aspects of CCA treatment, and to provide more information in a cost-effective manner, a key staff member was made available for radio call-in talk shows. This provided Weyerhaeuser with extensive publicity of its new product line and the CCA-treated product category in general.

Georgia-Pacific used a fastest-roofer contest to get publicity for its line of roofing products (Fig. 7-3). Preliminary contests were held at local distribution centers, attracting crowds and local media. Winners of the local contests were then eligible to compete in the national contest, again attracting more attention and media coverage.

PUSH-PULL STRATEGIES

There are two diametrically opposed promotional strategies. The pull strategy uses a carefully applied promotional mix aimed directly at final consumers to encourage them to "pull" the product off the retail store shelf. In other words, the promotional program is geared directly at the consumer and the consumer then requests that the retailer purchase more of the item in question. Sometimes this strategy is paired with a co-op advertising program where the manufacturer provides the retailer with ad layouts and graphics for the manufacturer's products which can be inserted into the retailer's advertisements. The manufacturer also shares the cost of the advertisement. Various formulas are used to calculate the cost sharing, and co-op advertising is a major tool for many big forest products producers.

The opposite of this is a push strategy, which can be effective where the retail salesperson can influence product and brand selection of the final consumer. In this instance, the manufacturer uses a promotional mix heavy on personal selling to "push" the product to the wholesaler or retailer and subsequently encourage the sales personnel at the wholesale or retail level to recommend that particular product to their customers.

WHAT DOES THE WOOD PRODUCTS INDUSTRY REALLY DO?

Historically the solid-wood products industry has relied heavily on personal selling and sales promotion as the major tools in its promotional mix. On the other hand, the paper side of the forest products industry has been a much stronger user of national advertising as part of its promotional mix. These differences result from the differences in the ultimate consumers of many of these products. For example, the potential purchasers of Kleenex tissues might be all persons in the United States over the age of 10, whereas the potential number of purchasers for a 4 × 8-foot sheet of oriented strandboard would certainly be much smaller and the predominant purchasers of such a product would be limited to home builders, repair and remodeling contractors, other construction contractors, and do-it-yourself customers. The paper industry clearly needs to use a promotional mix appropriate to reaching a broader spectrum of customers.

According to Duerr (1988) the solid-wood products industry has advertised to

Preliminary Competitions Underway for "World's Fastest Roofer" Contest

Professional roofers all over the country are sharpening up their skills in preparation for the third annual World's Fastest Roofer Contest, sponsored by Georgia-Pacific to mark National Roofing Month this spring.

Preliminary competitions are being held at G-P branches nationwide to determine finalists who will compete in Atlanta in May. The first roof-off for the 1989 contest was held at the Birmingham, Ala., branch in October, drawing an enthusiastic crowd and exceptionally well qualified contestants.

"If the Birmingham roof-off sets the standard for the others, we are going to have some very competitive contests around the country," says Don Glass, vice president of the gypsum and roofing division. Birmingham winner Darwin Thomas bested 15 other contestants by nailing one square (100 square feet) of Tough-Glass® shingles in 15 minutes, 50 seconds. His prize: A $325 cash award from Sillavan Lumber, co-sponsor of the Birmingham contest, and a trip to Atlanta for the national contest.

Winners of the national contest will nail down $10,000 worth of all-expense-paid luxury vacations: first prize is a vacation for two in Hawaii; second prize is a Caribbean cruise for two; and third prize is a trip for two to Las Vegas.

Branches that participated in 1987 and 1988 say the contest is an excellent way to raise visibility—and sales—for G-P roofing products. It's not too late to hold a local roof-off, according to Barbara Squires, G-P public relations manager for building products. Support is available from Atlanta for branches sponsoring local contests. For more information contact Squires at (404) 521-4741.

FIGURE 7-3
Georgia-Pacific's world's fastest roofer contest.

accomplish four main objectives. First, industry advertising has occurred which is aimed at differentiating forest products from alternative building materials such as metal or masonry. Second, public relations advertising by the large integrated firms has occurred to bring their names, activities, and products to the favorable attention of a wide variety of consumers. Third, promotional advertising has occurred which attempts to differentiate the individual producer on the basis of speedy service, wide variety, product quality, and sometimes even price. A fourth form of advertising that Duerr calls ''informational advertising'' can be found in many trade magazines, which provide information concerning prices and quantities available to potential buyers.

Trade associations have played a major role in promoting forest products. The biggest volumes of forest products are in the commodity grades, where the physical product of different manufacturers is very similar, and sometimes identical. Thus the competition between manufacturers is usually over factors such as service, delivery, credit reputation, etc., and not over the physical product itself. But the physical products produced by these groups of manufacturers do compete against products of a different kind produced by non-forest-products producers or in some instances other types of forest products producers. Some examples are wood studs versus aluminum studs, wood light-frame versus concrete or steel construction products, vinyl versus wood or aluminum siding, CCA-treated wood versus concrete foundations, hardwood lumber versus particleboard or MDF (medium-density fiberboard), and the list could go on. Trade associations have served as focal points to promote, in general, the product types produced by their members. For example, the Southern Forest Products Association has a major promotion campaign for southern pine lumber products, the Hardwood Manufacturers Association promotes hardwood lumber, the American Plywood Association promotes structural panels, and so on.

BIBLIOGRAPHY

Duerr, W. A. 1988. *Forestry Economics as Problem Solving.* Orange Student Book Store, Inc. Syracuse, NY.

Hutt, M. D., and T. W. Speh. 1981. *Industrial Marketing Management.* Dryden Press. Chicago, IL.

McCarthy, E. J., and W. D. Perreault, Jr. 1987. *Basic Marketing.* 9th edition. Irwin. Homewood, IL.

Rich, S. U. 1986. ''Recent Shifts in Competitive Strategies in the U.S. Forest Products Industry and the Increased Importance of Key Marketing Functions.'' *Forest Products Journal* 36(7/8):34–44.

Ray, M. L. 1982. *Advertising and Communication Management.* Prentice-Hall, Inc. Englewood Cliffs, NJ.

Shaw, R. J., and R. J. Semenik. 1985. *Marketing.* 5th edition. Southwestern Publishing Co. Cincinnati, OH.

Sheth, J. N. 1989. *Multivariate Statistics Shortcourse.* University of Illinois. Chicago, IL.

PART THREE

THE NATURE OF COMPETITION

Probably the most important management fundamental that is being ignored today is staying close to the customer to satisfy his needs and anticipate his wants. In too many companies, the customer has become a bloody nuisance whose unpredictable behavior damages carefully made strategic plans, whose activities mess up computer operations, and who stubbornly insists that purchased products should work.

Lew Young
Editor-in-Chief
Business Week

CHAPTER 8

USING ECONOMISTS' TOOLS

The most productive and profitable marketing programs are developed based on a clear understanding of market and competitive conditions and trends, and on an unswerving orientation to satisfying the customer needs and wants.

D. F. Healy

Marketing has been defined by some writers as basically a conflict or competition between corporations as well as the satisfying of human wants and needs. It follows, then, that a good marketing manager must be somewhat of an expert on competition and on the markets in which this competition occurs. Economists have for many years been students of market analysis and have developed several tools which can be useful to others in attempting to better understand the competitive nature of many markets.

SUPPLY AND DEMAND

The economists' basic tools for market analysis revolve around the notion of supply and demand. It is obvious that most customers have limited incomes and, therefore, must balance their needs against the prices of goods. The question is then opened: How do you measure customers' needs (desires) for a particular good? Economists answer this question in terms of the extra utility a customer can obtain by buying more of a good, or how much utility is lost if a customer has less of a good. In reality, we tend to measure such needs in terms of price and quantity.

Demand

The relationship between the price of a good and the corresponding quantity demanded at a given time is called a ''demand schedule'' or ''demand curve.'' Most demand curves are downsloping, meaning that, if prices are lowered, customers demand more goods. For example, in Fig. 8-1 we see that when price is lowered from $3.00 per unit to $1.00 per unit, the quantity demanded increases from 1 unit to 3 units. It is clear that the slope of this demand curve is important and can tell us many things about the marketplace. Demand curves (or, more properly, portions of demand curves) can be classified as inelastic, elastic, and unitarily elastic. For inelastic demand curves, the small decrease in demand due to a higher price will not be enough to decrease total revenue for the seller. In fact, total revenue will increase. For a demand curve exhibiting elastic demand, if the price is dropped the quantity demanded will increase enough to increase total revenue. For

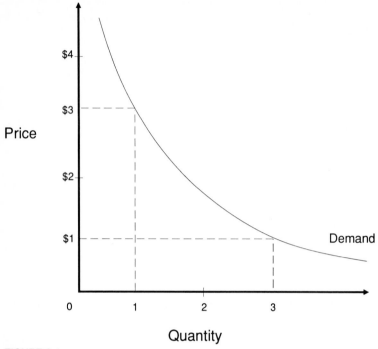

FIGURE 8-1
A typical demand curve showing the increase in quantity with lowered price.

unitarily elastic demand curves, price changes have a 1 : 1 impact on demand, leaving total revenue unchanged.

The key points to remember are that with elastic demand total revenue will *decrease* if the price is raised while with inelastic demand total revenue will *increase* if the price is raised. Clearly, marketing managers must know something about the slope of the demand curve for the products they market.

Demand elasticities are impacted by the availability of substitute goods and the urgency of need for the good. For example, in Fig. 8-2 there are two demand curves illustrated, one elastic and the other inelastic. For a product with many substitutes, such as ready-to-assemble (RTA) furniture, the curve is typically elastic. By lowering the price (assuming your competitors don't do the same thing) the RTA furniture becomes relatively more attractive to customers than its substitutes. This causes a disproportionate increase in the quantity of RTA furniture demanded, which results in an increase in the total revenue generated.

For a product with very few substitutes such as high-end solid-cherry furniture, the demand curve will typically be inelastic. Reducing price will not result in a disproportion-ate increase in quantity demanded and therefore total revenue will decrease. However, an increase in price will typically have a disproportionately small decline in the quantity of the good demanded, thereby increasing total revenue.

Of course, firms do not have totally flexible pricing at all levels of demand. This is

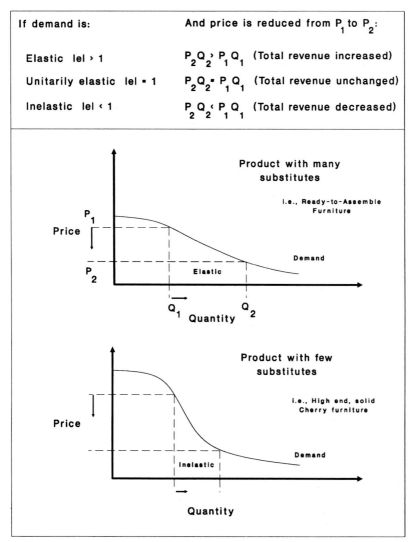

If demand is:	And price is reduced from P_1 to P_2:
Elastic lel › 1	$P_2Q_2 › P_1Q_1$ (Total revenue increased)
Unitarily elastic lel ▪ 1	$P_2Q_2 ▪ P_1Q_1$ (Total revenue unchanged)
Inelastic lel ‹ 1	$P_2Q_2 ‹ P_1Q_1$ (Total revenue decreased)

FIGURE 8-2
Elastic and inelastic demand curves.

because the firms producing the products or goods obviously incur a cost to do so. The higher the price, typically the more goods suppliers are willing to provide.

Supply

This relationship between the price of the good and the quantity of the good that suppliers are willing to provide at a point in time is called a "supply curve" (Fig. 8-3). The elasticity of both the supply and demand curves and their interaction helps to explain the

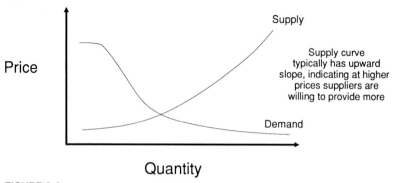

Price

Quantity

FIGURE 8-3
A typical supply curve.

nature of competition a marketing manager is likely to face. For example, an extremely inelastic demand curve means that the manager will have much choice in strategic planning, especially in the area of pricing, because for products with inelastic demand consumers typically like them and see few substitutes. Consequently, they are willing to pay higher prices before cutting back on the quantities they purchase. Of course, economic theory tells us that higher prices will likely boost supply, increasing competition. A smart marketing manager knows the impact of pricing on supply and potential competition.

MARKET STRUCTURES

Of course, many other factors also affect the nature of competition. Included among these are the number and size of competitors and the uniqueness of each firm's marketing mix. All of these factors combined make up the competitive environment or market structure for a given product or industry. Marketing managers must be able to determine the market structure they are operating in to be in a position to make effective decisions. Economists have defined several generic market structures which deserve some consideration. These generic market structures basically fall into four categories: pure competition, oligopoly, monopolistic competition, and monopoly. These four market structures provide a very useful framework for marketers of wood and paper products to develop a theoretical understanding of the marketplace without long periods of job experience. To understand market structure analysis, two key pieces of information are needed: the number of buyers versus the number of sellers and the type of product being exchanged.

Pure Competition

The pure competitive market has several important characteristics. The products are nearly homogeneous, with one product being essentially a perfect substitute for another product. There are many buyers and sellers who have a strong knowledge of the market. This is especially true in the softwood dimension market where even the largest lumber producer has only a small share of the market and market knowledge is widely available through various news and price information sheets such as Crow's, Random Lengths, and Madison's. For purely competitive markets, the ease of entry for buyers and sellers into the market is also important. New firms can easily start and new customers can easily come into a purely competitive market structure.

In a market which is nearly representative of pure competition, no single producer or buyer can effect a significant change in the market. For a given producer, the quantity it produces has virtually no impact on price, because its part of the industry's output is small. This results in basically a highly elastic or nearly flat demand curve for a single firm operating under pure competition (Fig. 8-4). Firms operating in purely competitive markets are typically price takers, accepting the prices for their goods which are set by the marketplace. This market is very characteristic of commodity markets for standardized grades of building materials and certain paper products.

Oligopoly

In an oligopoly, typically a few competitors offer similar things to what many times is a relatively large number of buyers. Oligopolies have a number of important characteristics. They usually produce homogeneous products, but not always. They are comprised of a few large producers relative to market size. However, sometimes an oligopoly can have a few smaller producers who follow the lead of the larger producers. Normally oligopolies have substantial barriers to entry resulting from economies of scale, expensive or patented technologies, and advertising/brand names. And last, most oligopolies recognize the mutual interdependence which exists among the firms of the oligopoly.

Oligopolies are an important market structure to understand for the forest products marketing manager. Typical oligopolies in the forest products industry include softwood plywood, other wood composite structural panels, disposable diapers, and newsprint. Many paper products are produced and marketed in oligopolistic settings.

In a competitive market, prices tend to fall to the minimum point on the industry's

FIGURE 8-4
A demand curve for a single firm operating under pure competition.

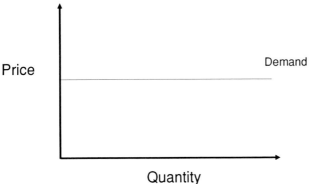

average cost curve due to competition. Firms operating in oligopolistic markets recognize this and thus their mutual interdependence. Many times, firms in oligopolistic markets will refuse to follow vigorous price competition. In the past, this has resulted in charges of collusion or price fixing against several oligopolistic segments of the wood products industry.

Oligopoly Pricing In an oligopolistic market, the legal line between price leadership and price fixing can be quite narrow. For successful collective management of pricing in an oligopoly, the following characteristics are required: a small number of firms, an easy-to-detect real price, high barriers to entry, stable markets, and limited antitrust actions. If the number of firms in an oligopoly is large, with a hard-to-detect real price, low barriers to entry, unstable markets, and vigorous antitrust actions, the opportunities for collective management of pricing are very slim.

To better explain the strong incentive to collude for pricing purposes in an oligopolistic market, let's turn our attention to Fig. 8-5. You may recall from basic economics that the most profitable level for a firm to produce is where marginal cost equals marginal revenue. Thus, for an oligopoly under complete collusion the ideal point to produce is where marginal revenue equals marginal cost, which is illustrated in Fig. 8-5 at price P_2 and quantity Q_2. If the oligopolists were to allow total competition, prices would tend to fall to the lowest point on the long-run average cost curve (P_1), which would result in lower profits.

Even though members of an oligopoly have a strong incentive to collude on pricing matters, each member of the oligopoly also has a very strong incentive to cheat on any such agreement. Undetected price cuts by a member of an oligopoly would do two things: (1) it would attract new customers who could now afford the lower price and (2) it would attract customers from other firms in the oligopoly. This results in two conclusions about price cutting in an oligopoly. First, demand for the price-cutting firm will be more elastic than industry-wide demand and the price which maximizes a single firm's profit will be lower than the price which maximizes the industry's profit.

Pricing in oligopolistic markets often tends to be inflexible. Many times even as basic costs of the operation change, prices tend to remain relatively constant. For those more interested in this phenomenon, most intermediate economics textbooks explain it through the so-called kinked demand curve concept. We'll leave that explanation to the economists for the moment; however, it is important to remember a couple of key points. For any oligopolistic firm increasing its price, its demand curve will be very elastic because other firms will, in most cases, maintain their lower prices. This means total revenues will fall for the firm increasing its price. For a firm decreasing its price, its demand curve will be very inelastic because other firms will typically respond by lowering their prices also. Again, this will result in lower total revenue for the firm lowering its price. This explains somewhat the reason that oligopolistic pricing tends to be relatively stable.

Of course, a fair question to ask at this point is: what is the legal mechanism by which oligopolistic markets can effect price changes? This occurs many times through the emergence of a price leader firm from among the member firms of the oligopoly. This price leader will normally set a price for all other firms to follow, which will maximize profits or sometimes generate a target rate of return.

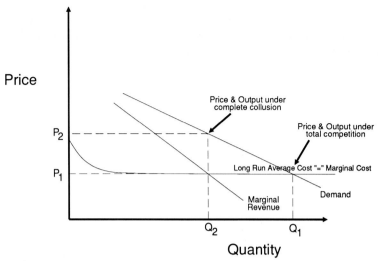

FIGURE 8-5
Range of price and output for an oligopolistic industry. (*Adapted from Gwartney, 1976.*)

The price leader must take the responsibility seriously. If followers cannot make a reasonable profit at the set price, they will likely try secret price cuts to expand sales. This will normally destabilize pricing in an oligopoly as other firms lose sales and cut their prices, resulting in a price war. This is seen frequently in the airline industry and in the automobile industry. The price leader must be a powerful firm in its market and usually walks a tightrope between the industry's interests, the antitrust laws, and its own interests.

Managerial Implications   Our discussion up to this point has been of markets exhibiting characteristics of oligopolies and pure competition. These markets can be difficult ones for marketing managers. Many of the tools of marketing are less effective in these markets. Pricing is controlled by the market or by a price leader. Product differentiation is possible, but more difficult in these markets as most products are relatively homogeneous. Differentiation tends to be on service variables in these markets.

With marketing tools limited, is it any wonder that forest products firms have historically concentrated on lowering production costs? It was the clear action they could take to increase profits, especially when they were operating in competitive and oligopolistic markets. However, another strategy would be to move their products into a different market structure which would give their managers a wider range of strategic options.

Monopolistic Competition

This market structure is called "monopolistic competition." "Monopolistic" implies that each firm tries to control its own market. And "competition" implies that there are still substitute goods or services. The key characteristics of a market operating under monopolistic competition are that products or services are heterogeneous in the eyes of some customers and that sellers feel there is competition for customers. Additionally, monopolistic competition usually implies low barriers to entry, large numbers of buyers and sellers, and small firms relative to the total market.

Because the goods offered by firms in this market structure are not identical, an individual firm faces a downsloping demand curve, not a flat one as under pure competition. Usually the demand curve is highly elastic because competitors may offer similar products and low barriers to entry can allow many rivals (Gwartney, 1976).

This market structure is typically very competitive and allows the use of a full range of marketing tools such as quality, promotion, service, and price to attract customers. Within the forest products industry, the better examples of this market structure can be found in retailing and secondary products such as furniture, cabinets, windows, doors, and many specialties. It is the market structure that many segments of the wood products industry are moving toward as they develop differentiated and specialty products and begin to serve niche markets.

FIGURE 8-6
A summary of basic market structures and their important dimensions. (*McCarthy and Perreault, 1987.*)

IMPORTANT DIMENSIONS \ TYPES OF SITUATIONS	PURE COMPETITION	OLIGOPOLY	MONOPOLISTIC COMPETITION	MONOPOLY
UNIQUENESS OF EACH FIRM'S PRODUCT	NONE	NONE	SOME	UNIQUE
NUMBER OF COMPETITORS	MANY	FEW	FEW TO MANY	NONE
SIZE OF COMPETITORS (COMPARED TO SIZE OF MARKET)	SMALL	LARGE	LARGE TO SMALL	NONE
ELASTICITY OF DEMAND FACING FIRM	COMPLETELY ELASTIC	KINKED DEMAND CURVE (ELASTIC AND INELASTIC)	EITHER	EITHER
ELASTICITY OF INDUSTRY DEMAND	EITHER	INELASTIC	EITHER	EITHER
CONTROL OF PRICE BY FIRM	NONE	SOME (WITH CARE)	SOME	COMPLETE

Monopoly

The last market structure in our list is monopoly. A monopoly is when one firm produces a unique product and has no competitors. Monopolies can be achieved in several ways. They can occur by patent, giving only one firm the right to produce a unique product. They can occur through legislation which allows a single firm to provide a given product or service. Or they can be achieved by individual firms which operate in niche markets too small to support other competitors or which produce products which other firms do not have the expertise to produce. A monopoly is a very rare market structure for the wood products industry and as such we will not give it further discussion.

SUMMARY

For a summary of the four market structures and their important dimensions, please see Fig. 8-6. The most desirable market structure is one which gives a company the ability to better control its own pricing and allows for marketing of products and not just selling.

Chap. 9 will discuss some tools and skills which will be useful to further analyze markets and allow for competitive strategy formulation.

BIBLIOGRAPHY

Chrysikopoulos, J., and B. Kirk. 1989. *Forest-Paper Outlook and Company Update Service.* Goldman Sachs Investment Research. New York.

Ellefson, P. V. 1979. *Economic Structure of the U.S. Timber Industry.* Staff Paper Series No. 11. University of Minnesota, Department of Forestry. St. Paul, MN.

Gwartney, James D. 1976. *Economics.* Academic Press. New York.

Healy, D. F. 1987. *New Strategies for Growth: Paper Distribution Plans for the Future.* Paper and Plastics Education Research Foundation. Great Neck, NY.

Rich, S. U. 1970. *Marketing of Forest Products: Text and Cases.* McGraw-Hill Book Company. New York.

Ries, Al, and Jack Trout. 1986. *Marketing Warfare.* McGraw-Hill Book Company. New York.

Stigler, G. T. 1966. *The Theory of Price.* Macmillan Co. New York.

Worrell, A. C. 1967. *Economics of American Forestry.* John Wiley. New York.

CHAPTER **9**

GENERIC COMPETITIVE STRATEGIES

I have always thought that one man of tolerable abilities may work great changes and accomplish great affairs . . . if he first forms a good plan and . . . makes the execution of that same plan his sole study and business.

Benjamin Franklin

Marketing today must be much more than satisfying human wants and needs (traditional definition). Today, every company has become customer-oriented. Knowing what the customer wants isn't too helpful if a dozen other companies are already serving his or her wants and needs.

For future success, a company must also be competitor-oriented. It must look for weak points in the positions of its competitors and launch marketing attacks or develop strategies to take advantage of those weak points. One of the best examples of this concerned the old American Motors Corporation (now part of Chrysler). If American Motors had developed a product strategy based upon identifying and serving existing customer needs, the result would have been a lot of products very similar to those of General Motors or Ford, which spent millions of dollars researching this same market-place to identify those same customer needs (Ries and Trout, 1986).

Strangely enough, when American Motors ignored conventional wisdom the company was much more successful. The unconventional Jeep product line was a winner for American Motors. However, American Motors passenger cars were mostly losers. The problem for American Motors was not the customer. The problem was General Motors, Ford, Chrysler, and all the many imports.

The moral of this story is that business is more than simply serving customers and the needs that they can articulate. It is also outwitting, outflanking, and outfighting your competitors. In short, marketing can be viewed as a competition with other firms as your competitors and the customer as the prize to be won.

HOW DO YOU OUTFOX THE COMPETITION TO ACHIEVE A COMPETITIVE ADVANTAGE?

Uncertainty

Anyone who listens to the national news is inundated with examples of the extreme amount of uncertainty in the business environment for U.S. corporations.[1] Congress is

---

[1]Many of the ideas on the next few pages have been adapted from Stephen South (1981).

constantly changing statues which directly affect businesses in a wide variety of areas, including taxes, personnel relations, environmental concerns, work place safety, product liability, and many others. The Federal Reserve Board, or, as it is known in the news, the Fed, is fine-tuning the growth of money supply and the interest rates, which combine to create additional uncertainty for U.S. businesses. The actions of a company's competitors also can work to create considerable uncertainty for business managers.

We read about the intensity of global competition constantly. Some areas which get considerable news are automobiles, computers, and other electronic devices. It is clear to even the casual observer of the international scene that the Japanese automobile and electronics industries have become very competitive on a global basis during the last 20 years. However, this increase in global competitiveness is not limited to automobiles and exotic electronics. Growing competition is becoming evident in the field of forest products. For example, New Zealand has significant plantations of radiata pine which will be coming onto the marketplace during the next several decades. They anticipate a rough doubling of their annual harvest and plan to export most of the increase. Other nations such as Canada, Chile, and the Scandinavian countries are becoming tougher competitors in the global market for forest products as well.

Elements of Business Success

One goal of management must be to reduce the influence of these forces on the business. But first it is helpful to understand two fundamental ideas which contribute to business success: operating effectiveness and competitive position (Fig. 9-1). Operating effectiveness is concerned with increasing the effectiveness of the production process, which in turn lowers production costs. It is a production orientation to business as discussed in Chap. 1. Operating effectiveness can be improved with economies of scale, production experience, vertical integration, and stable manufacturing processes.

Competitive position relates more to the firm's strategic marketing versus its operating efficiencies. It involves the selection of appropriate markets and the development of new products, product differentiation, and alternative distribution channels, and can also involve different cost or price structures which serve to give a firm a superior competitive position vis-à-vis its competitors. It is the marketing concept in action.

The forest products industry, over the years, has relied heavily on operating effectiveness for business success (Rich, 1986). This resulted because competitive arenas were more isolated and economic market structures were such that competition tended to focus on the operations side of the business and competitive positions tended to be very similar between firms. However, of the two elements of business success, competitive position (i.e., marketing) has clearly been shown to be the most important. Forest products firms which strategically concentrate on their competitive position versus operating effectiveness are typically more profitable (Rich, 1986).

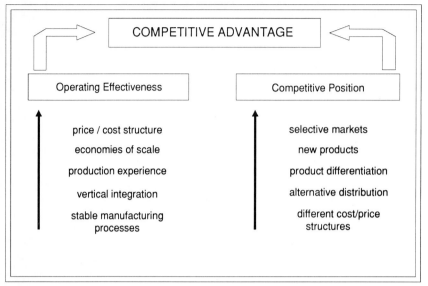

FIGURE 9-1
Fundamental elements contributing to competitive advantage. (*South, 1981.*)

Types of Competition

The idea of operating effectiveness and competitive position as elements of business success leads us to two kinds of competition which parallel the two elements of business success.

Reciprocal competition typically occurs between firms which compete on the basis of operating effectiveness. This generally occurs in industries with firms having similar strategic positions which are relying on operating differences to separate the successful firms. This type of competition occurs most frequently in the standardized industrial product markets, such as softwood dimension lumber, structural panels, or newsprint.

For firms which compete based upon their competitive position, the struggle occurs on a strategic basis with the competition occurring in the area of market segments, product offering, channels of distribution, branding, and such. Strategic competition allows firms to compete using a wide variety of market tools. Strategic competition most commonly occurs in markets which exhibit the characteristics of monopolistic competition or to a lesser extent in oligopolies which produce products not viewed as totally homogeneous.

To successfully compete in a strategic sense, managers must select competitive arenas for their firms which can be sheltered from changes in the business environment. They must also develop an advantageous position for their firm and its products as protection from the growing intensity of global competition.

Finding Sheltered Business Arenas

Numerous strategies are available which can assist in developing and finding sheltered business arenas. One strategy involves developing local operations in concert with the goals of the local governmental agency or country. Many local governments have a wide variety of goals which include jobs, trade policies, shared ownership, training of local people to assume managerial roles, and many others. Developing a business to match with these local goals can assist a firm in receiving favorable treatment, both politically and economically, from the local governmental bodies.

Working in areas where the products and manufacturing processes are technologically stable can also assist in developing shelter from competition. By working in areas where the technology is stable, firms can concentrate their efforts and energy on competing from a strategic standpoint rather than being leap frogged by the competition in terms of technology and then having to turn their attention to operational efficiencies.

The use of creative financial structures can minimize interest rate impacts and the fluctuation of foreign currencies. At one time in the United States, the use of local industrial development bonds was widespread, although they are less available today. They provided firms with typically below-market rates of interest for long terms. Local low-cost financing is also available in many Canadian provinces and in other countries.

Firms can choose to selectively participate in only those market arenas with sufficient stability for long-term profitability. This is a strategy which many wood products corporations are strongly pursuing. In general, the white paper markets have been much more stable than the markets for building products and many forest products corporations have added white paper manufacturing capacity in order to take advantage of this stability. For wood products manufacturers, the repair and remodeling/home center marketplace provides considerably more stability than does the market for new residential construction.

Patents and trade barriers are another way to develop a sheltered market arena. Japanese building codes historically have made it difficult for U.S. building products producers to export finished products to Japan. Another example has been the U.S. lumber industry's successful lobbying in the mid to late 1980s for a tariff on Canadian softwood construction lumber imported into the United States. Canada replaced the tariff with an internal tax of the same amount. However, the impact was the same: the price of Canadian lumber was raised for U.S. customers and a trade barrier was created.

Being first with a product in a given market can also result in advantages. Those first in the market with a product can develop brand loyalty and have the opportunity to develop manufacturing scale economies.

Another strategy closely related is that of product leadership. Product leadership can be defined as being the first to introduce a product in a specific size or class or grade. One example of this was Georgia-Pacific's introduction of plywood siding in 4 × 10-foot sheets. Plywood siding had previously been available only in 8-foot lengths.

Advantages can also be achieved through pricing strategies. A product can be priced high or low if performance characteristics warrant. Potlatch, for example, was very early in the market with its oriented strandboard, Oxboard. This product achieved strong brand loyalty, which enabled Potlatch to charge a premium price for its product (Sinclair and

Seward, 1988). Louisiana-Pacific, on the other hand, introduced an industrial particleboard brand named N.B.T. (Next Best Thing). They advertised this product as providing most of the properties of competing boards at a budget price.

Sheltered business arenas can also be developed through operating efficiencies by developing lower-cost structures. This is especially important in the area of industrial commodity products such as plywood sheathing and softwood dimension lumber. Lower costs can be developed through economies of scale, manufacturing experience, and sometimes vertical integration strategies.

Financial leverage has also been used over the years, especially by paper producers, to grow rapidly and increase their market share. They developed business arenas which were somewhat sheltered by controlling major shares of the market and through this were able to directly influence markets which are typically oligopolistic in nature.

Competitive Advantage

Competitive advantage can be elusive, but real. It can be achieved by concentrating on particular market segments or market niches. Offering products which differ from the competition can also provide a firm with advantages. Using alternative distribution channels and/or manufacturing processes can also help. And employing selected pricing along with fundamentally different cost structures can assist a firm in achieving competitive advantage in the marketplace.

For the advantage to be real, it must be tangible, measurable, and preservable for at least a short period. Competitive advantage offers the opportunity for sustained profitability rather than a situation where profits are simply competed away. As attractive as protected positions are, they do tend to break down over time and the strategically competitive firm is always on the lookout for new protected positions.

QUALITY AS A SOURCE OF COMPETITIVE ADVANTAGE

Products or services that fall below acceptable quality standards in the future will be treated more mercilessly. Consumers' standards have risen significantly, in step with new technological developments. And while most products or services can fairly be said to equal or surpass these standards, the ones that do not meet them will face grim prospects.

As the globalization of the American economy gathers speed, a renewed emphasis on quality will have the most severe practical consequences. It is not the key to marketing success so much as the necessary prerequisite for it.

Miller, 1988

In the wood products industry there has been a shift of corporate marketing strategy away from emphasizing commodity production—a production orientation—toward developing specialty products to better meet customer needs—the marketing concept. These new products are being produced with a renewed commitment to quality. This has encouraged the adoption of innovative process technologies to facilitate the manufacture of new and high-quality products designed for specific end users. Adopting new technologies and the development of new high-quality products are two of the leading trends in the forest products industry which can serve to increase a firm's competitive advantage (Rich, 1986).

Better quality can serve to enhance competitiveness in two ways, as shown in Fig. 9-2: first, through achieving a better market position (i.e., high sales and prices), and second, by generating cost savings with lower warranty costs, lower product liability, lower rework or scrap costs, and lower service costs.

A key element in the success of Japanese firms has been noted as product quality (Ouchi, 1981). Observers of successful U.S. firms also believe product quality to be an important component for success (Garvin, 1984b). The forest products industry is giving product quality more attention. During the mid to late 1980s product quality emerged as one of the top issues and problems affecting business in the wood building products industry.

The largest volume of wood products is sold primarily as commodities, with a standard set of voluntary grades widely accepted in North America. This has clearly created efficiencies in marketing activities and in building construction. However, some have argued that such a system of voluntary product standards tends to drive quality to the

FIGURE 9-2
Conceptual model of competitive advantage and quality. (*Garvin, 1984a.*)

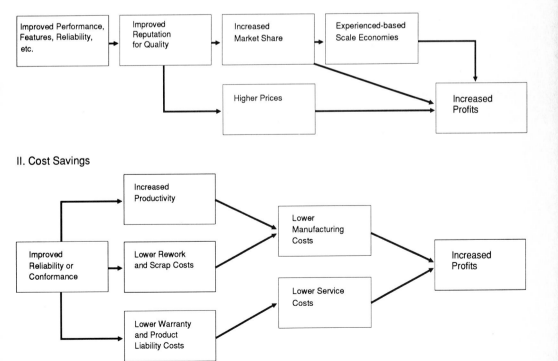

lowest common denominator (i.e., just high enough to pass the standard) (Keating, 1986). This has probably occurred in the wood products industry; however, it is recognized and a renewed emphasis on quality is occurring. A question to ask is: will people pay more for quality in commodity-like wood products? The answer is a qualified yes. In some markets there seems to be a quality premium (albeit most likely a small one) for some manufacturers' products (Sinclair and Seward, 1988).

The impact of product quality on market share and profitability has been debated. Conventional wisdom states that a strategy of high quality most often requires a perception of exclusivity which is usually inconsistent with high market share. However, some studies have shown a positive influence of product quality on market share (Phillips et al., 1983; Jacobson and Aaker, 1987). Some researchers have found no consistency in the quality of a firm's products and its profits while others have concluded that product quality is positively reflected in higher profits or return on investment (ROI) (e.g., a premium is placed on quality and customers are willing to pay the premium for quality) (Jacobson and Aaker, 1987).

Perhaps most surprising, the impact of product quality on cost is insignificant for most industries (Phillips et al., 1983; Jacobson and Aaker, 1987), although a very small positive relationship was found for raw and semifinished goods. Better quality can actually reduce costs by reducing scrap, rework, additional labor, work in process, warranty and liability claims, and inventory. It is unfortunate, but firms rarely calculate the real costs of poor quality.

For organizations that market a service (i.e., consulting foresters, campgrounds, outdoor recreation, etc.), quality may be even more critical to success. In one study, customers gave poor service as the primary reason they switched to the competition. Switching because of price was far down on the list (Sonnenberg, 1989). Cost, quality, and service should be viewed not as a series of tradeoffs but rather as complementary parts of a successful strategy (Haas, 1987).

UNDERSTANDING COMPETITION

The forces of competition can be visualized as shown in Fig. 9-3 (Porter, 1980). Competition can be thought of as being derived from five factors: potential new entrants, bargaining power of suppliers, bargaining power of buyers, threat of substitute products, and competition/rivalry among existing firms. Even with this view of competition, the goal of management is still the same—that is, to find a position where the company can best defend itself against these forces or can influence these forces in its favor.

Forces of Competition

New Entrants   New entrants to an industry can bring substantial financial resources to bear and can create new capacity and the subsequent desire to gain market share. Obviously, new entrants can cause considerable turmoil and shake up an existing marketplace. The degree to which new entrants pose a serious threat is related somewhat to the barriers of entry available to existing competitors. Barriers to entry can be developed in several areas: economies of scale, product differentiation, capital require-

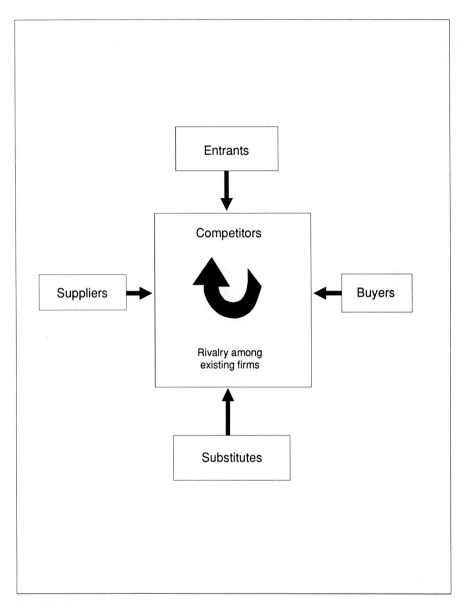

FIGURE 9-3
Porter's (1980) forces of competition.

ments, cost disadvantages independent of size, access to distribution channels, and government policy.

Economies of scale can force new entrants to come in on a very large scale or accept a cost disadvantage. Such scale economies may be present in production, research, and marketing, and many times in advertising and promotion.

Product differentiation can also create a barrier to entry by requiring new entrants to spend heavily on promotion and advertising to overcome customer loyalty to existing products. This is a substantial barrier to entry in many consumer product items such as soft drinks, cosmetics, and beer. Within the forest products industry, this barrier is also present in many brand name tissue product items. For example, the brand name Kleenex by Kimberly-Clark is a substantial barrier to entry for anyone wishing to compete in the consumer tissue market.

Substantial capital requirements to enter an industry can also create a barrier to entry. This is especially true if the captial is required for expenditures in areas which are nonrecoverable such as advertising or research and development. One of the major problems with new entrants to many industries is a lack of capital. This becomes especially acute for startup operations.

There are also cost advantages/disadvantages which are independent of company size. For example, forest products firms which own their own timberland at a historically low cost may have cost advantages which are simply not available to potential new entrants no matter what their size or attainable economies of scale. Other advantages can arise from learning curve effects, proprietary technology, government subsidies, or favorable locations. For other companies, cost advantages may be legally enforceable as in the form of a patent.

Government policy can also serve as a barrier for new entrants into an industry. Within the forest products industry, government policy on timber sales from pubicly owned lands and on environmental issues can clearly serve as barriers for new entrants.

Powerful Suppliers   Powerful suppliers can exert considerable bargaining power over firms in an industry. Powerful suppliers can raise prices or reduce the quality of goods available while the industry being supplied may not be able to recover the increased costs or compensate for the decreased quality. Supplier groups can be powerful when they are comprised of large companies which are more concentrated than the industry they sell to. If the product they supply is unique and/or if it has built-in switching costs, they can be powerful also. For example, the federal government is a very powerful timber supplier for many western sawmills and plywood operations. Many older mills need the large-size old-growth timber from public lands to operate efficiently, giving them a very high switching cost to go to smaller logs from private lands. Furthermore, the prices of timber can change with limited regard to the ability of the typical mill to recover increased stumpage costs in the marketplace.

If supplying firms can pose a believable threat to integrating forward into the industry's business, this can increase the power of the supplying firms. Also, if the industry is not an important customer of the supplier firm then the supplying firm's fortune is not tied closely to the industry's fortune, which gives the supplying firms more power.

Powerful Buyers   Buyers can be very powerful if they are concentrated or if they purchase in large volumes. This can be especially true for products which are purchased from the industry as standard or undifferentiated products. For example, a large building materials retailer chain purchases enormous quantities of standardized wood building products commodities. With such large purchases of these relatively standard products which are available from a large number of suppliers, it is free to play one company against another to get the best possible price.

Buyers can be very cost-sensitive and thus negotiate strongly for lower prices on the products they purchase which form a significant component of their own products and thus represent a significant piece of their costs. Also, if the buying firm is earning low profits this will create a great incentive to lower its purchasing costs. However, highly profitable buyers, when purchasing a product which is a very small piece of their final product, tend to be less price-sensitive.

In other instances, if the quality of the buyers' products is greatly affected by the industry's product, then buyers are typically less price-sensitive. Also, if the industry's product saves the buyers money, they tend to be less price-sensitive. On the other hand, if the product is unimportant to the quality of the buyers' products and does not save them money they may be very price-sensitive.

Buyers which can pose a strong threat to integrating backward to produce the products they are buying can exercise significant purchasing power over the industry.

A company's choice of suppliers to buy from or buyer groups to sell to is clearly a critical decision for the firm. The company can improve its strategic position by locating suppliers and buyers that have the least power to influence it adversely. One key element of this is trying to be selective concerning the buyers the company is willing to sell to. Not all firms have this luxury; however, if a firm sells to very powerful buyers it is typically an invitation to very low profits. One last note: firms that are low-cost producers in their industries or that have some other differential advantage can sometimes profitably sell to powerful buyers.

Substitute Products   Strong substitute products put limits on the profit potential of an industry. As the prices for the industry's products rise, substitutes become relatively more competitive and thus limit the industry's sales. Unless the industry can somehow improve the quality of its products or differentiate them from the substitute products, it faces a difficult situation.

With substitute products there are some trends which should create management concern. Substitutes which are improving their price performance tradeoff with the industry's products can cause significant problems. Another trend to worry about concerns substitutes which are produced by industries earning high profits. The deep pockets of the profitable substitute-producing industries allow them unusual flexibility in the marketplace, which can cause difficulty for their competitors.

Rivalry between Competitors   When firms are numerous or are of approximately equal size, competition can be very intense. If the market for the product produced is growing very slowly, then firms which are seeking to expand are constantly fighting others for market share, resulting in strong competition. When there is low differentiation between products and/or if switching costs between products are low, then it is easy for one competitor to raid another competitor's customer and create more intense competition.

If the industry producing the product has high fixed costs of operation, then competition can be intense because firms are reluctant to close plants in times of weak demand. This results in excess capacity, weak prices, and strong competition.

If capacity for the industry is typically added in large increments, then competition can be intense as new increments of capacity come on-line and create an excess supply in the market. This tends to result in more intense competition and again weaker pricing. This scenario is very commonly experienced in the paper industry.

High exit barriers to an industry also create intense competition because it is very expensive for firms to abandon production facilities. When rivals in the industry are diverse in strategies, this can also create more competition. Rivals with different strategies usually have different ideas about how to compete and tend to keep bumping into one another with their plans and strategies.

GENERIC STRATEGIES OF COMPETITION

Three generic strategies have been widely studied and written about. Any one strategy, if successfully implemented, may allow a firm to stake out a defended position in the marketplace. These strategies are overall cost leadership, differentiation, and focus (Porter, 1980). Let's examine each in turn.

Overall Cost Leadership

Developing an overall cost leadership position can produce a significant competitive advantage but will typically require a large capital expenditure and favorable access to raw materials. The achievement of overall cost leadership will typically require massive investments in large-scale efficient facilities and vigorous attention to many forms of cost reduction. Cost reduction will be in the form of controlling overall costs, controlling overhead, avoiding marginal accounts, minimizing research and development costs, minimizing service, and sometimes minimizing advertising.

Low cost relative to competitors is the central theme of this strategy. It is tough to implement and typically requires a large market share or some other special advantage. However, it can develop into a well-protected position because less efficient competitors usually suffer first in the face of competitive pressures.

Differentiation

The key idea with the differentiation strategy is to create something about the product that is perceived industry-wide as being unique. There are many bases for differentiation which were discussed in more detail in Chap. 4. However, to refresh your memory, some examples of ways to differentiate a product include quality, delivery, credit and terms, service, training, reputation/brand name, available technical information, price, etc.

Differentiation can provide insulation against competitors because of brand loyalty by customers and a resulting lower sensitivity to price. Differentiation is typically achieved through a mix of products and services, which is clearly a critical decision for the firm and requires a broad view of the concept of the product. This broad concept includes not only the actual physical product, but also the variety of image and service features that can have an impact on satisfying customer needs and meeting the competition.

In order to truly meet customer needs (i.e., differentiate the product) the firm must provide more than just the physical product. This supply relationship encompasses the entire set of benefits offered to the customer. In fact, with relatively standardized grading rules in North America for many commodity wood products, it is often easier to set a firm apart from its competitors along nonproduct dimensions.

Focus

Firms that pursue a focus strategy concentrate on a particular buyer group, segment of the product line, or geographic market. This strategy is built around serving a particular target market very well. The premise is that a firm is able to serve its narrow strategic market more effectively or efficiently than competitors that are competing more broadly. Serving a target market in this way can also create high switching costs which serve as a defense to the segment.

By effectively implementing a focus strategy, a firm can achieve differentiation by better meeting the market needs of its customers or lowering costs through specialization, or sometimes both. For most small- to medium-sized forest products firms, focus should be the strategy of choice.

Guerrilla Marketing   Guerrilla marketing is a special type of a focus strategy. It is helpful to remember that marketing is not something that is done only by large firms such as Coca-Cola or Ford. Marketing is also a way of thinking which can be very successfully employed by small firms. Guerrilla marketing is a particular type of marketing extremely well suited for small firms and, in particular, small wood products firms. Guerrilla marketing takes its name from the warfare analogy of guerrilla fighters. Just like the fighters, firms which successfully employ guerrilla marketing are small- to medium-sized, very flexible, very quick, and smart.

Guerrilla marketers view business as a competitive struggle. The struggle (battle) is for the marketplace and the customers go to the winner. Guerrilla marketers borrow many of their marketing tactics from guerrilla fighters. Guerrillas pick around the soft flanks of their stronger competitors, looking for an opening or niche in the market. They almost never attack stronger competitors head on in a direct battle for a market. To better understand guerrilla marketing, consider the following three rules of strategy (Ries and Trout, 1986):

Rule 1: Find a segment of the market small enough to defend   Just like good guerrilla fighters, guerrilla marketers look for a segment of the market which is small enough for them to be a major force. In other words, they focus on a small niche or specialty area of the market. This allows them to concentrate their energies upon a market small enough to eventually win and control.

The challenge here is to find a segment of the market small enough so that large competitors are not interested, yet still large enough for it to be profitable. The successful guerrilla tries to reduce the size of the battlefield (market) in order to achieve superiority in a localized area. This is a replay of the big fish in the small pond scenario.

Geography can be an ally in guerrilla marketing. For example, a firm may be the only one to offer a particular product or service in an area too small to attract additional competitors. A specialized product here will help. Even in a larger area, a company's business may be the only one which specializes in a given hard-to-find product or service. The plan is to subdivide the market into segments small enough that an individual firm can be a major player, and reduce the likelihood of strong competition from other firms.

Superior service can also be used to build a segment. Can the firm deliver its product faster than its competitors are able to? Does it treat all its customers like it would like to be treated itself? Does it offer credit, accept credit cards, provide delivery, offer operator training, or provide installation? All these things and many more make up the services a firm could provide to build a special segment or market niche for itself even in a crowded market.

A low-price strategy is not well suited to guerrilla marketers. There always seems to be someone who can make it or sell it for a lower price than you can! A low-price strategy is for the big firms with larger economies of scale and deeper pockets. Guerrillas are very tight with the dollar and watch their costs carefully, but they don't try to compete on low price. Guerrillas compete on better service and superior products, and sometimes by being the only game in town.

Rule 2: No matter how successful the company becomes, never act like the big firms Most guerrilla marketers must remain guerrilla marketers. To be successful, they must remain as lean as possible and have as many of their people as productive as possible. Formal organizational charts, job descriptions, and structured career paths are things which have little place in a firm operating under the principles of guerrilla marketing. Such entrapments will impede flexibility and quickness in trying to respond to changes in the marketplace.

This doesn't mean that guerrillas operate sweatshops; quite the contrary. These business firms are too small to afford nonproductive or disgruntled employees. They must have enlightened policies toward their employees to encourage their cooperation, goodwill, and hard work. Guerrilla employees must be flexible and be able to do a variety of tasks. Guerrilla employees may even be paid more than big-company employees, but they must be twice as loyal and work twice as hard.

A guerrilla firm needs to be able to take advantage of its talented people and small size to make quick decisions and to remain flexible in what may be a changing market.

Rule 3: Be prepared to abandon a market No market niche lasts forever! Look at the corner drugstore. Everyone needs friendly family-style service and everyone needs prescription medicines, vitamins, toothpaste, and shaving cream; right? Wrong! The discount chains, supermarkets, and large drugstore chains have just about put the local corner drugstore out of business. Few businesses last forever. Guerrillas always keep their eyes open for new market niche opportunities.

Remember, guerrillas do not attack stronger competitors head on, nor do they stand still and allow stronger competitors to attack them.

A guerrilla doesn't have the resources to squander in a marketplace where it has no chance of success. If a firm is providing a product or service and a much larger competitor moves into its marketplace that can provide the same item at the same quality and cheaper than the guerrilla shop can, it's time to consider moving on to a new opportunity. This is where the advantage of flexibility in a lean, motivated organization can really pay off. Rather than butt heads with a much stronger competitor, the guerrilla simply runs away to fight another day. That is, guerrillas search for a new market segment to concentrate on and focus their energies.

Some Tips   Here are a few additional points about guerrilla marketing to keep in mind.

Let People Know   Once a segment of the market is selected, potential customers must be told that a product or service is being offered. It's unlikely that the guerrilla marketer will call up a television network to place a series of advertisements during the Super Bowl half-time show. Rather, the guerrilla marketer may turn to local technology shows, trade association meetings, and selected sales calls on potential customers. More successful guerrilla marketers may expand their promotional efforts to include local trade journals, specialty magazines, or even direct mail.

Guerrillas carefully select their markets and customers, and they also carefully select the appropriate advertising methods to reach these targets. If their niches are based on a geographic strategy, then locally based advertisements are best. A unique-product-niche guerrilla may need to draw customers from a bigger market area and may choose to emphasize small advertisements in specialty magazines which are likely to be read by the target customer group. For example, in the case of guerrillas producing antique reproductions of solid mahogany furniture, readers of such magazines as *Colonial Home* or *Fine Homebuilding* may be potential customers.

Focus It   A critical aspect of such an advertising program is to choose a focus or theme and remain true to it throughout all of the advertising, regardless of the form. Choosing the advertising theme can be a tough decision. Decisions must be made about what will be emphasized and promoted about the business. Will it be a unique product, a special service, the superb quality of the product, speedy delivery, or any of a hundred different themes? Whatever is selected should communicate something special about the firm which separates it from its competitors. Commitment to the advertising/promotion program as a long-term investment is just as important as that new widget. The ad program must remain consistent to the special theme. It's not something to cut back on when times are rough.

Quality   One additional point for guerrilla marketers: they must understand that they have to offer a quality product or service for success. The most innovative strategy to find that particular segment coupled with the finest advertising in the world will not produce success if the product is of poor quality or design. The quality of whatever is being sold must be assured.

Generic Strategies Summary

Fig. 9-4 shows the three generic strategies in a matrix which has axes labeled ''strategic advantage'' and ''strategic target.'' We see from this figure that a differentiation strategy can be appropriate when the strategic market is industry-wide and there is some uniqueness in the product perceived by the customer. An overall cost leadership strategy can be appropriate when the strategic target is industry-wide and the firm has the opportunity/resources to achieve a low-cost position. For a firm to be successful in a focus strategy, it must focus on a particular segment of the market only and it may enhance this focus either through a product which is differentiated or by having a low-cost position, or both.

It is possible to modify these generic strategies and add an additional hybrid strategy combining low cost and differentiation (Day, 1984). A four-cell matrix, shown in Fig. 9-5, is used to position these strategies using customer price sensitivity as one axis and real or perceived relative differences in product offerings as the other axis.

Where customer price sensitivity is high and product differences are small, an overall cost leadership strategy is most appropriate. Where customer price sensitivity is high but significant differences are perceived in products, a differentiation strategy is recommended. Where customer price sensitivity is low and real differences in products are small, a hybrid strategy combining low cost and emphasizing differences between the products may be best. Where product differences are large and customer price sensitivity is low, a focus strategy is suggested.

FIGURE 9-4
Summary of Porter's (1980) three generic strategies.

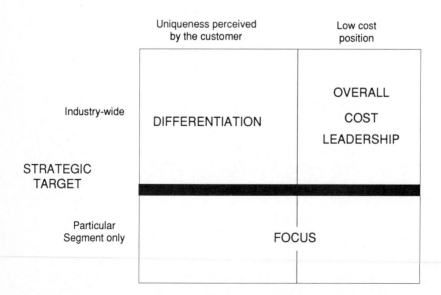

STRATEGIC ADVANTAGE

	Uniqueness perceived by the customer	Low cost position
Industry-wide	DIFFERENTIATION	OVERALL COST LEADERSHIP
Particular Segment only	FOCUS	

STRATEGIC TARGET

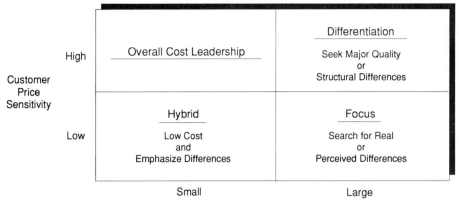

FIGURE 9-5
Day's (1984) generic strategies as a function of customer price sensitivity and product differences.

IMPLEMENTING STRATEGY

You may have heard the old cliché, a poor strategy effectively implemented is better than a good strategy inefficiently implemented. Even though this is generally acknowledged as a truism, there is precious little research or writing to assist in effective implementation of marketing strategy (Bonoma, 1984; Bonoma and Kosnik, 1990). Strictly by force of will, a clever leader of a guerrilla marketing firm may be able to secure effective strategy implementation. But for larger firms, ''It is invariably easier to think up clever marketing strategies than to make them work under company, competitor and customer constraints'' (Bonoma, 1984).

Consider a major forest products firm, which after many years of producing commodity products and offering them for sale through independent and company-owned wholesale distributors decided to change its stripes. A small, extremely talented group of people was hired to give the firm more of a marketing orientation. One of the early strategies was to take a commonly available lumber commodity product and package it with a variety of complementary specialty items, a computer-aided design system, and tons of sales literature, sales support, publicity, promotion, and advertising to turn the commodity product into a branded specialty product. There was considerable market demand for this product and a degree of customer dissatisfaction with the typical quality of this product available in the market at that time. In short, customers were unhappy with the product now on the market and a good opportunity existed to capture the upscale end of this market with a carefully branded product.

The strategy was carefully laid out and it was, by most accounts, a sound and potentially profitable strategy. But then the troubles started. This new branded product line had to be distributed through the company-owned wholesale distribution centers that were used to offering commodity products to price-sensitive customers. Now these distribution centers were told to sell this new higher-priced product using marketing tools they had never before used. Further, the distribution center managers traditionally had nearly total control of their operations and resented being told to carry in a different way a so-called new product for a higher price.

Other problems surfaced quickly as distributors and customers realized the newly branded commodity product was not of superior quality to other offerings of the same product by other manufacturers. The marketing strategy group tried to convince the commodity-oriented mill managers to produce a better-quality product. But this would have meant sorting out the better boards from the mills, which in turn would have resulted in the remaining mix of boards being of overall lower quality, thus antagonizing existing commodity customers.

Unfortunately, even though the strategy was good, implementing it in the face of the company's commodity culture was next to impossible. Such problems in marketing have two components, structural and human (Bonoma, 1984). In this case, the company's marketing structure was geared to commodity production and selling, a tough structural barrier to overcome. And the distribution center staffs (i.e., the human component) did not have the necessary skills or commitment to make the new strategy a success.

Changing a company's basic way of doing business is not easy and it is clear that this strategic move suffered from implementation problems. Another commodity-oriented forest products firm moved into a different product line consisting of nearly clear hardwood boards sold at a premium price to home centers. Do-it-yourself customers purchased the boards for home projects. This firm's strategy was laid out in a much less sophisticated fashion but it included a special processing facility to manufacture the boards and complementary items, a separate sales staff, and a distribution system separate from the company's main commodity business. It is easy to see how the implementation of this strategy avoided many of the potential structural and human problems encountered by the company in the previous example.

Strategy or Implementation

When is faulty strategy to blame for a failure and when is poor implementation the culprit? Obviously they affect each other. Fig. 9-6 may provide some help in diagnosing such problems. The first example falls into the lower left quadrant where a potentially good strategy failed due to an inability to implement. Of course, the cause of failure in many instances is difficult to determine because either a poor strategy or a good strategy can be masked by an inability to implement it. Poor implementation can easily cause management to doubt a good strategy.

With a poor strategy (upper right quadrant), good implementation may give management time to recognize the strategy failure and correct it, or good implementation may hasten failure. Some lower-level managers have stuck out their necks and violated current strategy policy to save a particular business situation because they recognized the strategy to be flawed.

Remember that poor execution of strategy can mask its appropriateness. As a result, managers should carefully examine their marketing practices leading to the strategy's implementation before concluding the strategy is flawed. Many problems which at first seem strategic are likely to lie instead in the execution of a strategy (Bonoma, 1984).

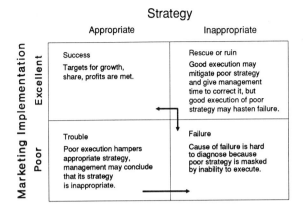

FIGURE 9-6
Diagnosing marketing strategy and
implementation problems.
(*Bonoma, 1984.*)

SUMMARY

Marketing fundamentals can be combined in many ways to develop competitive strategies. We have reviewed the notion of competitive advantage and why it is important. Generic strategies were also presented and the difficulty of strategy implementation was discussed. Before moving ahead to specific segments of the forest products industry, we will examine the trends which have buffeted this industry in Chap. 10.

BIBLIOGRAPHY

Bonoma, T. V. 1984. "Making Your Marketing Strategy Work." *Harvard Business Review* (March/April):69–76.

Bonoma, T. V., and T. J. Kosnik. 1990. *Marketing Management: Text and Cases.* Irwin. Homewood, IL.

Day, George S. 1984. *Strategic Market Planning: The Pursuit of Competitive Advantage.* West Publishing Co. St. Paul, MN.

Garvin, D. A. 1984a. "What Does Product Quality Really Mean?" *Sloan Management Review* (Fall):24–43.

Garvin, D. A. 1984b. "Product Quality: An Important Strategic Weapon." *Business Horizons* 27(May/June):40–44.

Haas, E. A. 1987. "Breakthrough Manufacturing." *Harvard Business Review* 65(2):75–81.

Jacobson, R., and D. A. Aaker. 1987. "The Strategic Role of Product Quality." *Journal of Marketing* 51(October):31–44.

Keating, B. 1986. "An Examination of Voluntary Product Standards." *Journal of Applied Business Research* 2(Winter):35–48.

Markell, Stephen A., Thomas H. Strickland, and Sue E. Neeley. 1988. "Explaining Profitability: Dispelling the Market Share." *Journal of Business Research* (16):189–196.

Miller, T. A. W. 1988. "31 Major Trends Shaping the Future of American Business." *The Public Pulse.* 2(1):1–8.

Ouchi, W. G. 1981. *Theory Z.* Avon Books. New York.

Phillips, L. W., D. R. Chang, and R. D. Buzzell. 1983. "Product Quality, Cost Position and Business Performance: A Test of Some Key Hypotheses." *Journal of Marketing* 47(Spring):26–43.

Porter, Michael E. 1980. *Competitive Strategy: Techniques for Analyzing Industries and Competitors*. The Free Press. New York.

Rich, Stuart U. 1986. "Recent Shifts in Competitive Strategies in the U.S. Forest Products Industry and the Increased Importance of Key Marketing Functions." *Forest Products Journal* 36(7/8):33–44.

Ries, Al, and Jack Trout. 1986. *Marketing Warfare*. McGraw-Hill Book Company. New York.

Shetty, Y. K. 1987. "Product Quality and Competitive Strategy." *Business Horizons* (May/June):46–49.

Sinclair, Steven A., and Kevin E. Seward. 1988. "Effectiveness of Branding a Commodity Product." *Industrial Marketing Management* 17:23–33.

Sonnenberg, F. K. 1989. "Service Quality: Forethought Not Afterthought." *Journal of Business Strategy* 10(5):54–57.

South, Stephen E. 1981. "Competitive Advantage: The Cornerstone of Strategic Thinking." *Journal of Business Strategy* 1(4):15–25.

CHAPTER 10

STRATEGIC TRENDS IN THE FOREST PRODUCTS INDUSTRY

The future is unpredictable, but it is not a random walk. In each market are strong likelihoods, built in dynamics, and even a few near-certainties.

George Day

Most recent strategic changes in the forest products industry have been dominated by the memories of the 1981–1982 recession. This recession in the wood products industry was caused by high mortgage rates and the subsequent collapse of the housing market. New housing starts in the United States, according to some estimates, dipped to below 1 million and only a few years earlier they had been above 2 million. Demand for paper also slowed, hurting paper producers as the economy struggled. This caused tremendous upheaval as the industry tried to cope with dramatic changes in its markets.

Many of the temporary changes were clearly unpleasant for the industry and its employees. For example, by the end of 1982:

- Forty percent of the workers of the Southern Forest Products Association member firms were laid off or working reduced hours.
- Fifty-five percent of the workers of the Western Wood Products Association member firms were laid off or working reduced hours.
- Twenty-five percent of British Columbia's sawmill work force were unemployed.
- Thirty-two percent of U.S. plywood workers were laid off (Corlett, 1988).

Such a trauma to the industry is not easily forgotten and in the years since then the North American forest products industry has undergone many changes. Obsolete operations have been closed or renovated. Companies have been restructured to cut costs and boost efficiency. New technologies have been employed to raise output, lower labor costs, and boost product quality. New nonhousing markets for wood products have been developed or expanded. And considerable industry consolidation has occurred.

MANUFACTURING COST REDUCTIONS

Manufacturing cost reductions have been achieved after a long and significant struggle. Several major firms have had long and acrimonious struggles with their unions to hold down or reduce wages. Many new cost-saving technologies have been adopted by firms. And western firms were released from costly government timber contracts and many got smarter about bidding on federal timber and about timber supply issues in general. The timber set-asides in the early 1990s for the spotted owl and other needs have completely turned the timber supply around and when export needs are considered it appears likely we may face a shortage of timber for domestic manufacturing.

Rich (1986) studied major forest products corporations in the time period 1976 through 1979 and again in 1984. He noted that there was an increase in the importance of technological strength of forest products firms between 1976 through 1979 and 1984. This serves to confirm the conventional wisdom that forest products firms are strongly investing in new technologies.

MARKETING FOCUS

Most major forest products firms and their associations have long realized the importance of better marketing for the industry. Most of the major associations in the United States and Canada have multimillion-dollar marketing/promotional programs under way for their member firms. Most of these programs are placing less emphasis on new residential construction and are increasing emphasis on repair and remodeling markets, light industrial and commercial construction markets, and export markets.

The major forest products firms have better organized their marketing efforts. Most major firms have national accounts systems in place. Through these systems they have a key account representative who is a liaison individual with each large national account they have. For example, Georgia-Pacific might have one person who is responsible for sales to Lowe's. Many of the firms have also undergone a reorganization to better address specific markets.

PRODUCT CHANGES

Most forest products firms have begun to place more emphasis on differentiated products over standard grades and sizes of commodities. Considerable emphasis is being placed on such differentiated products as machine stress rated lumber, lumber for wood trusses and trusses themselves, and radius edge decking.

Some firms are also placing considerable emphasis on specialty brand name products. They include engineered structural products such as Parallam by MacMillan Bloedel and Micro-lam by T. J. International. Louisiana-Pacific has pioneered a line of OSB-based horizontal siding products under the brand name Inner-Seal Siding. Weyerhaeuser moved into the specialty market with its ChoiceWood group, producing high-quality hardwood boards and moldings for sale in the home center market.

Firms also have a number of efforts under way to upgrade their commodity products out of the standardized industrial commodity classification toward a differentiated commodity. Such things as cut-to-size panels for the home center marketplace or industrial uses are now common. Many firms have also overlaid their panels, particularly industrial particleboard panels, to achieve more product differentiation. Some firms have developed new lumber grades to address special markets and other firms have begun to emphasize sanded grades of plywood.

The thrust in recent years for most firms has been to move from the production of standardized commodity items to differentiated and specialty products. Clearly, this is consistent with the information that firms are moving away from the dominant strategy of overall cost leadership into more market-oriented strategies such as differentiation and focus (Rich, 1986).

INDUSTRY CONSOLIDATION

One of the characteristics of mature industries is consolidation. The forest products industry is no different and major consolidations have become commonplace. In recent years, Champion International acquired St. Regis, James River acquired Crown Zellerbach, Georgia-Pacific acquired Superwood, U.S. Plywood, Great Northern Nekoosa, and Brunswick Pulp and Paper, International Paper acquired Hammermill Paper and Masonite, and Stone Container acquired Southwest Forest Industries. This list represents only a sampling of the larger merger and acquisition activities which have occurred. Many smaller purchases are being made all the time by forest products coporations.

It has yet to be seen if the major forest products firms can benefit from their increased size. As the wood products side of the industry moves more toward an oligopoly, the increased market share of major firms may allow them to have a stronger influence over the market as they do in the more oligopolistic paper segments.

Some studies have shown that medium-sized firms are most profitable, small companies the next most profitable, and the largest forest products firms the least profitable (Rich, 1986). Other studies have shown no direct impact of market share on profitability for forest products firms. A study of the largest softwood lumber and structural panel producers in North America concluded that market share had no significant direct impact on profitability (Cohen, 1989). Within a commodity-like industrial product market, there appears to be little profit advantage in building market share beyond a minimum efficient size. In the paper industry, market share appears to be irrelevant in terms of explaining firm profitability (Markell et al., 1988).

A PREFERENCE FOR WOOD

As the work and home environment we live in becomes increasingly part of the computer age, the term ''high-tech, warm touch'' has been spawned. This refers to people's need to touch the warmth of natural materials within our increasingly high-tech world and speaks well to consumers' continuing desires for real wood products. In a Virginia Tech study of over 1500 households, an overwhelming percentage of the respondents indicated a strong preference for real wood and wood veneers over wood-look plastics in furniture. Real wood will remain a material of preference for the majority of Americans. The warmth and patina of wood will continue to draw on people's emotions and influence their buying behavior.

The one exception to this may be in the use of tropical species. The strong concern over global warming and other environmental issues has placed the harvesting of most tropical species in a very negative light. In Europe, considerable consumer resistance has slowed the sales of products manufactured from tropical hardwoods. This same consumerism is spreading to the American marketplace. Some wood users are abandoning the use of wood from non-sustained-yield forests. This shows the signs of being a very controversial issue with a potentially negative outcome for some heavy users of tropical hardwoods.

FAVORABLE DEMOGRAPHICS

The dynamics of the American population also play a big role in determining the demand for forest products. The number of household units within the United States is growing at a rate much faster than the overall population. This is because of an increasing number of single-person households and a decline in the number of persons per household. Each household unit requires a certain basic level of housing and furnishings, creating a higher level of demand than would normally be expected based upon the general increase in population alone.

The prime age group which purchases the majority of forest products in the United States is the 24- to 54-year-old group. For the remainder of this century, this age group will be increasing as a percentage of the population. This is a result of the baby boomers passing through the population. This phenomenon of the population weighted heavily to traditional buying age groups should keep demand strong during the 1990s.

The U.S. marketplace, according to some economists, is becoming a two-tiered market. This is because of a decline in the U.S. middle class. High-paying factory jobs are no longer as prevalent as they once were, resulting in a polarization of the population with high-wage earners in the professional trades at one end and lower-salaried clerical and service workers at the other. This likely has good implications for producers of high-end and low-end products and may have negative implications for traditional producers of products designed for ''middle America.'' One only needs to witness the problems Sears has had over the last few years to recognize the problem in catering to the American middle class.

DEMAND FOR QUALITY

The baby boomers of the 1950s and 1960s grew into the yuppies of the 1980s and as this prime buying group in the population reached the 1990s some have called them ''couch potatoes.'' A less derogatory term coined for this phenomenon is ''cocooning.'' This increasingly affluent segment of the population has grown tired of the fast-paced yuppie lifestyle and has retreated to their homes in an endeavor to make them into ''cocoons'' as a shelter from the harried world. This, and other factors, have resulted in an increasing demand for upscale products. This is clearly evident in the housing market. The demand for upscale housing is at an all-time high. New buyers require massive master bedroom suites and gourmet kitchens in their new homes. The sales of wood flooring have strongly risen since the lows of the early 1980s and demand appears to be increasing.

Nowhere is this urge for quality more apparent than in the automotive marketplace. The Japanese car manufacturers, which seem famous for being able to gauge the pulse of American consumers, are continuing to tout their new luxury car lines. Honda was first with its Acura and more recently Toyota joined the fray with Lexus and Nissan with the Infiniti line. Of a less expensive nature, the sales of many other personal luxury items are on the rise also. Witness the increase in sales of expensive fountain pens and the mystique of Rolex watches.

All this has astonishing implications for manufacturers of high-quality items. Manufacturers of high-quality forest products should strive to take advantage of these trends and position themselves and their products to cater to this segment of the population, who will be demanding extraordinary quality in the products they purchase. However, manufacturers should not delude themselves into thinking that the only aspect of quality these upscale buyers care about will be the actual look and finish of the product itself. Buyers are increasingly concerned about the entire product. That is, they expect it to be delivered when delivery is promised. If they order a certain style and finish they expect that to be the style and finish which is delivered, etc. It's the entire package of the product and the related customer services which this new, more affluent group of the population will desire.

ENVIRONMENTAL CONCERNS

Another of the many trends impacting on the marketplace in the 1990s will be those political and environmental factors whirling around all manufacturing industries and particularly those using a natural resource. Many new health and safety issues will likely be thrust upon small manufacturers. We have already seen the wood dust issue and tighter emission controls on finishing systems. These and many other safety issues will make the job of the manger in the future much more difficult.

Some states are requiring that all items shipped into their borders which are flammable meet certain flammability standards. This has resulted primarily from the need to ensure safe homes and workplaces due to potentially toxic emissions from products as they burn during a fire. However, the vast array of regulations to enforce the state statutes has resulted in a nightmare for many small and large wood products manufacturers. Many forest products associations have lobbied heavily against such standards, but increasing regulation in this area is likely in the future.

As the concern over the environment continues to grow in the United States, the political reality is that we may have increasing difficulty in the future harvesting the trees we need to supply the raw materials for our many wood-using industries. To the extent that this difficulty grows, the prices of the raw materials available will also likely increase, thus resulting in increased problems and costs in procuring wood-based raw materials.

While concern over the environment may make some managers' jobs more difficult, it also presents many opportunities. Environmentally conscious companies may find advantages in the marketplace. Already hardwood lumber from the sustained-yield forests of North America is viewed as a more environmentally acceptable product in many European countries than tropical hardwoods (despite the fact that some tropical hardwoods are managed on a sustained-yield basis). Companies producing recycled paper have found strong markets for their products. Recycled pallets and other wood products are also widely accepted in many markets.

NEW TECHNOLOGY

More expensive and hard-to-get raw materials will, of course, rivet our attention even more strongly on maximum utilization. Equipment to enhance yield will be more important, as will better employee training. Better utilization of materials will assume increasing importance.

Another issue which will impact strongly in the future is the growing use of sophisticated equipment. This is clearly evident in the pulp and paper, softwood lumber, and structural panel areas. However, the advent of CNC (computerized numerically controlled) and NC (numerically controlled) equipment has also boosted the productivity of many furniture and cabinet manufacturers while at the same time greatly enhancing their flexibility and reducing setup times. A Virginia Tech survey of furniture manufacturers showed the purchase of CNC and NC equipment to be increasing rapidly during the early 1990s.

To the extent to which this trend continues, the small producer may find itself competing with larger, more heavily capitalized competitors that have gained lower per-unit cost and enhanced flexibility through the purchase of modern and sophisticated equipment.

THE FUTURE

What will the future bring to the forest products industry? Favorable demographics and a population that prefers wood are in the future. Also, there will be an increasing demand for quality products. But life and business will become more complicated as environmental and health issues impact more strongly on forest product businesses and their customers.

Companies will become larger and will invest in new technologies and in other manufacturing cost reduction methods. Products will become more differentiated and firms will become more marketing-focused.

BIBLIOGRAPHY

Cohen, David H. 1989. *The Adoption of Innovative Wood Processing Technologies in the Building Products Industry.* Unpublished Ph.D. dissertation. Department of Wood Science and Forest Products, Virginia Polytechnic Institute and State University. Blacksburg, VA.

Corlett, Mary Lou. 1988. *Forest Industries 1988–89 North American Factbook.* Miller Freeman Publications. San Francisco, CA.

Chrysikopoulos, John, and Bruck Kirk. 1989. *Forest–Paper Outlook and Company Update Service.* Goldman Sachs Investment Research. New York.

Markell, Stephen A., Thomas H. Strickland, and Sue E. Neeley. 1988. "Explaining Profitability: Dispelling the Market Share." *Journal of Business Research* 16:189–196.

McCahey, M. S. 1990. "California Stores Asked: Is Furniture Toxic?" *Furniture Today* (November 12):1, 22.

Porter, Michael E. 1980. *Competitive Strategy: Techniques for Analyzing Industries and Competitors.* The Free Press. New York.

Rich, Stuart U. 1986. "Recent Shifts in Competitive Strategies in the U.S. Forest Products Industry and the Increased Importance of Key Marketing Functions.'' *Forest Products Journal* 36(7/8):34–44.

Stureson, F. N. 1989. *Consumer Perceptions and Attitudes Regarding Ready-to-Assemble Furniture*. Unpublished M.S. thesis. Department of Wood Science and Forest Products, Virginia Polytechnic Institute and State University. Blacksburg, VA.

West, C. D. 1990. Competitive Determinants of Technology Diffusion in the Wood Household Furniture Industry. Unpublished Ph.D. dissertation. Department of Wood Science and Forest Products, Virginia Polytechnic Institute and State University. Blacksburg, VA.

PART **FOUR**

MAJOR FOREST
PRODUCTS INDUSTRY
SEGMENTS

The various segments of the forest products industry differ greatly. These differences are mirrored in the markets they serve, the products they produce, the capital intensity they employ, and the level of marketing tools they use. For our purposes we will divide the forest products industry into three major segments: (1) building products, which is dominated by structural panels and softwood lumber; (2) pulp and paper; and (3) hardwood lumber and secondary products, including such industries as furniture and pallets. We will examine each of these major industry segments in turn.

"*Well, Al, the sixties was <u>peace</u>. The seventies was <u>sex</u>. The eighties was <u>money</u>. Maybe the nineties will be <u>lumber</u>.*"

CHAPTER **11**

BUILDING PRODUCTS

Lumber is the most ubiquitous building product and represents a $10.5 billion market in the United States.

Louisiana-Pacific
1989 Annual Report

The production, distribution, and sale of wood building products at one time dominated most firms in the forest products industry. Their corporate policies were structured around the need to produce structural softwood lumber products. In more recent years, the production of pulp and paper products has assumed a more dominant role; however, the production and marketing of building products is still of immense concern to many companies.

Wood building products have long dominated in North American residential construction. The combination of their availability, versatility, and low prices has proven unbeatable for many years. Wood building products have had much less success competing for market share in the industrial and commercial construction markets.

In this chapter, we will first look closely at wood structural panels and softwood lumber. Then we will explore the markets for and marketing of wood building products in more depth. The impact of residential housing on demand for these products will be explored, along with other factors influencing the marketing of wood building products.

STRUCTURAL PANELS

The development of the wood structural panel has largely been of North American origin and is still dominated by North America. About half of the world's softwood plywood and 97 percent of its oriented strandboard are produced in North America (Widman, 1990). The invention of structural plywood has been credited to Carlson and Bailey. These men were employees of the Portland Manufacturing Company and in 1905 they constructed the first piece of structural plywood. The product gained slow commercial acceptance and was first used in such items as door panels, box crating, trunk stock, and furniture drawer bottoms. Early plywood grades were simple and consisted of good two sides, good one side, and drawer bottom stock.

Later in 1950, James d'A. Clark developed the first manufacturing facility to produce waferboard at Sand Point, Idaho. Rather than using veneers, this new product used small flat wafers of wood which were then glued together to form a structural panel. This initial manufacturing facility was unsuccessful and it was not until the 1960s that MacMillan Bloedel operated the first commercially successful waferboard plant at Hudson Bay,

Saskatchewan. The first successful U.S. waferboard producer was Blandin Wood Products of Grand Rapids, Minnesota (now part of Potlatch), in the early 1970s.

Waferboard technology soon gave way to the structurally superior oriented strandboard (commonly called OSB). In OSB, the small wafers of wood are still glued together; however, the wafers are now strands of wood which are oriented in layers at 90° to each other. This gives OSB properties similar to softwood plywood. Elmendorf Manufacturing Company in Clairemont, New Hampshire, opened the first OSB facility in the United States in the early 1980s. By the beginning of the 1990s, there was very little waferboard being manufactured in North America and most mills had converted to producing OSB.

By the early 1990s, two companies were dominant in structural panel production: Georgia-Pacific in plywood with approximately 5 billion square feet (3/8-inch basis)[1] of annual production and Lousiana-Pacific in OSB with approximately 2 billion square feet (3/8-inch basis). Georgia-Pacific is also a major OSB producer and Louisiana-Pacific a major plywood producer.

Structural Panels and the Product Life Cycle

As can be seen in Fig. 11-1, the growth of wood structural panels in the United States has been nothing short of phenomenal. However, this total growth can mask the changes that have been occurring within the structural panel market by product type. These changes can be examined by using the product life-cycle analogy. You might recall that in Chap. 4 the four stages of the product life cycle—introduction, growth, maturity, and decline— were discussed.

Looking again at Fig. 11-1, it is easy to see that the introductory stage for softwood plywood made from western species lasted until the end of World War II. After the war, production of western softwood plywood increased dramatically until the mid-1960s. During the early 1960s, the technology was developed to produce structural plywood from southern pine timber. Production was pioneered by Georgia-Pacific and the growth of southern pine plywood has remained strong through the time of this writing. The early growth of southern pine plywood benefited greatly by having lower rail transportation costs than western plywood to midwestern and eastern markets.

Two interesting things in Fig. 11-1 should be noted at this point: (1) when southern pine plywood entered into production, western plywood moved from a growth phase into a maturity phase and (2) because of the wide product acceptance of western plywood, and due to cost advantages in production and transportation, southern pine plywood skipped the introductory stage of the life cycle and went immediately to the growth stage.

In the late 1970s, U.S. production of waferboard and oriented strandboard began to gain momentum. The introductory phase for this product lasted until the early 1980s, and was largely served by Canadian producers, which typically export 45 to 60 percent of

---

[1]Structural panel production is reported based on the square feet of surface area. For example, a 4 × 8-foot sheet would have 32 square feet. Most structural panels come in a variety of thicknesses. Therefore, production volumes are converted to a standard 3/8 inch thick and surface measures adjusted accordingly.

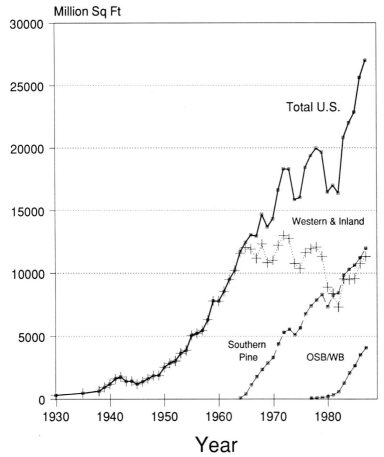

FIGURE 11-1
Product life cycles for U.S. wood structural panels.

their production to the United States. After the early 1980s, the product entered a strong growth phase, which at the time of this writing still continues.

It's important to note at this point that the shifts in the structural panel market were made possible by new technologies. But in most cases these technologies were available long before they were used to make panels commercially. The driving force behind the implementation of these technologies and the resulting shifts in the structural panel market has been timber supply and cost.

The strong acceptance of waferboard and oriented strandboard panels into the marketplace was made possible by a shift in the grading standards. This shift occurred in the 1980s as the structural panel industry went from specification-based standards for its products to a performance-based system of standards. The old product standards specified the species, grade of veneer, veneer thickness, and type of adhesive that was necessary to produce a plywood panel of a particular grade. While a new Com-ply panel or nonveneer panel (OSB/waferboard) might perform just as efficiently in use, it was obvious it could not meet the old product standards (i.e., it was no longer made of sheets of veneer). To solve this problem, the industry, led by the American Plywood Association, went to a system of performance-based standards. The standards for panels were based on performance criteria and not the type of manufacturing used to produce the panel.

Originally, the nonveneered panels had as much as a 30 percent production cost advantage over plywood and were sold in the marketplace at a discount to plywood prices. This lower selling price and much lower production cost coupled with the new grading standards allowed nonveneered structural panels to gain rapid acceptance in the structural panel marketplace. However, by the late 1980s to early 1990s, numerous new cost-reducing technologies for plywood had been developed, including high-moisture-content gluing, power backup rolls, spindleless lathes, and many others, which greatly reduced the cost of plywood production. In fact, some analysts believe that, in state-of-the-art plywood production facilities, production costs in the 1990s will be comparable to those for nonveneered panel production. Coupled with strong markets and available timber, this may partially explain the resurgence of western plywood production in the late 1980s (Fig. 11-1).

The strong growth of the nonveneered panel industry and the new technologies in the plywood side of the industry have combined to create the need for much larger mills. For instance, in 1978 there were 182 structural panel mills producing an average of 110 million square feet per mill. However, by 1988 the total number of mills had dropped to 169, but these mills produced an average of 161 million square feet per mill. This represented a 46 percent increase in the amount of production from an average structural panel facility (Anderson, 1989).

In 1989, North American structural panel production topped 31 billion square feet ($\frac{3}{8}$-inch basis), with U.S. production accounting for nearly 85 percent of the total. From 1990 to 1994, North American production is expected to climb 13 percent (Widman, 1990).

Role of the Plywood Association

The role of trade associations in the development of the North American system of light frame construction and the efficient production of products to service it has been significant. In structural panels, the dominant association was the Douglas Fir Plywood Association, now known as the American Plywood Association. This association assumed a leadership role in developing product standards and grades for plywood and later performance-based standards and grades which allowed the growth of the nonveneered panels. Appropriate testing and grade certification procedures allowed the association's grades to be widely accepted by building code officials and to become the marketplace standard by which all wood structural panels are evaluated (Fig. 11-2).

The development and acceptance of a single widely accepted standard set of grades in the United States created significant economies in the manufacture and marketing of structural panels and served as a major catalyst to the growth of the structural panel industry. Although the use of these standards and grades is in the strictest sense voluntary, their acceptance by building codes does present a formal legal barrier to certain markets for products which don't meet the standards. The American Plywood Association provides for its members a grade stamp and quality assurance program which has now been extended to other engineered wood composite products also. Although organizations such as Timber Products Inspection and others also offer grade stamp/quality assurance programs, the American Plywood Association dominates in the structural panel market.

In addition to developing traditional domestic residential construction markets through building code approval and quality assurance programs, the American Plywood Association is developing markets in the industrial area for shipping bins to replace steel drums and for truck trailer siding and for agricultural and commercial buildings. Strong development efforts have taken place in permanent wood foundations and export markets too. As is seen in Fig. 11-1, the growth of the wood structural panel market in the United States has been phenomenal, and the association played a strong role in that growth. However, other critical issues are also handled by the American Plywood Association and others for the industry. These issues include time supply, environmental concerns, and many others in which the industry needs a collective voice.

Regional Production Trends

As is easily seen in Fig. 11-1, the western states were essentially the only plywood-producing region up until the mid-1960s, at which time southern pine plywood from the southern and southeastern states began to come into the marketplace. The western region remained dominant in plywood production up until the early 1980s, at which time the volume of production by the southern pine plywood mills exceeded that by the western plywood mills. Additionally, many of the new oriented strandboard facilities were built in the inland and northern regions, which has resulted in more balance between the various regions than ever before.

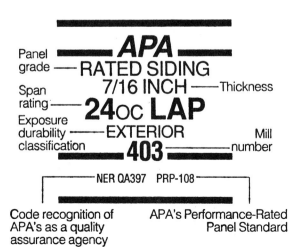

FIGURE 11-2
Example of an American Plywood Association grade stamp (trademark).

Major Markets

Residential Construction   New residential construction has been a very strong, sometimes even dominant, market for structural panels. In the 1990 to 1994 time frame, the American Plywood Association estimates that approximately 35 percent of the demand for wood structural panels will come from residential construction (Anderson, 1989). The housing market is comprised of single-family homes, multifamily homes, and mobile homes. Single-family homes consume over twice the amount of structural panels that multifamily homes do and mobile homes consume roughly 25 percent of the volume of structural panels that a single-family dwelling does. It is obvious that the mix of new housing starts can dramatically impact on the demand for structural panels.

Remodeling   The repair and remodeling market, or as it is sometimes called the R&R market, has been a growing market for structural panel use during the last decade or so. This particular market segment for the 1990 to 1994 time period is predicted to represent 21.5 percent of structural panel demand (Fig. 11-3). This market segment is divided into two major components, residential remodeling, which is approximately two-thirds of demand, and nonresidential remodeling, comprising the remaining one-third. There are by some estimates over 95 million housing units in the United States and 75 million of these units are estimated to be 20 years old or older. The increasing age of the U.S. housing stock serves as a powerful force driving demand for residential repair and remodeling. Nonresidential remodeling has been important as well. Older industrial and commercial buildings are continually remodeled to keep up with the times or better meet the needs of new tenants.

Industrial Markets   The industrial markets include a wide variety of segments which consume structural panels. The largest individual industrial market is for transportation equipment. Plywood panels are extensively used for truck and trailer liners, bodies and doors, for bus floors, and in boats, ships, and recreational vehicles. Another major segment includes the furniture and fixtures industry along with materials handling (such as pallets and crates), and other products which are manufactured for resale.

Nonresidential Construction   From 1990 to 1994, it is estimated that nonresidential construction will represent 12.5 percent of the structural panel demand (Fig. 11-3). This is a market with tremendous untapped potential for structural panel producers. Demand in this market can be impacted on by government policy as it relates to the economy, interest rates, and tax law changes. Probably the single biggest factor restricting demand is the poor market acceptance of wood products in many areas (especially east of the Rockies) for nonresidential construction. This stems partially from architects', contractors', and construction workers' familiarity and experience with steel and concrete and their lack of familiarity with wood.

Do-It-Yourself Home Use   The do-it-yourself (DIY) homeowner is predicted to represent 6.7 percent of structural panel demand between 1990 and 1994 (Fig. 11-3). This has generally been a growing market for structural panels. If the DIY segment is

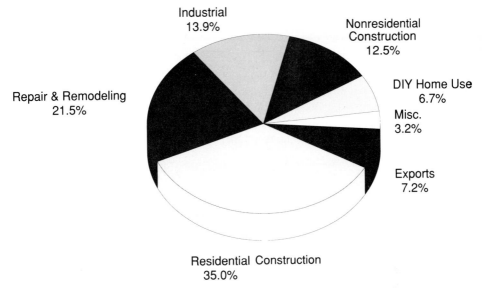

FIGURE 11-3
Average percentage of demand for structural panels by market segment from 1990 through 1994. (*Anderson, 1989.*)

combined with the residential portion of the repair and remodeling segment and the new residential construction segment, the true impact of the housing market for wood structural panels becomes clear very quickly.

Export Markets   One of the strongest growth opportunities for structural panels is in the overseas marketplace. Many associations, corporations, and government agencies have been working hard to secure building code approval in overseas markets for wood structural panels. This has resulted in substantial growth in the exportation of structural panels during the 1980s. For the time frame 1990 to 1994, exports are expected to continue their growth and will represent 7.2 percent of the demand for U.S.-produced structural panels (Anderson, 1989). When barriers to the Japanese market, such as Japanese building codes, are overcome, exports could expand even more.

Marketing of Structural Panels

The structural panel industry, dominated primarily by large-scale producers of commodity products, has traditionally relied on a production-oriented marketing approach. Analysts have attributed this to the low degree of product differentiation and the auction-like manner in which prices are set. As in an auction, prices are typically set between buyers and sellers in a fast-moving set of offers to sell at a given price by sellers and a set of corresponding offers to buy by potential buyers. Through this process prices are agreed upon. Weekly price newsletters like those by Crow's and Random Lengths report average prices for many standard items and many buyers and sellers use these prices as a starting point in their negotiations.

As with most wood building products, panels are mostly distributed by independent wholesalers and distributors. A Random Lengths survey of nonveneered panel producers showed that 61 percent of U.S. production goes through independently owned channels (Random Lengths, 1990). Another 25 percent of production moves through producer-owned distributors and 12 percent goes directly from producer to retailer (Fig. 11-4). Canadian distribution is similar but with a higher percentage moving through wholesalers/distributors and less going directly to retailers and end users.

Branding The primary focus of marketing for many forest products firms has been the generation of volume sales and the pursuit of cost-effectiveness in product distribution (Rich, 1979b). However, with the rapid emergence of waferboard and OSB the marketing philosophy of the industry appears to be changing. A majority of North American OSB/waferboard manufacturers have adopted a brand naming strategy for these products, deviating from their softwood plywood marketing policies, which generally included little or no emphasis on branding.

This is not to say that forest products firms have failed to use branding strategies. Many firms have, especially in the paper markets. Some names almost constitute market franchises such as Kleenex tissue, ScotTowels, and Hammermill bond. However, the wood products side of the industry has made much less frequent use of branding as a marketing strategy. Very few commodity grades of wood products, such as softwood dimension lumber or plywood, have been branded. Although Weyerhaeuser as early as the 1920s was using a brand name, 4-Square, on their softwood dimension lumber, few other companies followed suit.

FIGURE 11-4
Distribution of nonveneered structural panels. (*Random Lengths, 1990.*)

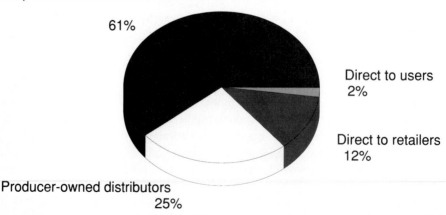

Independent wholesalers/distributors

61%

Direct to users
2%

Direct to retailers
12%

Producer-owned distributors
25%

TABLE 11-1
REASONS CITED BY MANUFACTURERS FOR IMPLEMENTING A BRAND NAMING
STRATEGY FOR THEIR OSB/WAFERBOARD PRODUCTS

Reasons	Frequency,* %
To differentiate the product from those of competitors	57.9
To better identify the product	26.3
To emphasize the product as a specialty product because the company was the first producer of a product or in a regional market	26.3
To develop a stable/loyal customer base	10.5

*Respondents may be represented in more than one category.
Source: Sinclair and Seward, 1988.

Why Use a Brand Name? In the late 1980s North American producers of OSB/waferboard were asked why they had introduced a separate brand name for their OSB/waferboard panel line. The rationale for choosing to/not to introduce a branding strategy varied considerably among manufacturers (Table 11-1) and was categorized into several primary areas.

The first and most frequently reported rationale (58 percent of the producers) was that the brand name served to differentiate the company's product line from those of its competitors. Product identification was perceived by 26 percent of the producers as being important in the development of their branding strategy. These manufacturers contended that the brand name served as an identification tool, increasing customer awareness while enabling the customer to relate specific product attributes and various product lines to the respective manufacturers. Sixty-eight percent of the producers agreed that their branding strategy provided a higher degree of protection from competitors than selling their product line generically.

Another factor, cited by 26 percent of the manufacturers as a basis for implementing a branding strategy, was the claim of being the first oriented strandboard or waferboard producer on the market (or in a regional market). These manufacturers considered these products a specialty line and wanted to stress that by introducing an individual brand name. Tied to this concept, one firm entering the market indicated it chose to introduce an individual branding strategy because other producers had set a branding precedent.

Interest in capitalizing on a stable consumer base by creating demand for the firm's brand was also mentioned as a basis for introducing a separate brand name. Manufacturers asserted that by developing a positive brand image for their reconstituted panel products, a resulting loyal customer base (i.e., professional building contractors) could be cultivated.

Ideal Panel Attributes The long-term success of structural panels, to a large extent, will depend upon how well the product is aligned with consumer demands. These demands were measured by surveying building materials retailers. These retailers provided a list of ideal structural panel attributes and a list of attributes which most influence their purchase decision in general.

As an ideal attribute, strength/stiffness was mentioned most frequently (41 percent)

(Table 11-2). Other important structural panel characteristics included low price (37 percent), surface/thickness uniformity (28 percent), dimensional stability (25 percent), and panel durability (19 percent).

Retailers also assessed the relative importance specific attributes had on their OSB/ waferboard purchasing decision (Table 11-3). Price and product availability were perceived by retailers as being of greatest importance when considering which particular OSB/waferboard brand to purchase. Service support, durability, and strength/stiffness were also important, while promotional support and panel weight ranked lowest in terms of importance. If a specific target market such as concrete forms were being analyzed, these attributes would likely be ranked differently.

Brand Name or Company Name?   Saunders and Watt (1979) reviewed a problem similar to the one facing OSB/waferboard manufacturers, where consumer confusion resulted from the variety of different brand names being assigned to nearly identical product offerings in the manufactured fiber market. To overcome problems plaguing this segment of the textile industry, they supported a product life-cycle approach, suggesting that new products be advertised generically in introductory stages. This enables the consumer to initially gain insight into the basic product attributes offered by the new product line. The promotion of the company image was noted as a more profitable strategy than introducing a large number of brand names where little product differentiation exists. Others have stressed the importance of the corporate image and

TABLE 11-2
IDEAL STRUCTURAL PANEL ATTRIBUTES AS PERCEIVED BY BUILDING SUPPLY RETAILERS

Ideal attributes	Response percentage, total United States,* %
Strength/stiffness	41
Low-priced	37
Product uniformity (surface/thickness)	28
Dimensionally stable	25
Durable	19
Moisture-resistant (weatherability)	13
Maximize performance/quality	13
Eliminate delamination	9
Lightweight	8
Workable (i.e., nailing, cutting)	8
Favorable appearance	8
Eliminate core voids	7
Improve span ratings	7
Skid-resistant surface	5
Smooth surface	5
Other	14

*Respondents may be represented in more than one item response category.
Source: Seward and Sinclair, 1988.

TABLE 11-3
RETAILERS' RANKING OF PRODUCT AND MANUFACTURER ATTRIBUTE
IMPORTANCE TO THEIR PURCHASE DECISION

Attributes	Retailers ranking attribute as important, %
Product attributes	
Price	78
Durability	67
Strength/stiffness	62
Dimensional stability	62
Quality/performance	59
Surface uniformity	54
Impact resistance	24
Panel weight	15
Manufacturer attributes	
Product availability	83
Service support	72
Company reputation	52
Sales support/financial incentives	37
Brand name	28
Depth of product line	27
Promotional support	22

Source: Seward and Sinclair, 1988.

insisted that a company's reputation as a reliable supplier of quality products reduces the need to develop and promote individual brand images (Hill et al., 1975).

Naming and promoting the new product (OSB/waferboard) generically during the introductory stage did not develop into an option. Early producers were few in number and the new products were available in only relatively small quantities. The new products had a distinct appearance and producers tried to capitalize on this by using a branding strategy to position the new products as specialties. Later producers followed the precedents already set but volume jumped dramatically, resulting in increased availability and the lessening of the specialty nature of the product.

Does Branding Work?   The emphasis on price and availability at the retail level suggests that OSB/waferboard products are sold/purchased on a commodity basis. Retailers are generally not concerned with the brand names when deciding which particular OSB/waferboard product line to purchase. However, some producers believe they benefit by marketing their OSB/waferboard product offering using a brand naming strategy. They believe that brand naming gives more protection from competitors than a generic name strategy, that brand naming identifies their product and differentiates it from those of competitors, and that brand naming promotes product recognition. In general, the most elusive brand naming benefit is the ability to command a higher price.

However, a few producers have made their branding strategy work. That is, their product line has achieved strong brand recognition and brand preference even to the point that customers are willing to pay a premium price for basically a commodity product line. Why have these producers succeeded while others clearly have not? Not all the factors are known, but what is clear is that these producers' products are perceived as the highest-quality products (Sinclair and Seward, 1988).

SOFTWOOD LUMBER

The production of lumber was one of the first industries of the early colonists of the United States. Early on, the sawmill industry was dominated by local firms serving local needs. Later the advent of large population centers and dwindling local resources forced lumber products to be shipped in from greater distances.

The production of softwood lumber in the United States reached its peak at about 45 billion board feet in 1909. In recent years, production has been as low as 23.8 billion board feet in 1982 and as high as 38.2 billion board feet in 1987.

The softwood lumber industry has historically been viewed as a fragmented industry which exhibits signs of maturity. Thousands of sawmills operate to produce the products demanded by our society today. In 1988, the top ten U.S. lumber producers manufactured 13.1 billion board feet of softwood lumber or only approximately 36 percent of the U.S. total (Corlett, 1989).

Softwood Lumber Industry Trends

Numerous trends have rocked the softwood lumber industry in recent years. One plague has been excess capacity. While the absolute capacity of the lumber industry is difficult, if not impossible, to measure, most analysts agree that in recent history the industry has had the capacity to produce more lumber than the market demanded at recent price levels. One source indicates that the industry could support the construction of 3 million homes per year while the typical market demand is only about half of that (Tillman, 1985). This overhang in capacity has also served to keep price increases on basic commodity grades of lumber products in check.

In the future, it appears likely that production of softwood lumber will not be limited by lack of market demand but rather by shortfalls in the timber supply. U.S. consumption of softwood lumber is predicted to increase substantially into the early 1990s. The combination of increasing demand, coupled with an array of new environmental and other factors restricting timber-harvesting activities, has been predicted to generate a shortfall in lumber supply (Table 11-4) (Widman, 1989).

Another trend impacting the softwood lumber industry (and structural panels) has been an increasing scale of production, which has caused ever declining numbers of mills and employees but increased production and productivity at the remaining mills. Perhaps the southern pine industry is as good an example of this phenomenon as any. In 1950, the southern pine lumber industry was producing approximately 9.9 billion board feet of southern pine lumber in 15,000 to 20,000 sawmills. Nearly 60 percent of the volume was in boards and finished grades and nearly 70 percent of total production went to residential construction. By 1961, production had dropped to 5.6 billion board feet in only 5000 to

6000 sawmills. Moving ahead to 1970, there were 7.7 billion board feet of southern pine lumber produced by 800 to 1200 sawmills. And the product mix had changed dramatically by this time also, with only 17 percent of the volume in boards and finished grades and nearly 82 percent in dimension lumber. In 1986, there were 11.9 billion board feet of southern pine lumber produced by fewer than 500 sawmills. In the space of 36 years, production was 20 percent higher and the number of sawmills had dropped from 15,000 to 20,000 to below 500 (Lindberg, 1986).

Temple-Inland, Inc., in its first-quarter 1989 report, spoke directly about the tremendous increases in productivity in southern pine sawmills. In 1984, it noted its Diboll sawmill required 5½ labor-hours to produce 1000 board feet of lumber. By 1988, that had been reduced to below 4½ labor-hours. However, the new sawmills under construction in early 1989 will require less than 3 labor-hours to produce 1000 board feet of lumber. In addition, the lumber yield from logs was increased by 6 percent at the older mills and by an even larger margin at the newer sawmills.

In order to compete with this new level of productivity in the industry, new and ever more expensive technologies are required. These new technologies are clearly raising barriers to entry into the industry to become an efficient producer. An article in *The Wall Street Journal* has indicated that advances in technology are reshaping the lumber industry (Bayless, 1986). The *Journal* article indicated that for many older mills the use of new technology is a clear threat, because many of them cannot afford the $15 million and up price tag required to modernize. The squeeze on raw materials supplies, especially logs from government timberlands, has added another level of uncertainty for independent sawmills considering making large investments (Bayless, 1986).

Domestic U.S. producers have, to a certain extent, abandoned portions of the commodity softwood lumber market to Canadian mills and other producers where possible. Their goal has been to move into more specialty product areas which can sustain higher margins with less cutthroat price competition.

TABLE 11-4
SOFTWOOD LUMBER CAPACITY AND CONSUMPTION PRELIMINARY FORECAST*
(MILLIONS OF BOARD FEET)

Year	Production Canada	United States	Consumption Canada	United States	North American excess (shortfall)
1989	24,600	35,300	6890	45,500	1514
1990	22,400	32,950	6650	42,200	698
1991	23,900	33,580	7670	46,300	(2906)
1992	24,800	34,700	7880	47,200	(2553)
1993	25,100	35,800	8000	51,400	(5961)

*Exports not shown.
Source: Widman, 1989.

An operator control station in a
modern softwood mill.

The Importance of Canada

Canada is a major lumber-producing country. As shown in Fig. 11-5, the production of softwood lumber in North America is split between Canada and the United States, with Canada producing 42 percent of the total and the United States approximately 58 percent. In recent years, Canada has supplied roughly 30 percent of the U.S. demand for softwood construction lumber. North American production and consumption of softwood lumber for 1988 are shown in Table 11-5. In that year, U.S. lumber producers were able to produce only 70.2 percent of the lumber demanded in the United States, and imports of Canadian lumber accounted for the remaining 29.8 percent (Widman, 1989).

The increase in imported Canadian softwood lumber during the 1980s caused considerable concern among many U.S. lumber-producing companies. Much of this concern stems from the differences in U.S. and Canadian timberland ownership patterns and timber sale practices. For example, in British Columbia, the major lumber-producing province in Canada, 94 percent of the standing timber is owned by the provincial government. Cutting rights are allocated to companies on a long-term basis and stumpage prices are calculated based on a formula which reflects market conditions and the costs of doing business. However, in the United States a much lower percentage of the timber is publicly owned and cutting rights are sold on a tract-by-tract basis to the highest bidder.

As a result, stumpage prices in the United States vary more widely based on market conditions and the overall supply picture than in Canada (Widman, 1990). This and other factors resulted in U.S. charges of unfair trade practices and in late 1986 Canada agreed to impose a 15 percent export duty on softwood lumber shipped to the United States. Several Canadian provinces opted to convert the 15 percent duty into provincial stumpage charges, which were then pumped back into forest management. At the time of this writing, the long-term impact of this surcharge is unknown and trade negotiations are ongoing.

Regional Production

Many species of softwoods in the United States are converted to softwood lumber. As shown in Fig. 11-6, the U.S. has three main timber-producing regions: southern, west coast, and inland. The southern region, which in the most recent years has been the largest lumber-producing region, is located in the southern and southeastern states. The dominant product of the region is southern pine. Southern pine is a species group comprised of several varieties of southern pine; however, the four main species are

FIGURE 11-5
1989 North American softwood lumber production by region. (*Widman, 1990.*)

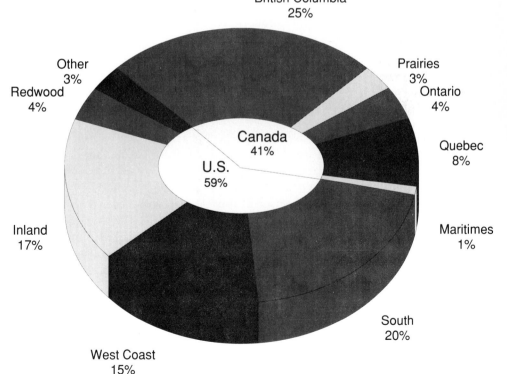

British Columbia
25%

Other
3%

Redwood
4%

Prairies
3%

Ontario
4%

Canada
41%

U.S.
59%

Quebec
8%

Inland
17%

Maritimes
1%

West Coast
15%

South
20%

TABLE 11-5
1990 CONSUMPTION AND PRODUCTION OF SOFTWOOD LUMBER IN THE UNITED
STATES AND CANADA

	Canada (thousand board feet)		United States (thousand board feet)	
Domestic production		22,810		36,825
Domestic consumption	6,950		46,300	
Exports offshore	3,967		2,414	
Exports to the United States	12,400			
Exports to Canada			486	
Imports from the United States		486		
Imports from Canada				12,400
Total	23,317	23,296	49,200	49,225
Adjustment for inventory		+21		−25
Total	23,317	23,317	49,200	49,200

Source: Widman, 1991.

shortleaf, longleaf, loblolly, and slash pine. The main market for southern pine over the last decade has been for preservatively treated wood products. By the early 1990s, over half of southern pine production was being treated.

The west coast region also plays a major role in timber production, with the main species being Douglas fir, true fir, and hemlock. West coast production is centered in the states of Washington, Oregon, and northern California.

The inland region, running from Idaho through the Dakotas and south to Colorado, produces large quantities of softwood lumber also. In most years recently, production in the inland region has exceeded that on the west coast. The dominant species produced are Ponderosa pine, lodgepole pine, inland varieties of Douglas fir, and various other cedars and pines.

The other two remaining U.S. categories are California redwood and other softwoods. Other softwoods include such species as red pine and eastern white pine, as well as smaller amounts of balsam fir and eastern spruce.

In Canada, British Columbia dominates lumber production, producing by itself 25 percent of the total North American output. British Columbian production is in two clear categories, coastal and interior. Coastal production has traditionally been marketed offshore and interior production has gone to U.S. and domestic markets. However, in recent years increasing volumes of interior wood have been exported offshore.

Major Markets

Residential Construction   Historically, housing starts have accounted for as much as 50 percent of the total U.S. softwood lumber consumption (Widman, 1989). As a result of this, forest industry analysts have traditionally paid very close attention to the current rate of housing starts and the forecasted housing starts as key indicators of the health of the lumber industry. As seen in Fig. 11-7, the percentage of the domestic softwood lumber market attributable to residential construction has varied in recent years from a high of 46 percent to a low of 33 percent.

When examining the market for softwood lumber in new residential construction, simply looking at the number of housing starts is insufficient. Several factors are at work. The construction marketplace has learned to use wood products much more efficiently to create housing. This is happening in two ways: (1) The industry from the early 1950s to the early 1970s built more mobile homes and multifamily dwellings at the expense of more lumber-intensive single-family homes. Since the 1970s the trend has been less clear (Fig. 11-8). (2) The industry decreased the lumber used per square foot of living space for all types of homes up through the late 1970s (Tillman, 1985). In Fig. 11-9, the increasing

FIGURE 11-6
U.S. lumber production forecast. (*Widman, 1989.*)

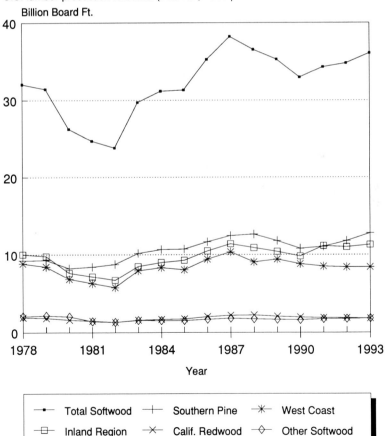

Billion Board Ft.

Legend:
- Total Softwood
- Southern Pine
- West Coast
- Inland Region
- Calif. Redwood
- Other Softwood

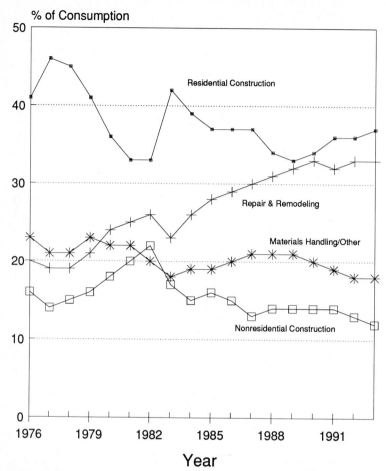

FIGURE 11-7
Domestic consumption of U.S. softwood lumber by end use. (*Widman, 1989.*)

efficiency of the construction industry is measured in the decline in board feet of lumber used per square foot of house constructed. This decline has resulted from many factors, including, but not limited to, a substitution of structural panels for softwood lumber, a decline in the use of wood floors, the advent of roof and floor trusses, slab construction, and the trend toward wider spacings of framing material. However, this trend bottomed out in the late 1970s and early 1980s as new homes began to feature more wood floors, wood doors, wood molding, wood cabinets, wood paneling, and other wood products. Recent data for single-family homes show a move toward more use of wood. As construction practices for mobile homes (HUD Code manufactured housing) became more like those for traditional housing, the wood use per square foot of dwelling also gradually increased beginning in the early 1970s.

Repair and Remodeling As shown in Fig. 11-10, the average percentage of softwood lumber consumed by the repair and remodeling market for the years 1989 through 1993 is estimated to be approximately 31 percent (Widman, 1989). The repair and remodeling market has shown strong growth over the last several decades. As shown in Fig. 11-7, the repair and remodeling market has grown from 19 percent of domestic softwood lumber consumption in the late 1970s to a forecasted 33 percent of domestic softwood lumber consumption by the early 1990s. Some analysts have predicted that the growth of softwood lumber consumed by the repair and remodeling market will eventually stabilize or at least grow at a slower pace (Widman, 1989).

Nonresidential Construction Historically, nonresidential construction has accounted for between 14 and 16 percent of the overall domestic consumption of softwood lumber. Consumption increased somewhat in 1981 and 1982 as high interest rates caused a tremendous decline in housing starts. Looking ahead to the future, Fig. 11-10 shows that nonresidential consumption will likely consume approximately 12.4 percent of softwood lumber from 1989 through 1993 (Widman, 1989). Nonresidential construction has consumed a fairly stable 6 to 7 billion board feet of softwood lumber (Widman, 1989). This market segment includes office buildings, retail stores, motels and hotels, churches, and public works projects. This segment tends to be less affected by recessions than new residential construction but on the other hand does not enjoy residential construction's cyclical peaks (Widman, 1989).

Materials Handling Another relatively stable market segment for the consumption of softwood lumber has been the materials handling and miscellaneous use segment.

New residential construction is a major market for softwood lumber and structural panels.

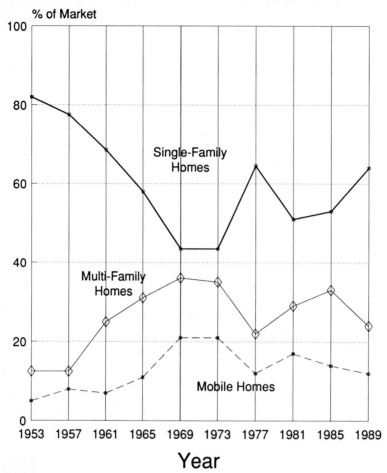

FIGURE 11-8
Distribution of the housing market by type of home. (*Adapted from Spelter and Phelps, 1984; new data added.*)

This segment has represented 18 to 23 percent of the domestic consumption of softwood lumber (Fig. 11-6). Slightly over half of this segment is consumed by such products as pallet stock, crates, and other types of packaging (Widman, 1989). Other uses of softwood lumber in this category include some rail ties and mine timbers along with industrial uses.

Exports   As is shown in Fig. 11-10, export markets were projected to comprise approximately 6 percent of the demand for softwood lumber in the period 1989 through 1993. Some analysts have projected that the average volume of softwood lumber exported from the United States will be approximately 3 to 3.5 billion board feet in the late 1980s and early 1990s. By far and away the largest customer for U.S. lumber exports is Japan, which in 1988 consumed almost 43 percent of U.S. softwood lumber exports.

Marketing Softwood Lumber

Much like primary producers of structural wood panel products, the producers of softwood lumber have typically been oriented toward production and not marketing. This is not surprising since the bulk of wood products sales by primary wood processors has, in the past, been largely characteristic of auction market sales of commodity products. The focus of marketing for many of these firms is chiefly the generation of volume sales and

FIGURE 11-9
Intensity of lumber use by type of housing structure, measured in board feet of lumber per square foot of housing structure. (*Adapted from Spelter and Phelps, 1984; new data added.*)

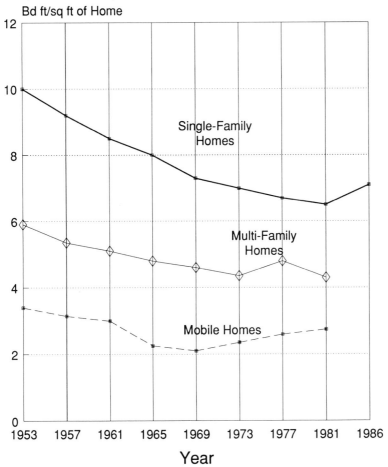

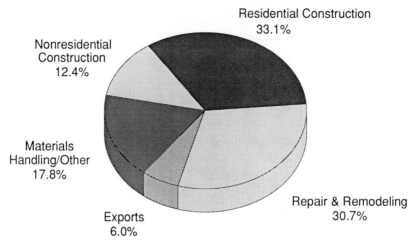

Residential Construction
33.1%

Nonresidential
Construction
12.4%

Materials
Handling/Other
17.8%

Exports
6.0%

Repair & Remodeling
30.7%

FIGURE 11-10
Average percentage of demand for softwood lumber by market segment from 1989
through 1993. (*Widman, 1989.*)

cost-effectiveness of product delivery (Rich, 1979b). Price, as with structural panels, is
largely set in an auction-like manner with weekly price newsletters like those published
by Crow's and Random Lengths providing the starting point for negotiations.

However, in view of recent economic trends which will alter the wholesale and retail
distribution of lumber and building materials, a shift toward a more substantial marketing
orientation by producers of softwood lumber may be necessary for survival in the
marketplace of the future. One study contacted retail lumberyards in fifteen north central
and northeastern states concerning their trade in softwood dimension lumber (Govett and
Sinclair, 1984).

Retail lumber dealers in the early to mid 1980s within the north central and
northeastern United States sold large volumes of lumber produced from local northern
species, of which Canadian spruce-pine-fir represented the most common species group
(Govett and Sinclair, 1984). Even within some geographic segments of the market area in
which Douglas fir and hem-fir had traditionally held the lion's share of the softwood
lumber market, spruce-pine-fir held a large market share in the mid-1980s.

Stud-size dimension lumber represented a segment of the total softwood lumber
market within which the spruce-pine-fir species group exhibited a particularly high
degree of market acceptance. These retail lumber dealers were virtually evenly divided in
their perception of whether their customers had a very strong preference for either west
coast– or southern-produced lumber over the northern dimension species. However,
almost 80 percent of the dealers felt their customers had a very strong preference for kiln-
dried lumber versus air-dried lumber.

With the exception of one neutral response, every responding lumberyard either
agreed or strongly agreed that do-it-yourself or walk-in-type customers purchased studs
almost totally on the basis of general appearance, straightness, and low price.

Table 11-6 shows how retail lumber dealers perceived large-order customers (building contractors) would independently rank the importance of nine characteristics of studs in their purchase decisions. Straightness, general appearance, and low price represented the characteristics of highest importance. This information continues to point toward the conclusion that softwood dimension lumber in standard grades is sold as a commodity product, with price remaining a critical component of the marketing mix but straightness and appearance apparently equally (if not more) important components.

Role of Trade Associations Like the structural panel market, trade associations have played a strong role in the promotion and quality assurance programs for softwood lumber. Structural lumber grades are relatively uniform between the United States and Canada and are set voluntarily by the American Lumber Standards Committee and the Canadian Lumber Standards Committee (see Table 11-7 for a sample of the various grades). However, the administration of grade-marking programs is largely performed by a wide array of mostly regional associations (Table 11-8). In the Pacific northwest, for example, the two major associations are the Western Wood Products Association and the Council of Forest Industries of British Columbia. Firms voluntarily pay a fee (usually based on production volume) for the right to use an association's grade stamp (trademark). In return, the association or inspection service has spot checks and training programs to ensure that the firms are grading properly.

Of course, in addition to their grading programs the associations generally provide a forum for the common promotion of several of a region's species or in some cases for a

Lumber exports are a significant factor in North American markets.

TABLE 11-6
PERCENTAGE OF LUMBERYARDS RATING VARIOUS
CHARACTERISTICS OF 2 × 4-INCH STUDS IMPORTANT TO LARGE-
ORDER CUSTOMERS

Characteristics	Importance rating, %
Straightness	97
General appearance	93
Low price	92
Absence of wane	86
Kiln-dried versus air-dried	78
Absence of stain and decay	76
Absence of insect holes	71
Strength	64

Source: Govett and Sinclair, 1984.

single species. These promotional efforts can take the form of national advertising for larger associations, buyers' guides (i.e., lists of members and the products they produce), lumber use guides, span tables, building plans, etc. Considerable effort has also been expended by the major associations on export market development.

In recent years, more association activity has been focused on timber supply and environmental issues. Also, associations serve as a very useful source of market data, with many larger associations having market data collection activities.

BUILDING PRODUCTS DISTRIBUTION

The various channel systems which currently distribute building products have evolved along with fluctuating market demands, the economic environment, and new technology. Where once the channel system was mostly composed of independent producers, middlemen, and retailers, there are now strong integrated members operating at all channel levels.

Building Products Market Prior to World War II

The building products market evolved primarily in response to the need for housing. As the industrial revolution asserted itself on the American continent, the crude pit sawing of lumber was replaced by sawmills which were operated by water power, then steam power, and eventually by electrical power (Wynn, 1946; Cox, 1974). Through improved technology in the sawmill, smaller-dimensioned lumber was produced which eventually became part of a light framing system. These small wood components provided satisfactory structural strength and warmth at a fraction of the volume of wood required for the log cabin.

As lumber demand increased, a distribution system developed. Local sawmills were quite prominent throughout the 1800s, but the sharp growth in the industrial centers of the east and midwest, coupled with the depletion of local timber resources, required lumber

TABLE 11-7
COMMON NORTH AMERICAN SOFTWOOD GRADES

Grade category	Metric size*	Nominal size*	Grade	Principal uses
Dimension lumber				
Light framing	38–89 mm thick 38–114 mm wide	2–4 in thick 2–5 in wide	Construction standard	Widely used for general framing purposes. Pieces are of good appearance but graded primarily for strength and serviceability.
			Utility	Widely used where a combination of good strength and economical construction is designed for such purposes as studding, blocking, plates, and bracing.
			Economy†	Temporary or low-cost construction where strength and appearance are not important.
Structural light framing	38–89 mm thick 38–114 mm wide	2–4 in thick 2–5 in wide	Select No. 1	Intended primarily for use where high strength, stiffness, and good appearance are desired.
			No. 2	For most general construction uses.
			No. 3	Appropriate for use in general construction where appearance is not a factor.

TABLE 11-7
COMMON NORTH AMERICAN SOFTWOOD GRADES (*continued*)

Grade category	Metric size*	Nominal size*	Grade	Principal uses
Stud	38–89 mm thick 38–114 mm wide	2–4 in thick 2–5 in wide	Stud	Special-purpose grade intended for all stud uses.
			Economy stud†	Temporary or low-cost construction where strength and appearance are not important.
Structural joists and planks	38–89 mm thick 140 and wider	2–4 in thick 6 in and wider	Select structural	Intended primarily for use where high strength, stiffness, and good appearance are desired.
			No. 1	
			No. 2	For most general construction uses.
			No. 3	Appropriate for use in general construction where appearance is not a factor.
			Economy†	Temporary or low-cost construction where strength and appearance are not important.
Appearance	38–89 mm thick 38 mm and wider	2–4 in thick 2 in and wider	Appearance	Intended for use in general housing and light construction where lumber permitting knots, but of high strength and fine appearance, is desired.

			Decking	
Decking	38–89 mm thick, 140 mm and wider	2–4 in thick, 6 in and wider	Select	For roof and floor decking where strength and fine appearance are required.
			Commercial	For roof and floor decking where strength is required but appearance is not so important.

			Timber	
Beams and stringers	144 mm and thicker, width more than 38 mm greater than thickness	5 in and thicker, width more than 2 in greater than thickness	Select structural	For use as heavy beams in buildings, bridges, docks, warehouses, and heavy construction where superior strength is required.
			No. 1	
			Standard† Utility†	For use in rough, general construction.
Posts and timbers	114 × 114 mm and larger, width not more than 38 mm greater than thickness	5 × 5 in and larger, width not more than 2 in greater than thickness	Select structural	For use as columns and posts in heavy construction such as warehouses, docks, and other large structures where superior strength is required.
			No. 1	
			Standard† Utility†	For use in rough, general construction.

*mm = millimeter; in = inch.
†Note: All grades are "stress-graded," meaning that working stresses have been assigned (and span tables calculated for dimension lumber) except those marked with a dagger.
Source: Adapted from Mullins and McKnight, 1981.

153

to be shipped to those markets from the south and upper midwest (Hidy et al., 1963; Morgan, 1982). At the outset, lumber boats carried cargo to the growing cities. Later, railroads became the primary carriers of wood products to the marketplace (Bernshon, 1943; Hidy et al., 1963; Cox, 1974; Morgan, 1982).

Technology also had a major impact on building products distribution through developments in lumber drying. Seasoned wood products proved to be superior structural members. They were easier to cut, lighter to carry, and stiffer from a framing-system standpoint; however, using them meant yarding lumber for several months while drying occurred. These lumber drying yards are believed to have evolved into the first distribution points in the industry (O'Dowd, 1984).

Lumberyards were later established in the large population areas, and then in the county seats, which were the commercial hubs of the rural areas. These early yards served

TABLE 11-8
MAJOR REGIONAL STRUCTURAL LUMBER TRADE
ASSOCIATIONS AND GRADING AGENCIES

West
California Lumber Inspection Service
California Redwood Association
Western Red Cedar Association
Western Wood Products Association
Council of Forest Industries of British Columbia
Interior Lumber Manufacturers Association
Cariboo Lumber Manufacturers Association
Alberta Forest Products Association
Pacific Lumber Inspection Bureau
West Coast Lumber Inspection Bureau
Redwood Inspection Service

East-northeast
Northeastern Lumber Manufacturers Association
Northern Hardwood and Pine Manufacturers Association
Canadian Lumberman's Association
Central Forest Products Association
MacDonald Inspection
Maritime Lumber Bureau
Ontario Lumber Manufacturers Association
Quebec Lumber Manufacturers Association

South-southeast
Southern Pine Inspection Bureau
Timber Products Inspection
Southern Forest Products Association*
Southeastern Lumber Manufacturers Association*

*Do not offer grading services.

Stud grade 2 × 4 inches from
Canada for sale at U.S. retailer.

primarily as wholesalers in this evolving channel system, acting as the sales arms of the producers among builders and contractors (Hidy et al., 1963; Cox, 1974; Twining, 1975; Maxwell and Baker, 1983).

By the early part of this century, the building products market was changing rapidly. The virgin timber of the south and the upper midwest had largely been cut, and the

An early lumberyard and forerunner of the modern home center.

industry was concentrating on the highly forested areas of the northwest. More effective distribution was needed, as producers were 2000 miles or more from their major markets. Lumber moved by boat from west coast ports down through the Panama Canal and up to the eastern markets (Bernshon, 1943; Cox, 1974; Twining, 1975). By the early 1920s this trade was expanding rapidly. Weyerhaeuser, for example, opened a major distribution yard in Baltimore in 1922 to handle west coast lumber shipments, and by the mid-1920s had a fleet of four steamships carrying lumber to supply a growing number of east coast yards (Hidy et al., 1963). The rail industry, by providing competitive service from east coast ports to inland markets and from the northwest to midwestern markets, also became a more significant factor in the lumber business.

Building Products Market After World War II

The four decades since 1950 have seen a marketing revolution in the distribution of lumber and building materials. Postwar demand for housing, further stimulated by increasing population trends in the 1960s and 1970s, placed periodic demands on the productive capacity of building materials producers. As investment funds flowed into manufacturing firms and more mortgage funds were required to finance the finished housing units, the total housing industry became one of the largest industries in the economy (Maxwell and Baker, 1983; O'Dowd, 1984).

Numerous factors have had a significant impact on the industry's growth pattern since World War II, including a scarce and costly timber supply, new product developments, improved manufacturing techniques, and increased transportation costs (Hidy et al., 1963; Cox, 1974; O'Dowd, 1984). Railroads provided strong competition to waterborne lumber shipments over long routes and trucks captured many of the shorter routes (U.S. International Trade Commission, 1982). By the mid to late 1970s imported lumber from eastern Canada, shipped mostly by truck, was displacing the U.S.-produced west coast lumber in the north central and northeastern markets. Rail shipments of British Columbian lumber to eastern reloading centers where it was transferred to trucks for further shipment also captured a large share of the southeastern market (U.S. International Trade Commission, 1982). In the early 1970s, nearly 60 percent of the lumber shipments from western U.S. mills went to northern and southern destinations; however, by the early 1980s this had dropped to 33 percent (U.S. International Trade Commission, 1982). This displacement of western lumber was partially the result of lower shipping costs for Canadian mills and increased competition from southern producers.

Also during this period after World War II, producers developed specialty products, advanced machinery began to appear in sawmills and factories, and the industry's main customer, the builder, grew into a completely different enterprise. Small builders grew into large mass builders, and in many cases the management changed from construction-oriented people to financial managers.

These large builders were very cost-conscious and were not as dependent on local dealers for land, financing, and materials as their small predecessors had been (Birkner, 1963). In their mass-production approach to tract housing, the large builders started to industrialize the home building function, creating a need for new types of precutting, prefabricating, installing, and erecting services which traditional channels did not provide (Rich, 1970).

In the distribution area, methods and systems were also changing. Specialty products and the needs of the large builders brought new demands and new opportunities for local wholesale distributors, who were able to inventory products for use in specific trading areas. The public demanded choices and building products had to cover a full range of colors, finishes, and textures to receive market acceptance (Rich, 1970; Anonymous, 1983a). Technology and production answered these needs and the role of the local wholesalers grew in importance; however, other intermediaries also were playing a role in this market, such as office wholesalers, reload centers, and distribution yards, to name a few.

Changes also occurred in the retailing of building products. Local retail lumber dealers now had to service very diverse segments in the retail building products market. The first segment consisted of the high-volume builders, who had increased demands for products, services, and credit. However, the different needs of smaller builders still had to be served. At the same time, the growing repair and remodeling market had to be dealt with. This market consisted of professionals and the expanding do-it-yourself market, which had new demands for information and sales assistance. These widely divergent needs resulted in the development of new types of retailers such as dealer-builders, construction supply dealers, full-service dealers, and the cash-and-carry dealer (Birkner, 1963).

On the merchandising side, product lines were exploding and the small corner display area had nowhere near the merchandising area required. By the 1990s, an average single-store firm stocked 10,000 to 15,000 different items in 10,000 to 22,000 square feet of indoor space while the larger multistore firms stocked 15,000 to 30,000 different items in 20,000 to 120,000 square feet of indoor space. Fig. 11-11 shows 1990 sales by product line for major home centers.

FIGURE 11-11
Distribution of sales by product line for major home centers/building materials retailers. (*Anonymous, 1990.*)

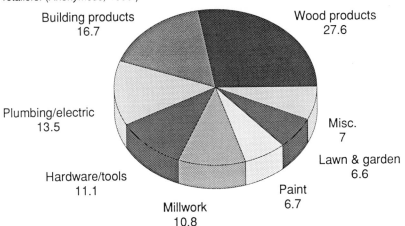

Building products
16.7

Wood products
27.6

Plumbing/electric
13.5

Misc.
7

Hardware/tools
11.1

Lawn & garden
6.6

Paint
6.7

Millwork
10.8

Do-it-yourself customer selecting a board to complete a home project.

Current Distribution Systems

Census data show that independent wholesalers are the dominant channels by which lumber and building materials are transferred from producer to retailer (Table 11-9).

Distributors, both independent and manufacturer-owned, are now called upon not only to handle the physical distribution of products and materials, but also to provide additional marketing and inventory control services (Rich, 1981; Lambert, 1984). Retailers are calling on wholesale distributors (and manufacturers) more frequently for help with in-store selling, product demonstrations, and training of store personnel. Increasingly, distributors are called upon to develop their own individualized merchandising programs for home center retailers as well to administer national manufacturers' programs at a local level (Anonymous, 1983a).

The emergence of strong networks of manufacturer-owned wholesale distribution centers has been a major force in the marketplace. The strongest network has been developed by Georgia-Pacific, although many other firms also have captive distribution centers. Most distribution networks work from a strong geographical base, and many function as full-line distributors for other manufacturers' products as well as their own.

TABLE 11-9
SALES OF LUMBER, PLYWOOD, AND MILLWORK THROUGH INDEPENDENT AND
MANUFACTURER-OWNED INTERMEDIARIES

	Total U.S. sales, %					
	1963	1967	1972	1977	1982	1987
Independent wholesalers	85	80	80	75	84	87
Manufacturers' sales branches and offices	15	20	20	25	16	13

Source: U.S. Department of Commerce, Bureau of the Census, 1989a.

One example of this strategy was seen in Weyerhaeuser's "First Choice" campaign. In this campaign, Weyerhaeuser stressed its expanded line of building materials, which included products such as asphalt roofing, fiberglass insulation, and other nonwood products typically not manufactured by Weyerhaeuser.

Commitment to strong captive distribution systems has been reiterated by several industry leaders, who addressed expansion of their distribution systems for building products as a "top priority for the 1980s" (Weyerhaeuser, 1981; Floweree, 1982). However, as the 1990s approached some wood products firms began to move back to their core business of manufacturing and away from distribution. Some of this move was triggered by the larger retailers beginning to buy directly from mills and the smaller retailers joining cooperative buying services that also purchased directly.

Wholesale Channels The paths and channels of distribution that building products may follow on their way toward the consumer are many (Fig. 11-12). The major links in this lumber distribution chain include sawmill sales offices, national lumber trading companies (office wholesalers), regional reload centers, large distribution yards, and local lumberyards and retailers (Leckey, 1989). It is common for a sawmill to operate a sales office which not only sells the production of its mill but also sells the production of many other mills. Sawmill sales personnel can have somewhat of an advantage being located at the sawmill site. This provides them with in-depth knowledge about what the mill can produce and is producing, which can assist them in making future sales decisions. Many times, these offices sell their products in rather large volumes to only a small number of customers. The sawmill sales office, in this way, can remain small while at the same time being able to move significant volumes of production. There is many times no need for traffic or credit experts and this operation's overhead can stay low (Leckey, 1989).

Another type of wholesaler is comprised of national trading companies, sometimes known as "office wholesalers." These wholesalers arrange for buyers and sellers of large quantities of lumber, yet rarely take title to the lumber themselves. These lumber trading offices handle lumber and plywood in very large quantities from producers and then market it to reloading centers, distribution centers, and larger retailers. Virtually all of this marketing is done over the telephone and many of these wholesale firms employ

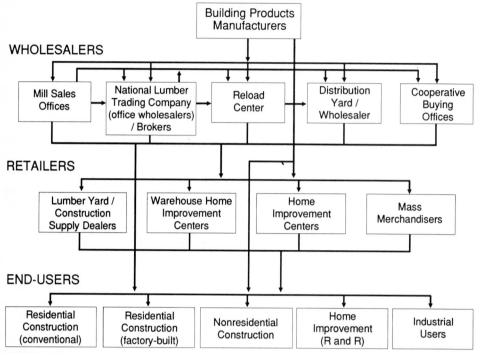

FIGURE 11-12
Simplified channels of distribution for wood building products. (*Adapted from Rich, 1982.*)

extensive credit and traffic departments. A number of the large producers of plywood and lumber operate national and regional trading centers for their firms (Leckey, 1989).

Reloading centers exist because of the differential in freight rates between various types of transportation, usually between railroad and truck. The reloading center typically accepts lumber, plywood, and other bulky wood products from the railroad and then reloads them onto trucks for shipment to their ultimate destination. Leckey (1989) gives the following example of the freight differential which supports reloading centers: the rail freight rate from western Canada to Niagara Falls, Ontario, Canada, for dry spruce lumber is approximately $80 per thousand board feet. However, if the lumber is shipped by rail across the U.S. border to Niagara Falls, New York (only 5 miles away), the rail cost jumps to $115 per thousand board feet. The additional cost is the result not of additional cost incurred by the railroad but rather of the way that rail freight tariffs are constructed. Other freight cost anomalies exist in other locations, which give reloading centers their advantage.

Large distribution yards operate much like reloading centers except on a smaller scale (Leckey, 1989). These full-service distribution yards typically purchase railcar and truckload lots for their own inventory. Most also carry a full assortment of other building products including roofing nails, concrete, siding, and many others. They purchase these goods in large lots and then ship truckloads of mixed items to retailers, contractors, and manufacturers in their geographic area.

Retail Channels   This wide array of players in the wholesale distribution channels sells products to an equally diverse mix of building materials retailers. These retailers range from traditional lumberyards or construction supply dealers to modern home improvement centers, warehouse home centers, and mass merchants.

The traditional lumberyard or construction supply dealer carries a full line of bulky building materials. It may also carry appliances, windows, doors, and many other items necessary to finish a home. The primary customer group of the traditional lumberyards has been the professional contractor.

As the repair and remodeling market has expanded over the last decade, the traditional lumberyard has discovered that in order to maximize its sales potential it must appeal to the growing group of do-it-yourself homeowners who want to find a variety of items available in small quantities, but also expect to find instructional manuals and material return services. This has given rise to the development of a new kind of retailer, the home improvement center or what is many times just called a "home center." As shown in Fig. 11-13, the growth in sales by home center retailers has been nothing short of phenomenal. This retail group grew in sales from less than $30 billion in 1977 to over $75 billion in 1988. Table 11-10 provides an overview of the largest home center retailers of the late 1980s.

The home center retailer typically carries a wide assortment of products for repairing, renovating, and maintaining a home. The primary customer of the home center is usually the do-it-yourself homeowner, but many if not most home centers have a special contractors' desk for sales to professional contractors. These home center retailers are now household names with Lowe's and Payless Cashways being the leaders in this particular retail segment.

The growth of the marketplace for the do-it-yourself homeowner spawned another type of home center in the late 1970s, the warehouse home center. Warehouse retailing is a merchandising concept that has been used successfully in consumer goods for quite some time. Warehouse retailing of building products, however, has been refined and combined with off-price merchandising to fit the particular requirements of home center retailing. According to some, warehousing of building materials began in 1979 when The Home Depot opened a 70,000-square-foot outlet in Atlanta. Since that time, numerous competitors have entered the warehousing market. The current industry leaders are Home Depot, Builders Square (a division of K mart), and Home Club.

Warehouse retailers carry an enormous assortment of products for home repair, renovation, and maintenance. These retail stores are large, ranging from 80,000 to 140,000 square feet of inside sales area. The emphasis is on low price, big product assortment (25,000 to 30,000 different items), and an appeal to the do-it-yourself homeowner; however, some warehouse retailers also offer services for professional contractors.

Another category of building materials retailers is the mass merchants. These firms are usually considered to be outside the building materials retail industry, but many carry numerous items which are typically found in home centers. Such firms include Sears, K mart, Wal-Mart, and Target stores.

Year

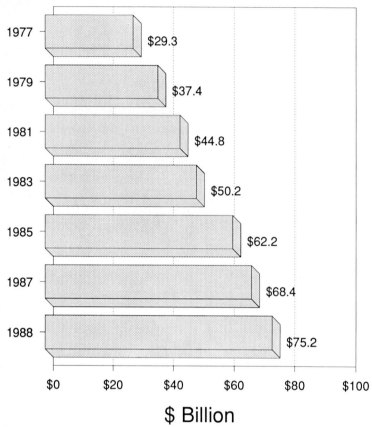

$ Billion

FIGURE 11-13
Home center industry sales. (*Franta and Johnson, 1988.*)

TABLE 11-10
TEN LARGEST HOME CENTER/BUILDING MATERIALS RETAILERS

Name	Estimated 1989 $ Volume	Number of stores
Home Depot	2.74 billion	119
Lowe's	2.67 billion	305
Builders Square (division of K mart)	2.0 billion	143
Payless Cashways	1.9 billion	195
Hechinger	1.24 billion	107
Grossman's	1.07 billion	157
Wickes Lumber	1.0 billion	196
Home Club	1.00 billion	58
84 Lumber	800 million	356
Wickes Cos.	675 million	119

Source: Anonymous, 1990.

Service is a major marketing tool for home center retailers.

Summary and Future Distribution Trends   Current channel systems used to distribute building products have evolved from an interim point for drying lumber into a complete network of both captive and independent middlemen operating at various levels in the channel system.

During the early part of this century the industry was concentrated in the northwest forests while the markets were in the growing east, technological advances in transportation expedited distribution systems, mass builders emerged, and wholesalers' responsibilities changed. Changes at the retail level meant catering to the very diverse segments of the retail building products market: high-volume builders, smaller builders, and the two segments of the repair and remodeling market, the DIY homeowner and the professional.

The repair and remodeling market, in particular, has brought about significant changes in the distribution of building products. One-stop home improvement centers have emerged, carrying a large variety of items available in small quantities as well as instructional manuals and material return service. These home centers cater to the DIY homeowner, professionals, and contractors. The DIY customer has become increasingly important, with the percentage of building materials retail sales to individual consumers growing strongly in the 1970s and 1980s.

The role of wholesalers will be more service-oriented in the future. A new sense of cooperation will likely be felt in the channel system, as manufacturers, wholesalers, and retailers combine efforts to meet consumer demands and provide more effective service. Manufacturer-owned wholesalers are a force in the marketplace because of their size, but independents account for the dominant amount of sales.

A vital component of the marketing function is distribution, without which the flow of products from producer to end user would come to a near standstill. Channel systems capitalize on the skills and resources of their individual members, tailoring their systems to meet the needs of their particular market segments. In the future, look for distribution systems to have a stronger awareness of and responsiveness to the changing demands of their target markets.

FACTORS DRIVING DEMAND FOR BUILDING PRODUCTS

Home building still represents the single largest market for softwood lumber and structural panels in the United States. The building products industry is less dependent now on new home building than it has been in the past due to efforts of individual firms and various trade associations to increase consumption of softwood lumber and structural panels in other markets. The most dramatic increase in consumption has been in the repair and remodeling markets, which to some extent increases the industry's dependence on housing as a market but diversifies that market away from being solely dependent on new construction.

Americans' traditional ties to the family home seem to be waning. A recent survey found that among people in the 45- to 54-year age range, 42 percent considered their home to be just another asset and only 19 percent felt it important to pass a particular home on to their children (Baker, 1989). This is mirrored somewhat in the shift toward more costly upscale housing. *The Wall Street Journal* has reported that the demand for new luxury homes is soaring (Celis, 1986). The National Association of Home Builders has estimated that by the early 1990s nearly eight out of ten home builders will be catering to upscale buyers (Celis, 1986). This ability to trade up in homes has been made possible by increasing affluence and the availability of low-cost fixed-rate mortgages (Celis, 1986).

Housing demand for wood products comes from two major areas: new home construction and the repair and remodeling of existing homes. We will first discuss the demand for new residential housing.

New Housing Demand

The importance of new housing construction cannot be overemphasized for the forest products industry but, in addition, it is a driving force of many segments of the U.S. economy in general. As a result of this, economists have developed very sophisticated computer models for estimating the future demand for housing. Some of these models include the ones developed by Marcin (1978) and Eckstein (1983). Other proprietary econometric models to estimate housing demand are in use by various organizations including Wharton Econometric Associates and the National Association of Home Builders, to name a few.

Spelter (1988) studied a number of these models to determine if a simplified procedure to estimate housing demand might work. He concluded that "ultimately . . . population and interest rates control most of the variation in housing activity." These two variables, population and interest rates, are the foundation of Spelter's model.

Spelter used the growth in the population of people most likely to form new households as the population variable in his model. He chose the 25- to 44-year-old group as the basis of his calculations. Projections of the numbers of people in this age class are available from the Bureau of the Census and the projections are quite accurate because we already know the number of people in younger age brackets and mortality and immigration rates are relatively stable.

As was expected, interest rates had a lagged effect on housing with lower rates resulting in higher projected housing starts. Population increases in the 25- to 44-year-old group also resulted in increased housing starts. Spelter's (1988) simplified model did an excellent job in comparison with the more complex models for estimating housing starts for single-family dwellings. However, it was somewhat less accurate than the more complicated formulas for predicting multifamily construction.

Many, many other factors can affect the demand for housing. And any user of a model must be prepared to make judgments as to the accuracy of any forecast. Spelter noted that variables not included in his model could have significant effects. Such variables include changes in government subsidies, which are independent of credit market forces, and the recent changes in tax laws with respect to multifamily construction, which have changed the economics of such investments.

Types of Housing

Housing is still largely dominated by the small to medium-sized builder. In the mid-1980s, the 100 largest home builders in the nation accounted for only about 15 percent of the new housing units erected (*Wall Street Journal*, 1984). Most of these units erected were the traditional stick-built structures which are constructed from individual pieces of building material on site. The term ''stick-built'' comes from the notion that they are built one stick of lumber at a time.

The alternative form of house construction, industrialized housing, had its start in the 1930s with the prefabricated house. In 1935, an experimental prefabricated house was built by the Forest Products Laboratory, with two more following in 1937 (Lawrence, 1968). The industry now produces three different types of structures: mobile homes, modular homes, and prefabricated homes. These types of units are described briefly below.

Mobile homes are now known by some as ''HUD Code housing.'' The Manufactured House Construction and Safety Standards Act of 1974 created mandatory national standards for manufactured housing. These HUD guidelines serve in place of the local building codes for mobile homes. As a result, HUD Code units (mobile homes) do not necessarily have to conform to local building codes. The factor that distinguishes mobile homes from other manufactured units, in addition to the code itself, is primarily that each unit is built with its own wheels and chassis, which are used for transporting the unit to the site where it will be used.

Modular housing, like mobile homes, is built almost entirely in factories. Ninety-five percent of the unit may be completed in the factory. The units are not transported on their own wheels; instead they are shipped, most often on trucks. The modular units are then assembled using cranes. Several subunits may be joined together to form one structure. These structures must meet local building codes.

Prefabricated housing includes precut units, kit homes, and panelized units. With precut units, the materials are cut in a factory and then shipped to the construction site. At the site, the precut materials are then used to build the unit. Kit units include log cabins, geodesic domes, and other structures. Here, the logs and other materials are shipped to the construction site. With panelized units, components of the unit are built in the factory and then shipped to the building site. The components typically include floor and roof trusses, interior and exterior wall panels, stairs, etc. Like stick-built and modular units, prefabricated housing must conform to local building codes.

Nearly all of the new homes built in the United States are built with some premanufactured parts, such as floor trusses, roof trusses, windows, and doors (Benoit, 1986). However, in the mid-1980s 34 percent of the new dwellings constructed were built from kits, modular units, or prefabricated housing (Benoit, 1986).

FIGURE 11-14
Shipments (in thousands of units) of modular and panelized units. (*Shapiro and Associates, 1987.*)

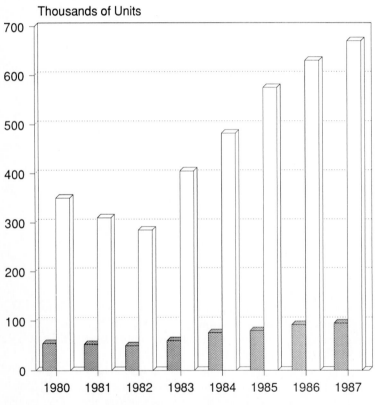

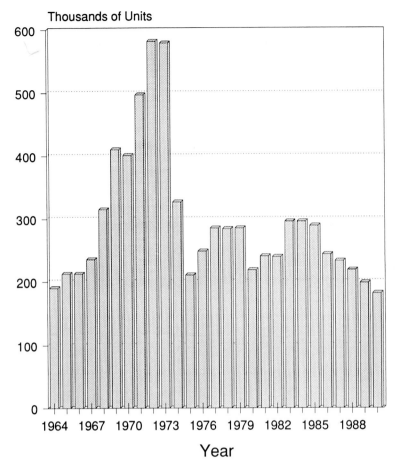

FIGURE 11-15
Annual shipments (thousands) of mobile homes from 1964 through 1989; 1990
data are estimated.

Sales of mobile homes have been declining to flat for several years. However, the demand for modular and prefabricated structures has increased. Fig. 11-14 shows the increasing shipments of modular and panelized units from 1980 to 1987.

The industrialized housing market can be described as fragmented. According to the National Association of Home Builders (NAHB), most producers of modular units ship to more than five states. However, the existence of state and local building codes makes it more difficult to ship to wide areas (NAHB National Research Foundation, 1987).

Mobile home producers have become the most centralized and capital-intensive segment of the U.S. home building industry. In 1983, 130 companies produced mobile homes in about 320 factories. The sales of these units amounted to almost $5.5 billion. The mobile home industry, in that year, consumed over $3 billion worth of building products, services, and other supplies (Manufactured Housing Institute, 1987). The top ten firms produced 54 percent of the units manufactured in 1983. However, there is relatively little forward integration because of the differing requirements of production, dealerships, and park management (Mathieu, 1986).

Fig. 11-15 shows the total number of shipments of HUD Code units from 1964 through 1990. There was rapid growth in shipments of HUD Code units, then called "trailers" or "mobile homes," until the early 1970s. This resulted in some overcapacity in this industry, one of the characteristics of a product in its mature stage. Partly because of overcapacity, price competition is very strong in this market.

It is difficult to estimate the exact number of companies producing modular housing. In 1986, estimates by industry specialists ranged from 137 producers to 195 producers (NAHB National Research Foundation, 1987). The National Research Center of the NAHB has identified 152 producers of modular housing, but considers this to be a conservative, or "lower-bound," estimate of the number of modular housing producers (NAHB National Research Foundation, 1987).

The market for modular housing is growing rapidly. It has been estimated that between 1980 and 1987 the number of modular housing units produced increased from 56,000 to 93,000 (Shapiro and Associates, 1987). Figure 11-14 provides more detail about the growth in shipments of modular housing units.

The prefabricated housing industry is very fragmented. This is largely because of the limited distances that producers can ship their products (Benoit, 1986). The market for

FIGURE 11-16
Residential remodeling expenditures (1978–1988). (*U.S. Department of Commerce, Bureau of the Census, 1988.*)

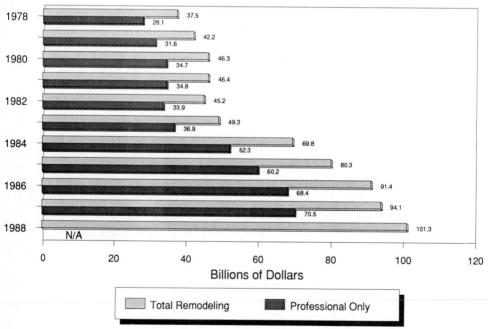

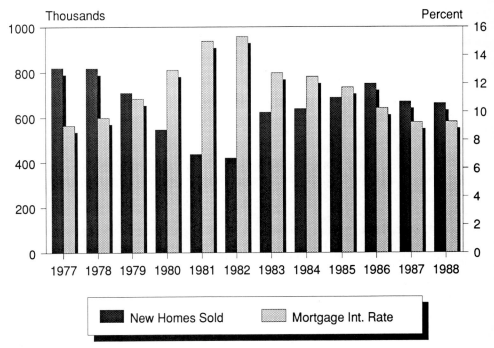

FIGURE 11-17
New single-family homes sold versus interest rates. (*Lowe's, 1989.*)

prefabricated housing is growing rapidly. Fig. 11-14 shows estimates of the portion of production accounted for by panelized units between 1980 and 1987. In addition to the growth in entirely prefabricated homes, there is also tremendous growth in prefabricated components used in stick-built homes. The majority of site-built homes use some component parts manufactured off site (Germer, 1987).

Repair and Remodeling

The residential repair and remodeling (R&R) industry experienced enormous growth in the 1970s and 1980s. Total residential R&R expenditures increased from $37.5 billion in 1978 to $101.3 billion in 1988 (Fig. 11-16). Professional remodeling alone increased approximately 11 percent annually, from $28.1 billion in 1978 to $70.5 billion in 1987 (Miller, 1989). Continued professional R&R growth appears likely through the 1990s but at a slightly slower pace (Miller, 1989).

Demand Factors   Many factors drive this market. The expansion of residential R&R through the 1980s was linked to high interest rates (Areddy, 1989; Cunniff, 1989), the increasing cost of new single-family houses (U.S. Department of Commerce, Bureau of the Census, 1989b) and the increasing volume of existing home sales (Anonymous, 1988).

Interest Rates   Interest rates have a distinct effect on new housing starts but have less influence on R&R (Shutt, 1987). As interest rates climbed in the late 1970s and early 1980s, the number of new housing starts fell. When interest rates peaked in 1982, at 15.14 percent, new housing sales hit a 20-year low (Fig. 11-17). The effect on R&R was much less devastating. With the exception of a slight decrease from 1981 to 1982 ($46.4 billion to $45.2 billion), the 10-year period from 1978 to 1988 saw expenditures for residential R&R consistently increase (Fig. 11-16).

Home Prices   New home prices also rose through the late 1970s and early 1980s. With the easing of interest rates after their 1982 peak, new housing starts rebounded to a level of approximately 1.7 million units in 1983. As interest rates eased, however, new home prices continued to rise (Fig. 11-18). From 1978 to 1988, the median price of a new single-family home increased 104 percent, from $55,700 to $113,500. These high prices made buying a new home more difficult financially.

During the same time period, existing home prices experienced similar yet less drastic increases. From 1978 to 1988, the median price of an existing single-family home increased 82 percent, from $48,700 to $88,600 (Fig. 11-18). On a comparative basis, the average price difference between the median-priced existing home and the median-priced new home was $7000 in 1978, broadening to $24,900 by 1988. As the price difference between new homes and existing homes increased, existing homes became relatively more affordable.

FIGURE 11-18
Median prices of new and existing single-family homes. (*Lowe's, 1989.*)

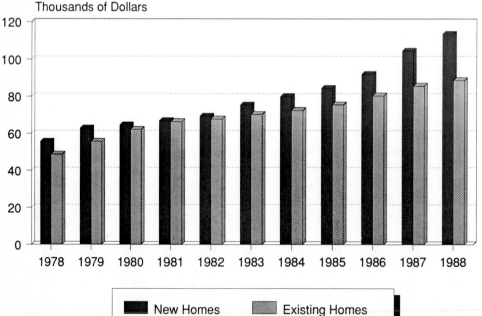

Millions

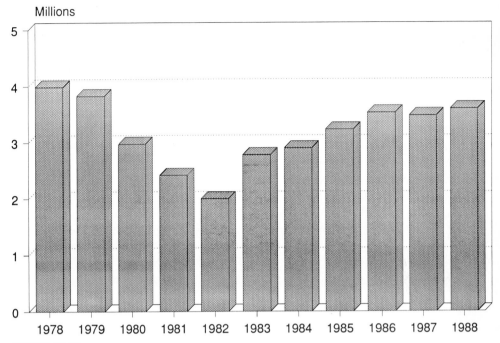

FIGURE 11-19
Volume of existing single-family home sales. (*Lowe's, 1989.*)

Existing Home Sales   Sales of existing homes are important in the remodeling market. Due to the fact that the most common periods for remodeling a house are shortly before selling and within 2 years of moving into a home (Shutt, 1987), the large volume of existing home sales clearly stimulated the growth of R&R. From 1982 to 1988, existing home sales rose from approximately 2 million to nearly 3.6 million (Fig. 11-19). The total volume of existing homes sold in 1987 was approximately 5½ times the volume of new homes sold in 1988 (Lowe's Companies, Inc., 1989).

Demographics   Various demographic trends also played a unique role in the growth of the R&R market. The first demographic trend related to R&R growth was the growth in the 35- to 54-year-old age group. This age group, closely associated with the baby boomers (ages 25 to 43 in 1989), has been growing and is expected to continue growing into the twenty-first century (Diez, 1988; Baker, 1989). Diez (1988) claimed that this growing middle age group may have helped stimulate the growth of R&R in the 1980s. From a nationwide survey of home improvement activities in May 1987, people in the age group 35 to 49 were shown to have participated in the greatest percentage (38 percent) of home improvements during the 12 previous months (Bloom, 1987). This same age group also represented the highest percentage of those who planned home improvements in the next 12 months.

A second demographic trend associated with the expansion of R&R was the increasing level of consumer affluence. U.S. disposable personal income (in current dollars) increased 56 percent from 1980 to 1987 (U.S. Bureau of Economic Analysis, 1988). This was aided by the influx of working women into the work force. From 1980 to 1987, 4.3 million married women entered the labor force (U.S. Department of Labor, Bureau of Labor Statistics, 1988). Dual-income families tend to do more home improvements, possibly because of their greater affluence (Garner, 1983; Seek, 1983; Parrott, 1988). In early 1989, dual-career families made up 60 percent of all families, and this was predicted to rise to 75 percent by 1995 (Herring, 1989).

Occupational Status   A positive relationship between occupational status and involvement in home improvement has been noted (Meeks and Firebaugh, 1974). Compared to a 15 percent growth in total employment, employment in professional and managerial occupations has been forecasted to increase 22 percent from 1984 to 1995 (Anonymous, 1986b). As the level of education completed is often associated with occupation, a higher level of education may be an indication of higher occupational status, and hence increased involvement in home improvement. In 1988, 26 percent of adult workers (25 to 64 years of age) were college graduates, up from 21 percent in 1978 (U.S. Department of Labor, Bureau of Labor Statistics, 1988).

Older Homes   Another factor driving the growing repair and remodeling market is the aging stock of houses in the United States. As the total inventory nears 100 million units, current statistics indicate that the overwhelming majority are over 20 years old. At 20 years, houses typically require major repairs such as roof replacement and new heating systems. Additionally, other remodeling activities are likely to occur in older homes to update and modernize such areas as kitchens and baths.

Repair and Remodeling Projects   Another trend affecting the allocation of expenditures for residential repair and remodeling was the shift from rental property to owner-occupied R&R. The growth in residential remodeling has historically occurred in maintenance and repairs (Baker, 1987). These maintenance and repair jobs are character-istics of the smaller fix-up jobs performed by landlords (Shutt, 1987). In the mid-1980s, consumer demand shifted away from maintenance and repairs and toward larger-ticket additions and alterations (which are more characteristic of owner-occupied R&R) (Baker, 1987).

Additions and alterations are the fastest-growing component of the remodeling market (Miller, 1989). Almost 90 percent of the R&R growth in 1986 was in additions and alterations (Baker, 1987). These larger-ticket projects require more time and skill, and are usually performed by professional remodelers. Characteristic of these large-ticket addi-tions and alterations are room additions and the complete remodeling of bathrooms and kitchens (Miller, 1989).

The bulk of kitchen and bath remodeling takes place within the first 2 years of home ownership (Higgins, 1987). After the first 2 years, the homeowner shifts attention to exterior projects (Higgins, 1987). Exterior projects are basically non-wood-related (with the exceptions of fences, decks, and some siding) and are commonly performed by the homeowner (Higgins, 1987). For those owning homes for 3 to 9 years, exterior projects represented 29 percent of their anticipated work, versus 22 percent for bathroom projects and 18.2 percent for kitchens. Even after owning their homes 10 to 20 years and 21 years or more, homeowners anticipated their greatest percentage of work to be in the exterior project group.

Long-Term Trends in Residential Repair and Remodeling   The expansion of residential repair and remodeling is expected to continue, although at a slightly slower pace, through the 1990s (Fig. 11-20). Seventy percent of all the homes which will be occupied by the year 2000 are already built (Lowe's Companies, Inc., 1987). This large portion of existing homes will need progressively more upkeep. Other long-term trends that may have an effect on the R&R market include a slow new-housing industry (Thompson, 1989), an increasing demand for larger and more upscale homes (Celis, 1986; Anonymous, 1989), and a growing consumer involvement in the professional remodeling process (Anonymous, 1987b; Baker, 1987; Shutt, 1987).

Housing starts are forecasted to stay in the 1.5 million range and may taper off to 1.4 million through the 1990s (Thompson, 1989). Since people are buying fewer new homes but continue to search for ways to meet their housing needs, they are likely to continue to repair and remodel their existing homes.

From a study conducted by the NAHB on 1800 homeowners who bought their homes after 1980, most would prefer to live in a home that was on average 30 percent larger than their current home (Anonymous, 1989). Second-time buyers were not willing to move into a smaller house but were willing to buy an unfinished house or one that could be

FIGURE 11-20
Professional remodeling forecast through 1993. (*Miller, 1989.*)

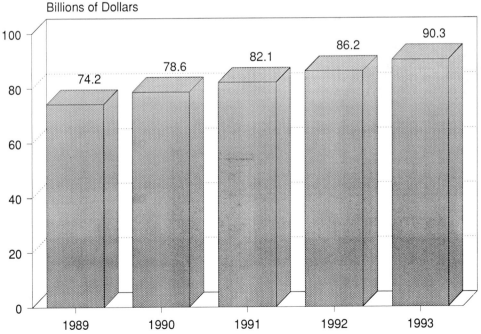

Deck made of preservatively treated softwood

expanded. This desire for larger homes may help stimulate a continuing demand for additions and alterations.

SUMMARY

The markets for building products are diverse but are clearly dominated by the housing market. Housing represents a two-part market, including new construction and the repair and remodeling of existing homes. Wood products "own" the residential construction market, but have had much less success in nonresidential construction. For small nonresidential buildings (5000 square feet or less) wood has as much as or more than one-half of the market (Spelter et al., 1987), but as buildings increase in size wood construction is less and less common.

Commodity-like wood building products are sold mostly on price, appearance, and availability. The channels of distribution include a wide range of midlevel intermediaries and numerous types of retailers. The wholesale and retail trade in building products is not dominated by firms owned by major manufacturers. Independent retailers such as Lowe's, Builders Square, and Payless Cashways have grown large enough to be a real factor in the markets for wood building products.

BIBLIOGRAPHY

Abt, R. C. 1987. "An Analysis of Regional Factor Demand in the U.S. Lumber Industry." *Forest Science* 33(1):164–173.

Anderson, R. G. 1989. *Regional Production and Distribution Patterns of the Structural Panel Industry*. E47. American Plywood Association, Tacoma, WA.

Anonymous. 1990. "1990 Giants Report." *Building Supply Home Centers* 158(2):48–84.

Anonymous. 1989. "What Do Homeowners Want?" *Home Center Magazine* 4086(5):33.

Anonymous. 1988. *The Current Housing Situation*. National Association of Home Builders, December 1988, Washington, DC. As reported in *Remodeling Trends, Facts and Figures on the Remodeling Market*. NAHB Remodelers Council. Washington, DC. Winter/Spring 1989.

Anonymous. 1987a. *1987 Profile of the Remodeler and His Industry*. National Association of Home Builders, Washington, DC. As reported in *Remodeling Trends, Facts and Figures on the Remodeling Market*. NAHB Remodelers Council. Washington, DC. Winter/Spring 1989.

Anonymous. 1987b. "NAHB Study Profiles the Remodeler." *National Home Center News* (March) (9):32–33.

Anonymous. 1986a. *Remodeling in the U.S.A. Qualified Remolder Magazine*—Research Report, Bureau of Market Research. Chicago, IL.

Anonymous. 1986b. "Demographic Forecasts: The Slowing Job Train." *American Demographics* 8(4):58.

Anonymous. 1984. "The 50 Billion Dollar Market." *Home Center Magazine* 4029(September):25–40.

Anonymous. 1983a. "Special Report: Marketing U.S.A." *Building Supply and Home Centers* 144(8):17–30.

Anonymous. 1983b. "Wholesaler Giants." *Building Supply News* 144(3):79.

Areddy, J. T. 1989. "New Construction Contracts Rose 8% in December but Declined for All of '88." *Wall Street Journal* (January 31):A4.

Baker, K. 1989a. "Market Outlook." *Building Supply and Home Centers* 157(1):36–37.

Baker, K. 1989b. "Market Outlook." *Building Supply and Home Centers* 157(October):32–33.

Baker, K. 1987. "Behind the Residential Remodeling Numbers." *Building Supply and Home Centers* 153(8):42.

Bayless, Alan. 1986. "Technology Reshapes Lumber Industry." *Wall Street Journal* (October 16):6.

Benoit, E. 1986. "Rise High the (Precut) Roof Beam." *Forbes* 137(2):120–124.

Bernshon, K. 1943. *Cutting Up the North*. Hancock House Publishers, Inc. Vancouver, B.C.

Birkner, E. 1963. "The Distribution Dilemma." *House & Home* (September):119–127.

Bloom, H. A. 1987. *The Home Improvement Market: Consumer Experiences and Expectations*. Newspaper Advertising Bureau, Inc. (July):3.

Buongiorno, J., and H. C. Lu. 1989. "Effects of Cost, Demand and Labor Productivity on the Prices of Forest Products in the United States, 1958–1984." *Forest Science* 35(2):349–363.

Celis, William. 1986. "Emphasis in Housing Market Shifts Toward Costlier Trade-Up Homes." *Wall Street Journal* (March 10):19.

Cleaves, D. A., and J. O'Laughlin. 1986a. "Analyzing Structure in the Wood Based Industry: Part 1—Identifying Competitive Strategy." *Forest Products Journal* 36(4):9–14.

Cleaves, D. A., and J. O'Laughlin. 1986b. "Analyzing Structure in the Wood Based Industry: Part 2—Categorizing Strategic Diversity." *Forest Products Journal* 36(5):11–17.

Corlett, M. L. 1989. *Forest Industries 1989–90 North American Factbook*. Miller Freeman Publications. San Francisco, CA.

Corlett, M. L. 1988. *Forest Industries 1988–89 North American Factbook.* Miller Freeman Publications. San Francisco, CA.

Cox, T. R. 1974. *Mills and Markets.* University of Washington Press. Seattle, WA.

Cunniff, J. 1989. "Interest Rates Hold Key to Market." *Roanoke Times & World News* (January 22):D83.

Diez, R. L. 1988. "Remodeling: Emphasis on a Growing Market." *Professional Builder* 15(2):15.

Eckstein, O. 1983. *The DRI Model of the U.S. Economy.* McGraw-Hill. New York.

Ellefson, P. V., and R. N. Stone. 1984. *U.S. Wood-Based Industry.* Praeger. New York.

Floweree, R. E. 1982. "Letter to Shareholders." *Georgia-Pacific Annual Report.* Atlanta, GA.

Franta, H., and M. L. Johnson. 1988. "The 1988 Market Profile." *Home Center Magazine* (September):67–87.

Garner, S. G. 1983. *Household Production of Housing Repairs.* Unpublished Ph.D. dissertation. University of Tennessee. Knoxville, TN.

Germer, J. 1987. "Kit Homes." *Progressive Builder* 12(May):9–16.

Govett, R. L., and S. A. Sinclair. 1984. "Marketing Research for Primary Processors of Northern Softwood Lumber." *Forest Products Journal* 34(5):13–20.

Greber, B. J., and D. E. White. 1982. "Technical Change and Productivity Growth in the Lumber and Wood Products Industry." *Forest Science* 28(1):135–147.

Herring, L. 1989. "Helping Spouses Helps Employers." *Personnel Journal* 68(1):35–37.

Hidy, R. W., F. E. Hill, and A. Nevins. 1963. *Timber and Men, The Weyerhaeuser Story.* Macmillan Co. New York.

Higgins, K. T. 1987. "Outside Projects Booming." *Building Supply Home Centers* 153(7):138–141.

Hill, R. M., R. S. Alexander, and J. S. Gross. 1975. *Industrial Marketing.* Irwin. Homewood, IL.

Lambert, A. 1984. "Support Programs Are Key to Sales Success." *Building Products Digest* 3(1):38–39.

Lawrence, J. D. 1968. "Manufactured Homes: Evolution and Market Potential for the Wood Products Industry." *Forest Products Journal* 18(9):57–59.

Leckey, D. 1989. *Trading Western Softwood Lumber.* Highland Press. Wilsonville, OR.

Lindberg, K. 1986. "The Story Behind Quality." *Market News* (August). Published by the Southern Forest Products Association. New Orleans, LA.

Lowe's Companies, Inc. 1989. *Annual Report.* North Wilkesboro, NC.

Lowe's Companies, Inc. 1987. *Annual Report.* North Wilkesboro, NC.

Manufactured Housing Institute. 1987. *Quick Facts About the Manufactured Housing Industry.* Manufactured Housing Institute. Arlington, VA.

Marcin, T. C. 1978. *Modeling Long Run Housing Demand by Type of Unit and Region.* USDA Forest Service Research Paper FP 308. U.S. Department of Agriculture. Washington, DC.

Mathieu, R. 1987. "The Prefabricated Housing Industries in the United States, Sweden and Japan." *Construction Review* 3(4):2–21.

Mathieu, R. 1986. "Manufactured Housing: The Industry in the Eighties." *Construction Review* 32(3):2–15.

Maxwell, R. S., and R. D. Baker. 1983. *Sawdust Empire.* Texas A&M University Press. College Station, TX.

Meeks, C., and F. Firebaugh. 1974. "Home Maintenance and Improvement Behavior of Home Owners." *Home Economics Research Journal* (3):114–129.

Merchant Magazine. 1989. (February):22.

Miller, P. H. 1989. "1989 State of the Remodeling Industry Report." *Remodeling Magazine.* A Hanley-Wood, Inc., Publication. Washington, DC.

Morgan, M. 1982. *The Mill on the Boot.* University of Washington Press. Seattle, WA.

Mullins, E. J., and T. S. McKnight. 1981. *Canadian Woods.* University of Toronto Press. Toronto, Canada.

NAHB National Research Foundation. 1987. *Modular Housing Industry: Structure and Regulation.* NAHB National Research Foundation. Washington, DC.

O'Dowd, F. E. 1984. Personal communication. National Building Material Distributors Association. Glenview, IL. February.

O'Laughlin, J., and P. V. Ellefson. 1981a. "U.S. Wood-Based Industry Structure: Part 1—Top 40 Companies." *Forest Products Journal* 31(10):55–62.

O'Laughlin, J., and P. V. Ellefson. 1981b. "U.S. Wood-Based Industry Structure: Part 2—New Diversified Entrants." *Forest Products Journal* 31(11):26–33.

O'Laughlin, J., and P. V. Ellefson. 1981c. "U.S. Wood-Based Industry Structure: Part 3—Strategic Group Analysis." *Forest Products Journal* 31(12):25–31.

Parrott, K. 1988. "The Relationship of Household Characteristics and the Home Remodeling Process." *Housing and Society* 15(1):56–69.

Random Lengths. 1990. *OSB/Waferboard Survey 1989–1990.* Random Lengths Publications. Eugene, OR.

Rich, S. U. 1982. *Langdale Lumber Company Case.* College of Business Administration, University of Oregon. Eugene, OR.

Rich, S. U. 1981. "Current Distribution Trends and the Independent Lumber Wholesaler." *Forest Products Journal* 31(12):12–13.

Rich, S. U. 1979a. "Corporate Strategy and the Business Investment Portfolio." *Forest Products Journal* 29(6):8.

Rich, S. U. 1979b. "Market Segmentation and Strategic Planning." *Forest Products Journal* 29(10):36.

Rich, S. U. 1970. *Marketing of Forest Products: Text and Cases.* McGraw-Hill. New York.

Row, Clark. 1964. *Changing Role of Retail Lumber Dealers in Lumber Marketing.* USDA Forest Service, So. For. Exp. Stat. SO-7. U.S. Department of Agriculture. Washington, DC.

Saunders, J. A., and F. A. W. Watt. 1979. "Do Brand Names Differentiate Identical Industrial Products?" *Industrial Marketing Management* 8(1):114–123.

Seek, N. H. 1983. "Adjusting Housing Consumption: Improve or Move." *Urban Studies* (20):455–469.

Seward, K. E., and S. A. Sinclair. 1988. "Retailers' Perceptions of Structural Panel Attributes and Market Segments." *Forest Products Journal* 38(4):25–31.

Shapiro, L. J., and Associates, Inc. 1987. *1987 Year-End Manufactured Housing Survey.* Unpublished document. L. J. Shapiro and Associates, Inc. Chicago, IL.

Shutt, C. A. 1987. "The Home Modernization Market: Where It Is, Where It's Going." *Building Supply Home Centers* 153(12):60–63.

Sinclair, S. A., and K. E. Seward. 1988. "Effectiveness of Branding a Commodity Product." *Industrial Marketing Management* 17:23–33.

Spelter, H. 1988. "A Simplified Two-Variable Formula for Projecting U.S. Monthly Housing Starts." *Forest Products Journal* 38(1):17–20.

Spelter, H., R. Maeglin, and S. LeVan. 1987. "Status of Wood Products Use in Non-Residential Construction." *Forest Products Journal* 37(1):7–12.

Spelter, H., and R. B. Phelps. 1984. "Changes in Postwar U.S. Lumber Consumption Patterns." *Forest Products Journal* 34(2):35–41.

Thompson, B. 1989. "The Battle for the Builder Is On." *Building Supply Home Centers* 156(5):68–72.

Tillman, D. A. 1985. *Forest Products: Advanced Technologies and Economic Analyses.* Academic Press. Orlando, FL.

Twining, C. E. 1975. *Downriver*. The State Historical Society of Wisconsin. Stevens Point, WI.

U.S. Bureau of Economic Analysis. 1988. *Survey of Current Business*. As cited in U.S. Statistical Abstracts, 1989.

U.S. Department of Commerce, Bureau of the Census. 1989a. *Preliminary Report Industry Series. 1987 Census of Wholesale Trade—Preliminary Statistics*. U.S. Department of Commerce. Washington, DC.

U.S. Department of Commerce, Bureau of the Census. 1989b. *Value of New Construction Put in Place*. C30. U.S. Department of Commerce. Washington, DC.

U.S. Department of Labor, Bureau of Labor Statistics. 1988. *News 88-423*. U.S. Department of Commerce. Washington, DC.

U.S. International Trade Commission. 1982. *Conditions Related to the Importation of Softwood Lumber Into the United States*. USITC Pub. No. 1241. U.S. International Trade Commission. Washington, DC.

Wall Street Journal. 1984. "The Largest 100 Home Builders." (April 25):33.

Weyerhaeuser Company. 1981. *Weyerhaeuser Annual Report*. Weyerhaeuser Company. Tacoma, WA.

Widman. 1991. *Markets '91: The Five-Year Outlook for North American Forest Products*. Widman Management Ltd./Miller Freeman Publications, Inc. San Francisco, CA.

Widman, 1990. *Markets 90–94: The Outlook for North American Forest Products*. Widman Management, Ltd. Miller Freeman Publications, Inc. San Francisco, CA.

Widman. 1989. *Markets 89–93: The Outlook for North American Forest Products*. Widman Management Limited/Miller Freeman Publications, Inc. San Francisco, CA.

Wynn, G. 1946. *Timber Colony*. University of Toronto Press. Toronto, Canada.

CHAPTER 12

PULP AND PAPER

In spite of Dagwood's need for sleep, wood pulps and the resulting collage of paper and other products produced from them are the most significant products produced by the forest products industry. More tons of product and dollars of revenue are generated through pulp- and paper-producing activities than by any other manufacturing system of the forest products industry (Tillman, 1985). Moreover, pulp mills also serve as markets for waste by-products of lumber and plywood manufacturing systems. In fact, the by-product utilizing capacity of the pulp industry is so important to the lumber industry that when operating rates of pulp mills decline, many times sawmills are forced to curtail their production because of the lack of market for their pulp chip by-products.

Used with permission, King Features Syndicate, Inc.

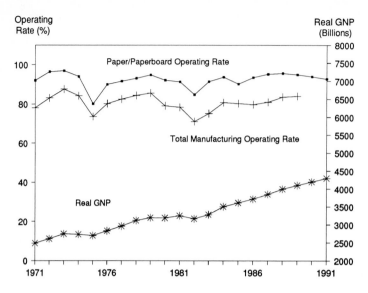

Note: Estimated for 1989-1991.

FIGURE 12-1
Paper-paperboard operating rates for 1971 through 1991.
(*Chrysikopoulos and Kirk, 1989a.*)

THE NATURE OF THE INDUSTRY

The pulp and paper industry is a highly capital-intensive industry. The capital intensity of the industry by almost any measure is far and above the capital intensity of any other segment of the forest products industry and among all manufacturing industries the pulp and paper industry ranks near the top in terms of capital investment per employee (Tillman, 1985). This high capital intensity creates a barrier to entry for new entrants that do not have access to sufficient capital. It also causes a relatively high portion of the total cost to be in the fixed costs of the facilities which in turn requires the pulp and paper industry to operate at relatively high operating rates to produce a profit. These high operating rates are illustrated in Fig. 12-1, which shows the historical pattern of paper-paperboard operating rates from 1971 through 1991 (1989, 1990, and 1991 are estimated). When the operating rates for the paper-paperboard industry are compared to the operating rates for total manufacturing in the U.S. economy, the paper-paperboard rates are consistently higher.

The growth in the demand for paper is strongly correlated to growth in real GNP (gross national product). This relationship is illustrated in Fig. 12-2, which shows the apparent U.S. consumption of paper-paperboard and the apparent consumption of paper-paperboard per billion dollars of real GNP for a 20-year period. Over this period, apparent consumption of paper-paperboard averaged approximately 20,914 tons per billion dollars of GNP (1982 dollars) from 1971 through 1991. As shown by Fig. 12-2, this has been relatively consistent.

While the paper industry is considerably more stable than most other segments of the

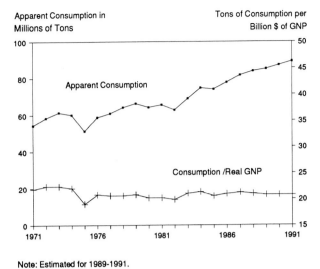

Note: Estimated for 1989-1991.

FIGURE 12-2
Paper-paperboard apparent consumption and apparent consumption per billion dollars of real GNP (1982 dollars). (*Chrysikopoulos and Kirk, 1989a.*)

forest products industry, it still experiences its business cycles. The close relationship of paper-paperboard to GNP can be seen in Fig. 12-1. During years of declining GNP, such as 1974 and 1975, and again in 1980 and 1982, we see the operating rates for paper-paperboard facilities dropping in concert with the GNP. In the late 1980s, the overall operating rate for the paper-paperboard industry was relatively high. Market pulp operating rates were extremely high, at 99.5 percent (Melnbardis, 1989). This extremely high operating rate resulted from a strong demand for market pulp and a subsequent steady rise in market pulp prices from $425 per metric ton in early 1986 to $800 per metric ton in the first quarter of 1989 (Melnbardis, 1989). Usually when operating rates reach so close to capacity and prices rise accordingly, the pulp and paper industry adds additional capacity to meet the increased demand. However, given the tremendous economies of scale in the industry, capacity additions are normally added in very large increments, which can cause the market to be temporarily flooded with extra capacity until demand catches up.

During the first quarter of 1989, pulp producers worldwide had more than fifty new projects on the drawing boards. Those projects in North America alone were valued at more than $4 billion (U.S.) (Melnbardis, 1989). It was estimated that these new projects could boost world market pulp capacity as much as 42 percent. As a result, operating rates for market pulp producers were predicted to fall in the early 1990s (Melnbardis, 1989) and they did so. While not necessarily to the same extent, most segments of the pulp and paper industry go through similar cycles, with demand being very strong, then additional capacity being added in very large increments to meet that demand. This floods the market with more product than it can use, which causes operating rates to fall and prices and profits to drop to a point where they remain until demand catches up with supply again. Then the cycle repeats itself. Some analysts have gone so far as to call this cycle of boom and bust a vicious circle.

Pricing

> Montreal-based CIP Inc. and Bowater Inc.-Darien, Connecticut planned to raise their U.S. newsprint prices $40 a metric ton, to $650 from $610. CIP led the pricing with a move to $655 a metric ton, but backed off to $650 to match Bowater, some publishers said. (Zehr, 1987)

The Wall Street Journal carried the previous quotation in the third quarter of 1987, noting that CIP and Bowater planned to raise their newsprint prices on January 1, 1988. What you see here is an example of price leadership within the paper industry. The price leaders, CIP and Bowater, were signaling to their customers and the rest of the industry that they intended to raise their prices. This method of pricing is very common in many strongly oligopolistic segments of the paper industry, including uncoated printing and writing papers, newsprint, unbleached kraft, and linerboard (Rich, 1983). Rich (1983) describes the method of price determination in these industry segments as one of target return pricing, tempered by marginal cost pricing. Price leadership can be shown to exist where one company is typically the first to announce price changes and other firms then follow. The leadership role can sometimes shift among the dominant producers; however, the leader's job is to carefully adjust pricing to changing costs and demand conditions without precipitating a price war. The leader should be able to make the announced price stick.

In the example noted, the industry operating rate for newsprint production in Canada had been running at about 97 percent, indicating a very strong market demand. CIP led the pricing action and attempted to move the price to $655 per metric ton. In this instance,

Large rolls of paper at site of manufacture.

however, CIP was unable to make the full increase stick when Bowater announced only a $40-per-ton increase to $650 per metric ton. Thus, CIP was forced to back off on its initial price to match Bowater's smaller price increase.

When operating rates are above 93 percent or so, this usually signals periods of relatively strong demand and a sellers' market (Rich, 1983; Zehr, 1987). When operating rates are in this area or higher, the price leaders will typically use a target return pricing scheme by setting a price which will allow them to earn a satisfactory rate of return on their capital investment (Rich, 1983). This return on investment is many times defined as the ratio of total income to total operating assets and may have an inflation premium built into it also.

During periods of weak demand, again defined by operating rates falling below 93 percent, target return pricing by price leaders is usually tempered by marginal cost pricing (Rich, 1983). Marginal cost pricing involves setting the price of a particular product such that the company can secure the largest marginal contribution to fixed overhead and profit. In periods of weak demand (i.e., low operating rates) many paper firms are willing to price their product below its average total cost all the way down to its marginal cost, in an effort to minimize losses (Rich, 1983).

Within many segments of the paper industry, the dominant producers have significant market shares, heavy capital investments, and new and efficient paper-making machines. When demand is strong, a price leader typically emerges from these dominant producers and leads prices to a fairly high level to achieve a target return on investment. This still leaves plenty of orders to keep all their paper-making machines running. Smaller mills are typically content to price at about the same level because market demand is strong enough to keep them also running well above their breakeven points. However, during periods of weak market demand the price level may actually be determined by smaller, marginal mills even though the price change announcements may still come from the dominant-firm price leaders. During periods of weak demand, the smaller, more inefficient mills try desperately to boost their market share sufficiently to keep their mills and paper machines running. They, many times, can do this only by lowering their prices below the prevailing industry level. If enough smaller mills are successful such that the large dominant mills begin to lose market share and see their operating rate drop over a period of several months, then the price leader among the dominant firms may take the industry price level down close to the marginal price of the smaller, more inefficient mills. The large modern producers may still make a profit at this level although they will typically fall short of attaining the target return goal that the higher price provided. When market demand strengthens again, the price leader will move the industry price up once more to a target rate of return level (Rich, 1983).

PAPER MARKET TRENDS

While paper markets are clearly subdividable into discrete product-market segments, it is useful to discuss overall trends in the paper markets before discussing each of the discrete product-market segments in turn.

There are many strong substitute products for paper, including plastics and electronic communications. Many are predicting a slight slowing of the growth in paper demand vis-à-vis the growth in GNP due to the increasingly strong performance of substitute products. While this relationship is not shown in any dramatic way in Fig. 12-2, it does bear watching. Worldwide papermaking has increased and this has resulted in increased international competition. This has been especially worrisome to U.S. producers of market pulp that have large eucalyptus pulp projects planned and completed in South America (Melnbardis, 1989). Other strong international competition is coming in the fine-paper market from European and Scandinavian producers, and in the newsprint and pulp market from Canada. Growing paper exports have been predicted for Japan and other Pacific Rim countries, assuming that the U.S. dollar stays at moderate levels. Many of these countries do not have the natural resource base to support their needs for paper and paper products and must import wastepaper and pulp chips to supply their domestic paper producers.

The legislative situation is causing considerable uncertainty for the industry, especially in the areas of recycled paper and environmental concerns. Mandated percentages of recycled paper in newsprint and other paper products have the potential to negatively impact on market pulp producers and paper producers in rural areas without an economical supply of wastepaper. Tighter environmental regulations are having a major impact on older mills which are difficult to retrofit, but almost no producers are escaping additional environmental costs.

Many producers are recognizing the need to narrow their product line to achieve efficiencies and market clout. Thus we see producers beginning to specialize in certain grades and types of papers and showing a willingness to sell facilities not related to meeting those company goals and needs. One example of this in the early to mid 1980s was International Paper's move out of the newsprint business.

Recent acquisitions and takeovers have continually emphasized the increased vulnerability of pulp and paper producers with poorly performing company assets. These assets can take many forms, such as poorly performing pulp and paper mills or timberlands which are not making a strong contribution to corporate profitability.

Operating rates for the mid to late 1980s have been strong and this has lead to many firms announcing capacity increases in a wide variety of pulp and paper grades. Many of these capacity increases are coming on-line in the early 1990s and will have a moderating influence on operating rates and also profits. Should the strong capacity increases come on-line at a time of weak or negative GNP growth, the industry could experience a period of extremely weak markets.

Competition in the Paper Industry

The pulp, paper, and converting industry is dominated by a relatively small number of giant integrated firms; however, there is a multitude of small firms. The giant firms compete almost exclusively on cost and the smaller firms typically opt to compete in a focused way. These smaller firms sometimes have a geographic focus, a customer focus, or another form of focus strategy (Bauerschmidt et al., 1986).

Competitive methods geared to customer service are important in the paper industry (Bauerschmidt et al., 1986). These include quick delivery, immediate response to customer orders, producing to order for special customers, and having higher quality standards than competitors. However, cost control is also very important, with higher production efficiency than competitors and tight control of overhead competitively important. Bauerschmidt et al. (1986) have noted that "efficient plants and people are not competitive options, but fundamental necessities."

The structure of the industry forces most firms into similar pricing and usually precludes aggressive price competition. Given this, efficient manufacturing facilities are a prerequisite and most competition is in the area of customer service.

Successful Papermakers

Clarence Brown, Vice President at Shearson Lehman, in an address to the Forest Products Research Society noted the common characteristics of successful papermakers (Brown, 1986):

1 Successful papermakers specialize in serving only a few markets exceptionally well. Too much product diversity can be a handicap.

2 While market share is important, successful papermakers do not rely on large size alone. They concentrate on adding value to their products, which provides their customers with unique performance characteristics. Customers in turn are willing to pay a little extra to obtain these products not readily available from other suppliers.

3 Successful papermakers are not afraid to be different. The difference may be in production, labor relations, finance, or other areas. They refuse to do it one way just because it was always done that way before. They are always moving ahead and are tough for the competition to catch.

4 Diverting resources (people and capital) into areas they know little about is carefully avoided. The number of ill-fated diversification ventures for papermakers is large and instructive, including Cuban electric utilities, oil and gas production, manufactured housing, home furniture, carpet manufacture, school bus manufacture, and many others.

5 Successful papermakers keep a well-proportioned balance sheet. They stay financially well managed. This means not debt-free but rather with a wise and judicious use of credit.

PRODUCTS

Up to this point we have discussed the pulp and paper industry as a single large industry. However, it does not produce a set of homogeneous products; rather, the products produced are very different in their production facilities, uses, pricing, distribution, and marketing. Many times the underlying demand factors can also vary from one product to another. Figure 12-3 provides a breakdown of the major paper grades. Within each of these grade-segments there are numerous subsegments which are beyond our ability to discuss individually. However, we will tackle each of these major grade-segments in order to better understand the diversity and uniqueness of the paper industry.

Newsprint

Approximately 75 percent of the newsprint consumed in the United States is consumed by daily newspapers (Chrysikopoulos and Kirk, 1989a). Other consumers of newsprint include weekly newspapers, government printers, and commercial printers.

The long-term consumption rate for newsprint roughly parallels the growth in real GNP (adjusted for inflation) which has, over time, averaged approximately 2.4 percent per year (Chrysikopoulos and Kirk, 1989a). As with other segments of the paper industry, when faced with very high operating rates such as occurred in the mid to late 1980s, the

FIGURE 12-3
Major paper grades.

NEWSPRINT:

       Newspapers
       90% mechanical pulp -- 10% chemical pulp

UNCOATED GROUNDWOOD:

       brighter than newsprint, used
       for catalogs & directories

COATED PUBLICATION PAPERS:

       magazines, catalogues, periodicals, etc.

UNCOATED FREE SHEET:

       books, bond paper, envelope, etc.
       Free Sheet means no more than 10% mechanical pulp

BRISTOLS:

       note cards, postcards, file folders, etc.
       paper is thicker than .006 inch

PACKAGING & CONTAINER BOARD:

       Kraft paper, linerboard, corrugated medium, etc.

TISSUE:

       restroom supplies, sanitary products, paper towels, etc.

MARKET PULP:

       sold to paper mills and chemical users of pulp

newsprint industry responds to such conditions with strong capacity additions which tend to drive operating rates down.

Over half of the newsprint consumed in the United States is imported from Canada, with lesser amounts coming from Scandinavian countries. U.S. overseas exports have generally been trending down as a result of strong domestic demand. A large portion of these exports has been to Japan and during the late 1980s Japan invested strongly in North American newsprint-manufacturing capacity.

North America is the world's low-cost producer of newsprint, producing nearly half of the world's supply (Widman, 1990). Canada alone was responsible for about one-third of the world's production and 60 percent of global exports in the late 1980s (Widman, 1990).

Prices of newsprint rose strongly in the late 1980s. However, price stability problems have been projected for the early 1990s as significant new U.S. and Canadian capacity comes on-line (Chrysikopoulos and Kirk, 1989a). Recent laws in some states requiring that newspapers contain a significant percentage of recycled fiber are adding additional uncertainty to this market. These laws may especially hurt Canadian mills with limited access to recycled fiber.

Uncoated Groundwood

Telephone directories, magazine and newspaper inserts, and some supercalendered magazine papers are made of uncoated groundwood. It has similar properties to newsprint but is considered of higher quality and may be thought of as a value-added newsprint. U.S. consumption of uncoated groundwood is increasing faster than real GNP growth.

About 80 percent of Canadian production is exported to the United States and the United States imports over 40 percent of its uncoated groundwood needs. Most U.S. imports are from Canada but some supercalendered grades are imported from Europe (Widman, 1990).

Coated Paper

Coated paper is used by commercial printing operations for inserts, catalogues, and annual reports. It is also used in magazine publishing and in the publishing of other periodicals. Coated paper largely falls into one of two types: coated groundwood papers and coated free-sheet papers. The consumption of coated papers during the 1980s increased roughly 2 to $2\frac{1}{2}$ times the real GNP growth rate (Chrysikopoulos and Kirk, 1989a). These high growth rates are the result of strong gains in magazine circulation and in catalogue merchandising. Demand growth for coated papers is predicted to continue at a significantly higher level than that of real GNP (Chrysikopoulos and Kirk, 1989a).

Imports have played a strong role in the coated paper marketplace. In 1989, imports represented approximately 7 percent of U.S. consumption (Widman, 1990). Western Europe is the dominant supplier and the United States imports mostly coated groundwood. Exports have accounted for approximately 1 percent of U.S. production and are expected to remain essentially unchanged in the near term. U.S. producers typically do not have established marketing organizations with primary responsibility for sales to offshore markets (Chrysikopoulos and Kirk, 1989a).

As a result of the strongly growing demand in the United States, there were strong increases in capacity during the late 1980s. European and Canadian capacity in coated paper is also growing at a high annual rate. These strong increases in capacity have caused pricing to be sensitive to new startups.

Uncoated Free Sheet

Uncoated free sheet is used primarily as office paper for forms, bond paper, envelope paper, copier paper, and computer printer paper (Fig. 12-4). Secondarily, uncoated free sheet can be used as printing paper. The term ''free sheet'' is derived from the fact that the paper is largely free of mechanical pulp.

As with other paper grades, the demand for uncoated free sheet moves in tandem with changes in real GNP. During the 1970s, the consumption of uncoated free sheet grew at about 1.5 times the real GNP growth rate (Chrysikopoulos and Kirk, 1989a). However, during the 1980s the consumption of uncoated free sheet accelerated to approximately twice the rate of real GNP growth, reflecting the tremendous growth in the use of copier machines and computer printers (Chrysikopoulos and Kirk, 1989a). The continuation of this trend is expected to maintain strong long-term growth rates for uncoated free sheet at slightly less than twice the annual growth in real GNP (Chrysikopoulos and Kirk, 1989a).

In 1988, imports represented approximately 5 percent of uncoated free-sheet consumption and by the early 1990s this is predicted to increase upward to 8 percent (Chrysikopoulos and Kirk, 1989a). About 40 percent of these imports are from Canada, which had a significant expansion of its uncoated free-sheet industry in the late 1980s and early 1990s. U.S. exports of uncoated free sheet are expected to remain, for the near term, at less than 1 percent of production (Chrysikopoulos and Kirk, 1989a).

This particular paper grade-segment has seen strong competitive pressures from new capacity additions (particularly Canadian) and a movement toward large commodity-type production mills specializing in a given commodity product (i.e., copy paper). These trends have periodically caused weak pricing in the industry.

Bristols

Bristols are thick, stiff papers usually 0.006 inch thick and thicker. They are everyday papers used in note or index cards, file folders, tags, and business reply cards. Bristols can come coated or uncoated and in many finishes and colors.

In the late 1980s, U.S. bristol production increased slightly to approximately 1100 tons per year or about 10 percent of the volume of uncoated free sheet (*Pulp and Paper,* 1989). The paperless office concept and electronic filing systems could hurt bristols in the long run, but they have caused no significant near-term decrease in demand.

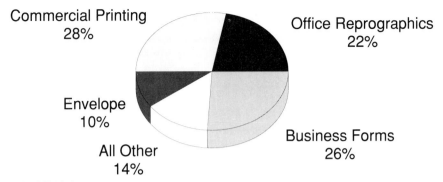

FIGURE 12-4
Markets for uncoated free sheet. (*Chrysikopoulos and Kirk, 1988.*)

Container Board and Kraft Paper

Container board is comprised of linerboard and corrugated medium. It is the basic raw material for producing containers and is the largest segment of the world's paper industry. Consumption is clearly a function of the demand for corrugated containers, which has basically a 1 : 1 relationship with real GNP. Approximately 85 percent of the U.S. total kraft linerboard output is consumed in the domestic manufacture of corrugated containers (Chrysikopoulos and Kirk, 1989a).

Linerboard represents one of the paper products of which U.S. producers are significant exporters. In 1989, approximately 2.1 millon tons of linerboard were exported, representing 13.5 percent of U.S. domestic production (Widman, 1990). Since

The use of paper for packaging is more widespread than may be thought.

1977, exports of linerboard have varied between 10 and 17 percent of domestic production.

Operating rates for linerboard have typically been high as a result of a smaller number of grades. Many capacity increases in the United States have resulted from the enhancement of existing mills rather than the development of new linerboard mills. Linerboard production is concentrated in only a few companies in North America, Europe, and to a lesser extent Japan. The use of recycled fiber programs is strong in some areas. As a result of a strong worldwide economy and small capacity increases, linerboard prices were up significantly during the mid to late 1980s.

Unbleached kraft paper is produced primarily for bags, sacks, shipping wrap, and other converting products. Bag and sack shipments account for about 62 percent of total shipments (*Pulp and Paper*, 1987). Consumption of unbleached kraft has been declining and production capacity has also declined, keeping operating rates in line. Grocery bag and sack paper shipments have been negatively impacted by stiff competition from plastic bags, resulting in an overall decline in shipments of unbleached kraft paper.

Tissue

About 98 percent of tissue output falls into the category of sanitary tissue products, which includes bathroom, facial, toweling, napkins, and base stock for other sanitary products (*Pulp and Paper*, 1988). The two major markets for tissue are the consumer market and the commercial and industrial market. Consumer tissue is typically purchased by individuals through retail outlets. Commercial and industrial tissues are bought by hotels, restaurants, offices, factories, and other businesses and government agencies.

During the late 1970s through the late 1980s, the tissue industry expanded at approximately 2.2 percent per year, which is slightly less than the long-term growth rate of real GNP (*Pulp and Paper*, 1988). This relatively slow growth in the U.S. marketplace has prompted Scott Paper and James River to make major investments in the faster-growing Western European tissue market. Overseas markets, in general, tend to be growing much faster than the U.S. domestic marketplace. There is minimal import pressure in tissue and limited exports as well. However, a number of U.S. firms, including Scott Paper and James River, are investing in tissue production operations in markets outside the United States.

The consumer tissue market is highly competitive, with retailers and consumers being very price-conscious. The industry went from 1981 to 1988 with virtually no price increases (*Pulp and Paper*, 1988). In 1988, the industry was able to average approximately a 5 percent price increase. Hidden pricing increases did occur in earlier years in the form of reduced sheet counts, etc.

Market Pulp

Sixty-five percent of the world's fiber needs for paper and paperboard production is met by virgin wood pulps, 25 percent by recycled paper, and 10 percent by other plant fibers and minerals. Recycled fiber is used extensively in Europe and some Pacific Rim nations where wood is expensive and in short supply. In other areas such as Latin America and

Large rolls of tissue positioned for rewinding to household size rolls.

Southeast Asia plant fibers such as bagasse (sugar cane) are used as pulp raw material (Widman, 1990).

Wood pulps are of several types including mechanical, chemical (sulfite and kraft), and chemi-thermomechanical. Mechanical pulps have high yields produced by grinding the fibers apart. No wood components are dissolved away as in the chemical pulps; however, this leaves the lignin on the fibers, which will cause paper produced from mechanical pulp to yellow over time.

Chemical pulping provides lower yields because it removes the lignin and other components while separating the wood fibers. This gives a longer, smoother fiber which is stronger.

Chemi-thermomechanical pulp is a process where chips are pretreated with chemicals and heat. Then pulp is mechanically produced. Thermomechanical pulp is also a mechanical pulp which uses heat in conjunction with the mechanical pulping process. Both chemi-thermomechanical and thermomechanical processes are higher-yielding processes than chemical or simple mechanical pulping.

Bleached northern softwood kraft pulp is considered the premier grade based on its longer fiber lengths and strength. Canada and the United States are the world's leading producers of this grade.

Most of the wood pulp produced (75 percent on a worldwide basis) is consumed at the site of production in making paper or paperboard. Only 25 percent is sold on the open market as "market pulp" (Widman, 1990), although some countries, especially Canada and the Nordic nations, sell more of their pulp as market pulp. Chemical pulps dominant the pulp market.

North America is the major world producer of chemical market pulp, as shown in Fig. 12-5, with approximately half of the world's production capacity. Market pulp is primarily used for paper manufacture but other significant markets for pulp as adsorbents in diapers and other sanitary products as well as feed stocks for chemical products exist.

The late 1980s saw a bull market in market pulp, with market pulp cost relative to paper prices hitting all-time highs (Kirk and Chrysikopoulos, 1989). Figure 12-6 shows the narrowing gap between pulp costs and paper prices, with pulp costs estimated to be at 82 percent of paper prices in 1989. This strong market resulted in many new capacity additions announced in the late 1980s (Melnbardis, 1989). Other analysts are worried that the strong capacity increases coupled with less fluff pulp use in diapers due to super adsorbents will hurt profitability in this segment in the 1990s. Additionally, mandated recycling programs could also lower demand for market pulp.

Recycled Fiber

Strong recycling efforts are under way in many, if not most, parts of the world. The heaviest use (on a percentage basis) of recycled paper fiber is in areas of low availability of virgin sources of fiber. In the late 1980s, 49 percent of the fiber used in Japanese paper production was recycled fiber. Some countries in Europe have wastepaper utilization rates of 40 to 60 percent. The United States in the mid to late 1980s was the world's largest consumer of wastepaper; however, this amounted to recycling only about one-quarter of

FIGURE 12-5
World chemical market pulp capacity, 1987. (*Kirk and Chrysikopoulos, 1989.*)

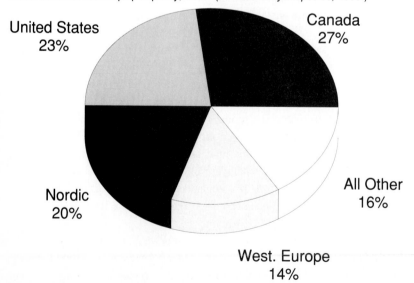

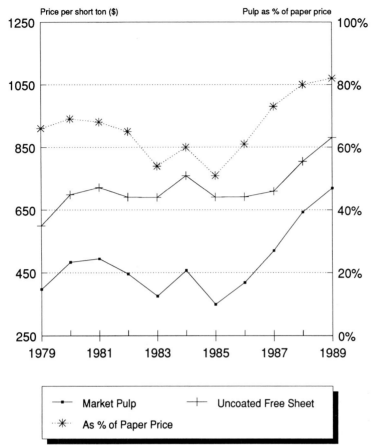

FIGURE 12-6
Relationship between U.S. market pulp costs and paper prices. (*Kirk and Chrysikopoulos, 1989.*)

its paper production (Widman, 1990). Recent goals set by the American Paper Institute call for much higher recovery recycling rates by the mid-1990s.

Many technical questions remain about recycling, including ones about paper quality, deinking, and the number of times a fiber can be recycled. However, in spite of these and other economic issues wastepaper recycling will no doubt increase strongly in the future and will be a major factor influencing many forest products markets.

INTERNATIONAL TRADE

Consistently through the mid to late 1980s, the United States ran a trade deficit in pulp, paper, and paperboard products. In 1989, approximately 80 percent of U.S. imports came from Canada but only 10 percent of U.S. exports went to Canada (Chrysikopoulos and Kirk, 1989b). Newsprint and market pulp are the dominant products imported. For the period 1985 to 1989, newsprint averaged 45 percent, market pulp 24 percent, and printing and writing paper 16 percent of the value of U.S. pulp, paper, and paperboard imports (Chrysikopoulos and Kirk, 1989b).

U.S. exports are mostly commodity product grades, especially pulp and linerboard. Export markets for higher-value grades and access to the mill-dominated merchant structure in Western Europe are limited. From 1985 to 1989 market pulp averaged 38 percent of the value of U.S. pulp, paper, and paperboard exports. Kraft linerboard averaged 12 percent, bleached kraft paperboard averaged 7 percent, industrial packaging papers averaged 5 percent, writing and printing papers averaged 4 percent, and a wide variety of other paper products contributed approximately 34 percent of the value of pulp, paper, and paperboard exports (Chrysikopoulos and Kirk, 1989b).

PAPER DISTRIBUTION

Historical Development

Paper wholesalers, or as they are usually referred to in the trade, merchants, dominated the distribution of paper prior to World War I, with the exception of newsprint and paper sold to magazine publishers. Paper producers were much smaller in those years and paper merchants typically assumed full responsibility for stocking, pricing, promoting, and selling paper products. As a result, paper merchants were able to dictate to mills the specifications of the paper they should manufacture. Mills then manufactured paper to meet the specifications of the merchants, who then sold the paper under their own merchant brands (Rich, 1970).

During the 1920s, as paper demand expanded, mills also increased in size and it became obvious that long runs of a standard paper line provided economies of scale in paper production. This encouraged the manufacturers to produce and advertise their own mill brands. Producers developed their own sales forces and branch sales offices in order to reach more paper merchants on a national basis and get their mill brands better known among printers and other customer groups (Rich, 1970). This situation has continued until the present day, with some larger merchants still stocking and merchandising their own merchant brands of paper, but with most papermakers trying to promote their own mill brands through independent paper merchants.

As time progressed, paper manufacturers began to take more and more of an interest in the distribution and marketing of the paper products they produced. Some paper manufacturers such as Mead and Champion International developed extensive networks of paper merchants. Others, especially in the fine writing and printing paper area, developed systems of independent paper merchants who received an exclusive right to distribute the paper manufacturer's product in a given geographical area. These franchises or exclusive distribution arrangements have been widely used by paper producers that do not own their own captive merchant system.

Current Distribution Channels

Current distribution channels vary widely depending on the paper product (Fig. 12-7). For example, most newsprint and paperboard go directly to end users and not through merchants, while consumer tissue products may go through a variety of wholesalers/merchants. Study Fig. 12-7 carefully to become familiar with the current distribution systems for paper.

Types of Paper Merchants   The paper merchant industry has evolved over the years and this evolution resulted in three basic types of paper merchants: industrial, fine paper, and dual. Industrial merchants sell a wide variety of industrial paper products including kraft paper, brown bags, disposable paper goods, tissues, and many other janitorial and cleaning supplies. Fine-paper merchants typically stock printing paper, copy paper, computer paper, and an assortment of other office and printing items. Dual merchants, as the name implies, tend to stock items that both industrial merchants and fine-paper merchants would carry. As merchants have increased in size in recent years, the trend has been for the larger industrial or fine-paper merchants to become dual merchants.

FIGURE 12-7
Simplified channels of distribution for paper products.

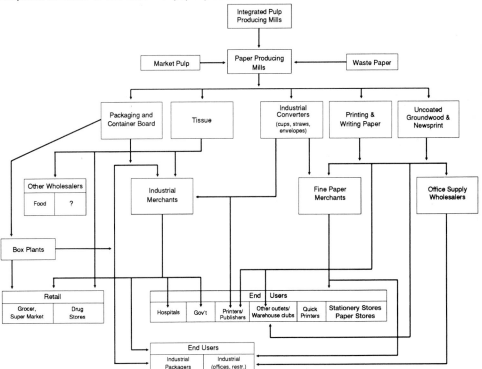

Paper merchants serve as stocking wholesalers while performing some office whole-saling functions as well. Fine-paper merchants many times can serve as office whole-salers for larger orders of copy paper and other products. The fine-paper merchant takes the order and transmits it to the manufacturer; the product is shipped mill-direct to the final customer. A substantial portion of the sales of a fine-paper merchant can be through this channel. Industrial paper merchants, on the other hand, due to selling a wider variety of products in typically smaller quantities, have less opportunity to move products directly from the mill to their customers. However, many industrial paper merchants do arrange for some factory-direct shipments. In some paper markets a dual system operates, with the manufacturer's sales force handling larger customers through mill-direct shipments and independent merchants handling smaller-volume customers. Of course, where the line is drawn between larger and smaller customers can be a point of conflict between manufacturers and merchants.

> The paper merchant of today is moving out of the shadows and into a position of high visibility in the paper industry. As a result, the traditional stereotype of the merchant as an "unsophis-ticated" and labor-intensive business is giving way to a more enlightened view. In fact, the distribution of paper and related disposable products is evolving from a fragmented collection of business establishments to a highly sophisticated, programmed distribution system. The metamorphosis is still underway, but the pace of change is rapidly accelerating. (Healy, 1987, pg. 12)

It's clear from the quotation above that the paper merchant of today is a rapidly changing business and that the practices of yesterday will not suffice tomorrow. The paper business has become very price-sensitive at the wholesale level. For commodity grades of paper, extreme price sensitivity can exist. Such grades include copy paper and computer paper. Within these price-sensitive commodity grades there is evidence of very little brand preference. Most paper is sold on the basis of price and service. As the customers of paper merchants have changed, the ability of merchants to render the services needed has begun to have more and more impact on their ability to sell their products.

PAPER CUSTOMER TRENDS

Printing Paper Customers

The mid-1980s through the 1990s have been and will be a period of substantial change in the composition and marketing practices of printing paper merchants. The so-called quick-copy center has become a major factor for fine-paper merchants. Healy (1987) expects a shakeout in the quick-copy industry. This is likely to occur as the better-managed copy centers incorporate more and more advanced technology and marketing into their operations. The less able copy centers will see themselves forced out of business because there are very few professional offices or businesses that cannot afford their own copying equipment.

The problem for fine-paper merchants in this scenario is how to provide adequate service to the former customers of the quick-copy centers. Where there once were only a few larger-volume customers, in the future there will be many small-volume customers.

One solution to the problem of many small customers has been the cash-and-carry retail paper store. Healy (1987) noted that in 1987 all types of merchants said they had a 50/50 chance of opening a retail store in the next 5 years. The home office explosion has also created another large segment of low-volume users which can be served by paper stores or more likely by catalog sales.

This trend of more in-house printing/copying may also affect the small sheet-fed commercial printers who serve smaller clients (Healy, 1987). These commercial printers are finding that with the advent of desktop publishing, sophisticated software, and laser printers, their services are now in less demand. In the area of larger commercial printers, new technologies for printing and graphic arts are rapidly impacting on the demand and use of paper. One area in which this has been strongly evident is color printing. As color printing has increased, the demand for coated papers has risen consistently (Healy, 1987).

Another major trend for the fine-paper merchant has been the increase in web printing. New technologies have dropped the economies of scale in web printing such that web printers are able to economically run shorter press runs from the web press, which requires a continuous ribbon of paper fed from a roll. The costs of web printing are approaching the costs of the former sheet-fed printing equipment (Healy, 1987). Advances are likely to continue in this area and paper merchants may be forced to develop roll stock programs for web printers or lose out on a significant portion of the commercial printing business.

These changes all have significant implications for the fine-paper merchant. Large-scale commercial printers are likely to be even more demanding of the technical expertise and service abilities of the merchant. The shift from small printing and quick-copy operations to in-house operations will require many new capabilities by the merchant. Additional customer services of all kinds will be expected, coupled with the marketing and logistical challenges of efficiently serving an increasing number of smaller accounts.

Industrial Paper Customers

Fine-paper merchants carry a rather homogeneous product line. However, this is not true of the industrial paper merchant, which carries a variety of separate and distinct products to serve diverse market segments (Fig. 12-8). These products range all the way from sanitary papers/tissues to food service disposables to packaging equipment and supplies to janitorial supplies. In order to provide full-line service to customer groups, industrial paper merchants have begun to specialize in one or more of these product areas.

By specializing, the merchant's goal is to develop a one-stop shopping, full-service relationship with customers in a given market segment in the hope that this will increase customer loyalty (Healy, 1987). Each specialized segment of the market requires the industrial paper merchant to offer a unique set of products and services to satisfy the customer. Merchants become almost consultants to customers. They go well beyond simply being providers of commodity products and help solve problems and provide training and other value-added activities.

In order to examine the customers for industrial paper merchants, it is necessary to look at these customers on a segment-by-segment basis. As is easily seen in Fig. 12-8, the packaging equipment and supplies segment is very important to industrial paper

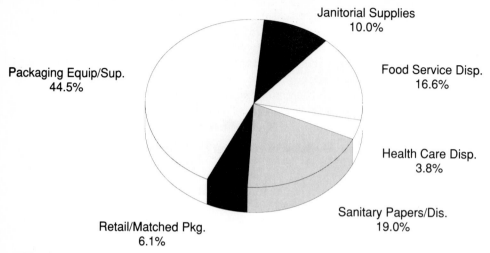

FIGURE 12-8
Typical industrial merchant sales mix in the 1980s. (*Healy, 1987.*)

merchants. The health of this segment is closely tied to the overall U.S. economy. As manufacturing picks up, more products are shipped and more packaging equipment and supplies are required.

The food service disposables market is an increasingly important market for industrial and also dual merchants (Healy, 1987). This market, as a whole, has been growing strongly as a result of several factors. More fast food, which is typically served in disposable containers, is being consumed. Furthermore, it is being consumed from a wide array of outlets. One type of outlet that has been increasing in numbers is the deli, which has been taking away business from supermarkets. Another is the food service section of convenience stores. In addition, the feeding of employees in offices and plants, health care facilities, and schools and colleges is becoming more oriented toward using disposable products. For merchants to capitalize on this trend they are required to carry a complete line of disposable products including cups, napkins, lids, trays, plastic utensils, etc.

The sanitary papers market can be typically divided into several segments including facial tissue, bathroom tissue, paper towels, and paper napkins. This segment has experienced strong growth over the years for industrial and dual paper merchants. Growth has typically come from new accounts in office buildings, health care institutions, and also manufacturing facilities. Additional growth has resulted from the affiliation of the sanitary papers segment with janitorial supply needs and the food service disposables market segment. With expanding office construction and an increasing need for nursing homes and extended care facilities, the sanitary papers market is expected to continue to grow. Many paper producers have developed innovative systems for dispensing larger rolls of bathroom tissue and paper towels to reduce labor in office buildings and health care facilities. Even more value-added products and strategies will benefit this market segment in the future.

As a complementary product line to food service disposables and sanitary papers, the janitorial supplies segment is becoming increasingly popular for industrial paper merchants. Janitorial supplies exhibt a superior profitability versus some of the other paper goods and also have a high stock turnover rate (Healy, 1987). To effectively serve this market segment, industrial merchants are required to carry a variety of floor finishes, cleaners, powered floor equipment, mops, paper goods, chemicals and soaps, brushes, and other items. The trend in some areas is to provide an entire system of products which can be used for the total janitorial needs of the purchasing client. This may require the industrial merchant to provide training and other services as part of its product program.

The future health of the industrial paper merchant seems to be closely related to several important customer segments. Those customers in health care and food services are purchasing increasing amounts of disposable products. The industrial paper merchant must compete for these customers with other classes of wholesalers as well as other paper merchants. The industrial merchant must learn to service these customers better and more effectively than do the alternative sources of supply.

As many of the customers of industrial paper merchants become larger, they may wish to centralize their purchasing needs and establish working relationships with a smaller number of suppliers (Healy, 1987). The performance of these suppliers will become relatively easy to measure and the penalties for nonperformance could be harsh.

Another trend affecting the industrial paper merchant is the aversion to carrying inventory (Healy, 1987). No channel participant wishes to carry inventory. Many customers are becoming oriented to a just-in-time approach and are selecting suppliers that can provide products when they need them at the correct time. It appears for the next several years that computers and telecommunication advances are likely to cause dramatic changes in the order entry and inventory management practices of all members of the channel of distribution. Direct computer linkages from customer to industrial merchant and from industrial merchant to manufacturing mill will likely become commonplace.

Overall Market Trends for Paper Merchants

It should be clear by now that the paper merchant business is becoming increasingly sophisticated. This sophistication is occurring at all levels from ordering products, to managing inventory, to supplying customers.

Paper retail stores are growing at a rapid rate and are increasing their ability to serve small-order sales. These retail stores are likely to develop a significant share of the office paper market. Many of these stores are owned by paper merchants and some are independent.

The distribution of paper is consolidating from a fragmented collection of independent merchants to a system dominated by larger firms. Mergers and acquisitions are continuing at a rapid pace. It appears that size alone may become an important factor for a paper merchant. Healy (1987, p. 46) notes that ''at some point agility and cunning may not be enough for the smaller independent.'' The survival of the smaller independent local paper merchant will largely be due to the firm's ability to identify and take advantage of small profitable market niches.

The traditional concept of inside sales personnel in the paper merchant industry is changing toward the use of telemarketing and catalogs (Healy, 1987). Telemarketing is a more proactive posture in that sales personnel are trained to call assigned accounts on a regular basis to request orders and also to make sales presentations. In a fully developed operation, the sales personnel also receive incentive compensation based on sales and achievement of gross margin goals (Healy, 1987).

New technological developments in printing and the ever expanding sophistication of product offerings in other areas have required a more specialized sales force than in the past. The benefits to be derived from trained, specialized sales personnel have become obvious (Healy, 1987). These personnel are better able to serve as technical sales consultants to specific industry segments.

RELATIONSHIPS BETWEEN PRODUCERS AND MERCHANTS

Developing and sustaining strong working relationships (partnerships) between firms in business marketing is increasingly viewed as a means for firms to reduce real costs and/or add value. This need for a channel partnership viewpoint in industrial distribution has been occasionally called for by academic writers (Webster, 1976; Narus et al., 1984). One study concluded that the firms best able to implement their business plans viewed the maintenance of a partnership relationship with their industrial distributors as a key part of their business strategies (Bonoma, 1984).

When fine-writing and printing paper producers were asked to rate the importance of several factors to their marketing strategy and ultimate success, they ranked a strong partnership working arrangement with independent paper merchants as the most important success factor (Sinclair, 1990). In fact, the top three rated success factors all related to producers' relationships with paper merchants (Table 12-1). This probably should not be

TABLE 12-1
PERCENTAGE OF FINE-WRITING AND PRINTING PAPER PRODUCERS RATING THE
FOLLOWING ATTRIBUTES AS IMPORTANT TO THEIR MARKETING STRATEGIES AND
ULTIMATE SUCCESS

Attributes	Percent
Strong partnership working arrangements with independent paper merchants	80
Receiving market information from paper merchants	77
Exclusive or franchise distribution system for your products through independent paper merchants	68
Captive/company-owned paper merchants	38
Export markets for your paper	25
Captive/company-owned paper retail/office supply stores	23

Source: Sinclair, 1990.

surprising as many successful paper producers have developed exclusive or franchise distribution systems using independent paper merchants. Many of these systems have been in place for a number of years.

Neither export markets for paper or captive/company-owned paper retail or office supply stores were viewed as being important to paper producers' ultimate success. However, a group of producers do own strong captive/company paper merchant systems and did feel that their captive merchants would be important to their success in the future.

Partnership Role

At least from the producer's perspective, the majority of relationships between the producer and paper merchant appear to be based on a partnership concept, with many of the producers working hard at maintaining and enhancing the partnership relationship (Sinclair, 1990). This is consistent with the viewpoint of the merchants as expressed by Healy (1987): "Merchants envision a number of changes in their relationship with manufacturers and suppliers. What once may have been an adversarial system seems to be giving way to a more performance-oriented set of interrelationships."

Partnership Advantage   For a successful partnership, it is crucial that each partner provide its partner firm with some advantage relative to other, potential partners. This notion has been called "partnership advantage" and has received only limited attention in the marketing literature (Sethuraman et al., 1988). To better define partnership advantage, one study asked producers to rate the importance of nine characteristics of partnership advantage. Market penetration, financial stability, and local reputation were given the highest mean importance (Table 12-2).

On the other hand, paper merchants expect the producers will become more demanding in (1) credit and terms of sale, (2) the marketing effort expended by the merchant, and

TABLE 12-2
PERCENTAGE OF FINE-WRITING AND PRINTING PAPER
PRODUCERS RATING THE FOLLOWING CHARACTERISTICS OF
AN INDEPENDENT PAPER MERCHANT TO BE IMPORTANT

Characteristics	Percent
Market penetration ability	93
Financial stability	93
Local reputation	93
Knowledge of the local market	90
Prompt payment of bills	90
Management capabilities	90
Ability to handle customer problems	87
Amount of inventory carried	77
Technical sales capability	77

Source: Sinclair, 1990.

(3) the minimum allowable order size (Healy, 1987). This is consistent with the data in Table 12-2, which show marketing skills and financial concerns as the most important characteristics.

Changing Relationships

Consolidations and acquisitions are resulting in fewer independent paper merchants. As a result of fewer independent paper merchants to work with, the relationship between producers and the remaining merchants will, by necessity, become more partnership in nature. Such partnership activities include more sharing of market information, better cooperation to penetrate specialty markets, and mutual goal setting (Healy, 1987; Sinclair, 1990).

SUMMARY

The pulp and paper industry remains a dominant part of the forest products industry and for most major integrated firms comprises over 50 percent of sales. Paper sales are generally influenced by the overall economic health, with some grades such as newsprint having roughly a 1 : 1 relationship with real GNP growth and others like coated paper showing demand growing at 2 to $2\frac{1}{2}$ times the growth in real GNP. Paper products have developed a distinct market system different from building products. A key issue in this system, as in the system for building products, is the relationship between producer and merchant (wholesaler).

Many issues in the relationship between producer and merchant remain unresolved. Some of these issues include growing captive merchant systems, direct mill sale policies, and the long-standing exclusive/franchise merchant system.

Fine-writing and printing paper producers generally believe they practice a partnership relationship with independent paper merchants and view this relationship as a critical factor for their business success and marketing strategies. For a paper merchant to be a good distributor, producers expect the merchants to have market penetration ability, local reputation, knowledge of the local market, and financial strength. Producers want merchants that will be reliable partners with a commitment to the relationship. Consolidations and acquisitions in the paper merchant industry will likely continue making the remaining larger merchants even more important.

Look back at the quote from Healy on page 196. It is clear that the paper merchant is rapidly changing and that the practices of yesterday will not suffice tomorrow. Manufacturers and merchants that can help lead this change and develop true partnerships will have a competitive advantage in the marketplace.

BIBLIOGRAPHY

Bauerschmidt, A., D. Sullivan, and J. Weber. 1986. "How Companies Compete in the USA." *Pulp Paper International* 28(5):48–51.

Bonoma, T. V. 1984. "Making Your Marketing Strategy Work." *Harvard Business Review* (March–April):69–76.

Brown, C. W. 1986. "A Financial Analyst's View of the North American Pulp and Paper Industry." *Proceedings 47351, North American Wood/Fiber Supplies and Markets: Strategies for Managing Change.* Forest Products Research Society. Madison, WI. pages 28–31.

Buongiorno, J., and H. C. Lu. 1989. "Effects of Cost, Demand and Labor Productivity on the Prices of Forest Products in the United States, 1958–1984." *Forest Science* 35(2):349–363.

Chrysikopoulos, J., and B. Kirk. 1989a. "Forest-Paper Outlook and Company Update Service." *Investment Research, Goldman Sachs* (March 21). New York.

Chrysikopoulos, J., and B. Kirk. 1989b. "Paper-Paperboard Industry." *Investment Research, Goldman Sachs* (June 16). New York.

Chrysikopoulos, J., and B. Kirk. 1988. "Boise Cascade Corporation." *Investment Research, Goldman Sachs* (December 6). New York.

Dagenais, M. D. 1976. "The Determination of Newsprint Prices." *Canadian Journal of Economics* 59:442–461.

Healy, D. F. 1987. *New Strategies for Growth: Paper Distribution Plans for the Future.* Paper and Plastics Education Research Foundation. Great Neck, NY.

Kirk, B., and J. Chrysikopoulos. 1989. "Market Pulp: Nearing a Crossroad." *Research Brief, Goldman Sachs* (March 10). New York.

Melnbardis, R. 1989. "Wood Pulp Makers Run Expansion Spree." *Wall Street Journal* (March 27):B8.

Narus, J., N. M. Reddy, and G. L. Pinchak. 1984. "Problems Facing Distributors." *Industrial Marketing Management* 3:139–148.

O'Laughlin, J., and P. V. Ellefson. 1981a. "U.S. Wood-Based Industry Structure: Part 1—Top 40 Companies." *Forest Products Journal* 31(10):55–62.

O'Laughlin, J., and P. V. Ellefson. 1981b. "U.S. Wood-Based Industry Structure: Part 2—New Diversified Entrants." *Forest Products Journal* 31(11):26–33.

O'Laughlin, J., and P. V. Ellefson. 1981c. "U.S. Wood-Based Industry Structure: Part 3—Strategic Group Analysis." *Forest Products Journal* 31(12):25–31.

Pulp and Paper. 1989. "U.S. Paper Industry Will Be near Peak Performance Level This Year." *Pulp and Paper* (January):48–56.

Pulp and Paper. 1988. "Tissue: Producers Seek Relief from Rising Pulp Prices; Volume Should Remain Strong through 1988." *Pulp and Paper* (February):13.

Pulp and Paper. 1987. "Kraft Paper: Demand Still Declining, But Strong Liner Demand and Shutdowns Keep Market Tight." *Pulp and Paper* (November):13.

Quicke, H. E., J. P. Caulfield, and P. A. Duffy. 1990. "The Production Structure of the U.S. Paper Industry." *Forest Products Journal* 40(9):44–48.

Rich, S. U. 1986. "Recent Shifts in Competitive Strategies in the U.S. Forest Products Industry and the Increased Importance of Key Marketing Functions." *Forest Products Journal* 36(7/8):34–44.

Rich, S. U. 1983. "Price Leadership in the Paper Industry." *Industrial Marketing Management* (12):101–104.

Rich, S. U. 1970. *Marketing of Forest Products.* McGraw-Hill Book Company, New York.

Sethuraman, R., J. C. Anderson, and J. A. Narus. 1988. "Partnership Advantage and Its Determinants in Distributor and Manufacturer Working Relationships." *Journal of Business Research* 17:327–347.

Sinclair, S. A. 1990. "Paper Manufacturers' Relationship with Paper Wholesalers: A Partnership?" *Forest Products Journal* 40(9):24–28.

Tillman, D. A. 1985. *Forest Products: Advanced Technologies and Economic Analyses.* Academic Press, Inc., Orlando, FL.

Webster, F., Jr. 1976. "The Role of the Industrial Distributor in Marketing Strategy." *Journal of Marketing* (July):10–16.

Widman. 1990. *Markets 90–94: The Outlook for North American Forest Products.* Widman Management, Ltd./Miller Freeman Publications, Inc. San Francisco, CA.

Zehr, L. 1987. "Two Newsprint Makers Plan Price Rise of 6.6% for U.S. Firms Effective Jan. 1." *Wall Street Journal* (September 8):13.

CHAPTER 13

HARDWOOD LUMBER AND
SECONDARY PRODUCTS[1]

We have a definite lack of product specialization in our hardwood industry. . . . Too many of us are attempting to be everything to every hardwood consumer and we pay the price. Why not specialize?

James Gundy

Within this chapter, we will discuss the role and importance of the hardwood lumber industry and the major industrial users of hardwood lumber. The similarities and differences between hardwood lumber and that of the more commercially important commodity softwood products will be noted.

HARDWOOD LUMBER

In contrast to the more commercially important softwoods, only modest attention has been given to the marketing and supply and demand relationships of hardwood lumber. As a result of this lack of attention, scientific and economic literature describing the markets for hardwood lumber is less available than for the more widely studied softwoods.

One example of this lack of information has been the problem of knowing how much hardwood lumber is actually produced in the United States. The U.S. Department of Commerce's current industrial reports have shown hardwood lumber production ranging from 5.1 to 7.7 billion board feet in the 30 years between 1957 and 1987. However, other analysts of the hardwood lumber market have concluded that since 1970 hardwood lumber usage figures have been higher than reported production figures plus imports (Cardellichio and Binkley, 1984). Luppold and Dempsey (1989) have examined the U.S. Department of Commerce numbers and have developed revised estimates of U.S. hardwood lumber production as shown in Fig. 13-1. This figure shows a low in modern hardwood lumber production of 7.1 billion board feet in 1981 and a high of almost 11 billion board feet in 1987. Hardwood lumber has been estimated to account for about one-third of the value of all domestically produced lumber, both hardwood and softwood (Luppold and Dempsey, 1989).

One reason for the extreme difficulty in even estimating hardwood lumber production is the fragmented nature of the industry. Some mills are part of major forest products

---

[1]This chapter is co-authored by Dr. Robert J. Bush, Assistant Professor of Forest Products Marketing, Virginia Polytechnic Institute and State University, Blacksburg, VA.

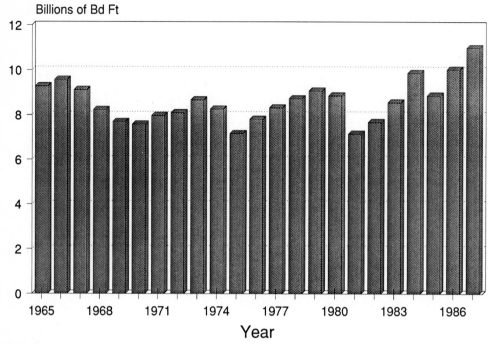

FIGURE 13-1
Estimated hardwood lumber production in the central and eastern United States. (*Luppold and Dempsey, 1989.*)

corporations, but most are independent operations. With the exception of west coast alder mills and a few eastern hardwood mills, most hardwood sawmills are small operations which draw resources from a relatively small area and this makes them very difficult to study. One study did show the median eastern and southeastern hardwood sawmill has only slightly more than 23 employees, produces not quite 4 million board feet per year of lumber, and has a productivity per employee per year of approximately 175 thousand board feet (Bush and Sinclair, 1989). In fact, the three largest hardwood lumber producers in 1987 controlled only 4.1 percent of the hardwood production in the United States and the top ten controlled approximately 7.7 percent (Corlett, 1989; Luppold, 1989).

The typical small size of a hardwood sawmill does not lend itself well to the adoption of new and more efficient lumber production technologies. Advanced technologies such as computer-controlled/assisted processing equipment and more sophisticated types of headrigs have not been found to be associated with higher levels of labor productivity in hardwood sawmills (Bush and Sinclair, 1989). This, of course, does not preclude the existence of other benefits such as enhanced product quality and better grade recovery. In contrast to softwood-producing sawmills, hardwood sawmills exhibit no apparent economies of scale in labor productivity (Bush and Sinclair, 1989).

Location of the Industry

The hardwood lumber industry is concentrated close to the major hardwood forests in the eastern and central United States. Of the total 1988 hardwood production, 97 percent was produced in the eastern United States. For the purposes of commercial trade, this area is often separated into three regions: northern, southern, and Appalachian.

Since 1965, the proportion of total hardwood lumber production accounted for by the northern region has increased and southern production has decreased (Luppold and Dempsey, 1989). Luppold and Dempsey (1989) attribute this charge to higher sawtimber quality in the northern region, access to markets for low-quality lumber products, and the emphasis on and availability of softwoods in the southern region.

A fourth region, western, is also often defined. While this region produces only a small proportion of total hardwood lumber production, the western hardwood lumber industry is important in some geographic markets. In addition, several of the largest U.S. hardwood lumber producers are located within this region.

Hardwood Lumber as a Product

While a conceptually simple product, hardwood lumber is manufactured in a surprising variety of forms. Commercially traded hardwood lumber varies by species, level of process, thickness, and length. Because wood is a natural material, hardwood lumber also

Most hardwood sawmills are still small when compared to softwood mills.

varies in the number and types of defects that are present. To account for this variation and provide a level of product standardization, grading systems have been developed.

In the United States, hardwood lumber is graded into three basic categories: factory lumber, dimension parts, and finished market products. Factory lumber grades differ from dimension and finished market products in that the former are based on the proportion of the lumber that can be cut into smaller usable pieces. Dimension parts and finished market products are graded assuming each piece will be used as is rather than cut into smaller pieces.

The majority of hardwood lumber is sold under the factory grade rules developed by the National Hardwood Lumber Association (NHLA). Standard NHLA lumber grades and the general requirements of each are provided in Table 13-1.

Buyers of hardwood lumber use NHLA grades as a starting point but may have additional requirements such as more stringent minimum dimensions, set width (all one specific width), set length, or limitations on color and defect that are not addressed by the grade rules. In addition, NHLA rules contain modifications to suit the markets for certain species (e.g., red alder).

Hardwood lumber is traditionally sold by the thousand board feet in random-length, random-width loads. Standard lumber thickness varies from $3/8$ to 6 inches and standard lengths range from 4 to 16 feet in 1-foot increments. However, certain lumber grades may place additional limits on the allowable ranges of length and width.

Eastern hardwood lumber is produced from a wide variety of species. Among the most important are red and white oak, ash, yellow poplar, soft and hard maple, gum, cherry, and walnut. Hardwood lumber is sold green (i.e., with a relatively high moisture content) or kiln-dried. Kiln drying adds value to the lumber and may reduce shipping costs (since the

TABLE 13-1
GENERAL REQUIREMENTS FOR NATIONAL HARDWOOD LUMBER ASSOCIATION
FACTORY LUMBER GRADES

Grade	Cuttings	Width	Length
Firsts	$91^{2}/3\%$ clear	6 inches +	8–16 feet (max. 15% 8 and 9 foot)
Seconds	$83^{1}/3\%$ clear	6 inches +	8–16 feet (max. 15% 8 and 9 foot)
Selects	$91^{2}/3\%$ clear	4 inches +	6–16 feet (max. 5% 6 and 7 foot)
No. 1 Common	$66^{2}/3\%$ clear	3 inches +	4–16 feet (max. 5% 4 and 5 foot)
No. 2 Common	50% clear	3 inches +	4–16 feet (max. 10% 4 and 5 foot)
No. 3A Common	$33^{1}/3\%$ sound	3 inches +	4–16 feet (max. 25% 4 and 5 foot)
No. 3B Common	25% sound	3 inches +	4–16 feet (max. 25% 4 and 5 foot)

weight of the lumber is reduced). However, some buyers prefer to purchase green lumber and dry it themselves. Hardwood lumber is also sold rough (i.e, the surfaces of the boards are unfinished) or surfaced. As with drying, surfacing adds value but is not desired by all types of buyers.

Western hardwood lumber is generally limited to red alder, big leaf maple, and black oak. Red alder, a fine-textured wood similar to yellow poplar and soft maple, is used by manufacturers of furniture (including upholstered), cabinets, pallets, and millwork. Unlike eastern hardwood lumber, red alder is typically sold kiln-dried and surfaced. While there is a strong demand for western hardwood lumber exports to Pacific Rim countries, relatively little is shipped to the eastern United States.

Major High-Grade Market Segments for Hardwood Lumber

Furniture   The furniture industry is a major and relatively stable market for hardwood lumber (Fig. 13-2). Furniture (including the dimension it buys) is the second largest industrial market for hardwood lumber. Only the pallet industry uses more. However, the wood household furniture industry is the largest single user of high-grade hardwood lumber and is one of the largest users of hardwood veneer. Usually only No. 1 Common and better grades are used in exposed solid wood parts of furniture and kitchen cabinets. Since these higher grades of lumber command high prices the furniture industry is very important economically to the hardwood lumber industry (Reynolds and Gatchell, 1982).

Finally, the furniture industry is important to the hardwood lumber industry because during the last several decades this has been the most stable market for hardwood lumber (Luppold, 1987). The market for furniture is driven by many factors, with two of the most

FIGURE 13-2
Hardwood lumber markets, 1987. (*Luppold, 1989.*)

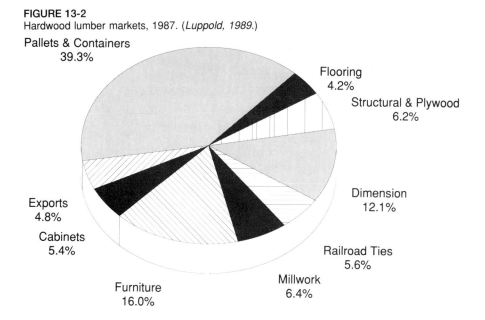

Pallets & Containers
39.3%

Flooring
4.2%

Structural & Plywood
6.2%

Exports
4.8%

Cabinets
5.4%

Dimension
12.1%

Railroad Ties
5.6%

Furniture
16.0%

Millwork
6.4%

important being the need to furnish new households and the replacement of worn-out furniture. In addition, disposable personal income and general unemployment rates influence the amount of new furniture purchased in the United States (U.S. Department of Commerce, 1985).

Furniture Industry Issues That Impact the Hardwood Lumber Market The furniture industry is facing many issues that impact on the hardwood lumber market. First, the loss of market share by domestic furniture manufacturers to foreign furniture manufacturers has been significant. Foreign-made furniture has become increasingly important in U.S. markets (U.S. Department of Commerce, 1985). Second, the substitution of other materials in furniture has resulted in a smaller amount of hardwood lumber being used in the average piece of furniture. Third, the furniture industry is vulnerable to economic cycles (Spelter et al., 1978). Finally, changes in the demographic characteristics of the United States influence furniture consumption.

The loss of domestic market share to foreign furniture producers has had an important impact on U.S. furniture producers. Foreign competition has had an advantage because of the generally lower cost of labor overseas. In addition, the strong dollar in the early 1980s gave foreign producers further cost advantages. Under these favorable conditions many foreign manufacturers were able to upgrade their production facilities, and thus erode one of the traditional strengths of U.S. furniture producers.

Lumber consumption rose in relation to furniture production between 1928 and 1947, but decreased after that. By 1960, the "average piece" of furniture had only 58 percent of the hardwood lumber used in 1947. This decline was largely because of changes in styles and substitution of other materials (Robinson, 1965). The use of particleboard, medium-density fiberboard (MDF), hardboard, and plywood have been important in reducing the amount of hardwood lumber used in the average piece of furniture (Spelter and Phelps, 1984).

Style changes have also influenced the species used in furniture. For example, the use of oak in furniture has increased dramatically (Morrissey, 1985). In addition, the use of softwood lumber in household furniture has increased at the expense of hardwood lumber (Luppold, 1987).

The furniture industry is very susceptible to changes in the economy and to business cycles. Furniture sales are usually tied to changes in housing construction and sales. When housing construction and sales are good, furniture sales are also good. Therefore, changes in the economy, interest rates, and other factors that influence the number of housing starts have an important influence on furniture sales, and thus on hardwood lumber markets.

An additional important factor in furniture consumption is the change in U.S. demographics. The largest purchasers of furniture are in households headed by people 35 to 54 years old. Between now and the end of the century the 35- to 54-year-old age group is expected to grow at a much faster rate than the total U.S. population. This strong growth in the prime furniture-buying group should be beneficial for the furniture market (Epperson and Wacker, 1988).

Although there has been a decrease in the amount of lumber used per unit in the production of furniture, there has also been an increase over the past three decades in the total amount of furniture produced. This has resulted in an upward trend in the consumption of hardwood lumber for furniture production (Cardellichio and Binkley, 1984).

Hardwood Dimension Products   The hardwood dimension[2] market is an important segment of the market for hardwood lumber (Fig. 13-2). The furniture industry is the largest user of hardwood dimension materials, although its share of the dimension parts market has declined because of imported parts and furniture. The second largest use of hardwood dimension products is in cabinets (National Dimension Manufacturers Association, 1985).

One of the problems faced by the U.S. hardwood dimension industry is that it is tied to factors that cause demand to be cyclic. Factors that influence demand include (1) the general economic condition of the United States, (2) the demand for furniture and cabinets, and other items made from wood components, (3) the mixtures of materials used to manufacture furniture and cabinets, (4) the amount of parts manufacturing done by the furniture industry versus the furniture industry's willingness to purchase parts from outside manufacturers, and (5) the amount of competition from foreign manufacturers of furniture and wood parts (National Dimension Manufacturers Association, 1985). These factors combine to cause the number and value of orders placed for hardwood dimension products to vary greatly from year to year.

Changes in the mixtures of materials used to manufacture furniture and cabinets have an important impact on the hardwood dimension products industry. When manufacturers of furniture and cabinets switch to other materials, this decreases the demand for hardwood dimension products.

In recent years, foreign producers of hardwood dimension stock and furniture have gained significant market share in the United States. The total value of hardwood dimension stock imports was estimated at over 23 percent of the value of domestically produced hardwood dimension stock in the mid-1980s (National Dimension Manufacturers Association, 1985). From the late 1970s to the early 1980s, there was a 154 percent increase in imports of wood and upholstered furniture (National Dimension Manufacturers Association, 1985). By 1989, furniture imports exceeded $3 billion (McKee, 1990a).

Foreign competitors are strongest in wood household furniture. They take advantage of lower labor costs and the increased popularity of knock-down/ready-to-assemble (RTA) furniture, and they have improved the quality of their products. In addition, foreign competition has successfully adapted to U.S. style and finish preferences. Foreign competition has been less successful in upholstered household furniture because of its bulky nature and shipping difficulties (National Dimension Manufacturers Association, 1985).

Factors that influence the U.S. hardwood dimension industry's battle with imports include price, quality, marketing and distribution, and level of manufacturing technology. Price has been a major reason foreign competition has done so well in the U.S. market. Advantages of lower labor costs and better exchange rates have allowed many

---

[2]In this context, "dimension" is a term used to mean partially to fully machined parts for furniture, cabinets, and other goods.

foreign manufacturers to sell their products more cheaply than U.S. manufacturers can for goods of similar quality (National Dimension Manufacturers Association, 1985). U.S. manufacturers no longer have advantages in production technologies, equipment, or the quality of their products. Foreign competitors have access to excellent equipment and have upgraded the quality of their products.

U.S. producers still have a limited competitive advantage in furniture and dimension product marketing and distribution. U.S. producers have a competitive advantage because of better control over transportation, better inventories, and their long-term relationships with buyers. However, foreign producers are improving their marketing of knock-down/RTA furniture and are diminishing U.S. manufacturers' shipping advantages.

Millwork   The millwork industry is comprised of firms primarily producing windows, window parts, doors, door parts, and wood moldings. It buys mostly higher grades of lumber and competes with the export and furniture markets for material. The majority of millwork firms are small and many use much more softwood than hardwood lumber.

Hardwood millwork in housing increased from 235 board feet per unit in 1950 to 340 board feet per unit in 1976 (Spelter and Phelps, 1984). By the late 1980s, hardwood millwork demand continued quite strong due, according to some analysts, to the growing do-it-yourself (DIY) and professional remodeling markets (Tomasko, 1988). Luppold (1989) called hardwood millwork the "surprise industry" of the 1980s due to its strong growth, but noted that market data on end use markets for millwork did not exist.

Threats to the use of hardwood lumber in millwork production are the substitution of softwood lumber and the use of other wood-based and nonwood materials. Softwoods have a dominating share of the millwork market (McKeever and Martens, 1983) and metal and plastic parts for windows and doors are commonplace. Newer threats include the molded wood fiber products wrapped by veneer or vinyl and the reconstituted or finger-jointed wood core printed and/or embossed with a wood grain look.

Cabinets   In contrast to many wood-based industries, the kitchen cabinet industry has grown dramatically during the last several Censuses of Manufactures. From 1972 to 1982 the number of manufacturers grew by 65 percent (U.S. Department of Commerce, Bureau of Census, 1985). The 1987 census data show a doubling of hardwood lumber use by the cabinet industry from 1982 to 1987 (Luppold, 1989).

Demand for kitchen cabinets and bathroom vanities is closely linked to housing starts and residential repair and remodeling activity (Ackerman, 1987). Demand for hardwood lumber by the cabinet industry is influenced by the relative level of price competition. Increased price-based competition between manufacturers will encourage the use of lower-cost substitute materials for hardwood lumber.

The use of substitute materials and decreased housing starts can be viewed as threats to the lumber-oriented cabinet industry. Lower housing starts may hurt cabinet consumption; however, the trend toward larger, more upscale homes may result in more cabinet use per housing unit. The industry already consumes large quantities of plywood, particleboard, and medium-density fiberboard as lumber substitutes (McKeever and

Martens, 1983). Ackerman (1987) reported that laminate-faced cabinet panels have displaced solid wood in 15 to 20 percent of the kitchen cabinet production.

Exports   The total amount of hardwood exported from the United States has grown tremendously. From the mid-1970s to the mid-1980s world demand for hardwood logs, lumber, and veneer from the United States almost tripled. This growth has been largely due to a growth in exports to Europe and the Pacific Rim (Araman, 1987). Figure 13-3 shows how the total volume of hardwood lumber imports and exports have changed since 1950.

In 1960, exports accounted for only about 2 percent of the hardwood lumber produced, but by 1980, exports had doubled to 4 percent (Cardellichio and Binkley, 1984). The total value of hardwood lumber exports also grew rapidly. In 1979, hardwood lumber exports represented about 10 percent of the value of total hardwood lumber shipments. By 1986, exported hardwood lumber accounted for about 20 percent of the value of all hardwood lumber shipments. Clearly hardwood exports are increasing, and are becoming a more and more important part of the hardwood lumber market.

Growth of U.S. hardwood exports is likely to be strongly influenced by several factors including (1) the price of U.S. hardwoods relative to that of other suppliers, (2) the amount of high-quality material available for export, (3) changes in the species de-

FIGURE 13-3
Hardwood lumber imports and exports from 1950 to 1988. (*Luppold, 1989; Nolley, 1989.*)

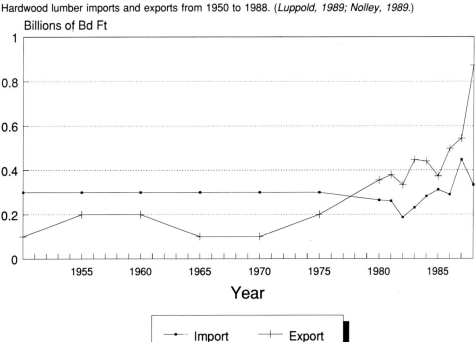

manded, (4) the sales efforts made by U.S. marketers and their competitors, and (5) the demand other countries have for hardwoods (U.S. Department of Commerce, 1985). Worldwide concern over the loss of tropical forests has caused many to seek out temperate hardwoods as substitutes for tropical species, further boosting demand for U.S. hardwoods.

Major Low-Grade Market Segments for Hardwood Lumber

Pallets   The use of hardwood lumber for pallets has risen sharply, largely due to the increase in mechanization of shipping in the United States. Since the mid-1930s the only downturn in the production of pallets in the United States occurred during the economic recessions of 1974–1975 and 1981–1982. The number of pallets used in the United States rose from about 9 million in 1948 to over 250 million in 1980 (Spelter and Phelps, 1984). By 1985, pallet production in the United States had increased to 450 million per year (McCurdy and Ewers, 1987) and by the early 1990s it exceeded 500 million per year. The amount of lumber (softwood and hardwood) used to make these pallets rose from about 220 million board feet in 1948 to approximately 3.5 billion board feet in 1980 and then to 4.4 billion board feet in 1987 (Spelter and Phelps, 1984; McCurdy and Ewers, 1987; Luppold, 1989). The growth in the amount of hardwood lumber consumed in pallets has resulted from the increased use of pallets. This has occurred even though the actual amount of wood used in a typical pallet has declined (Cardellichio and Binkley, 1984; McCurdy and Ewers, 1985).

Although some softwood lumber is used in constructing pallets, the majority of lumber used in making pallets is hardwood. It was estimated in 1982 that 83 percent of the lumber used in pallets was hardwoods. Approximately half of the hardwood lumber used to make pallets was oak. Oak was used more often for the heavier nonexpendable pallets, while other woods were used for lighter expendable pallets (McCurdy and Ewers, 1985).

Pallet use is expected to continue to grow well into the future. Improved materials handling methods will be needed to hold down labor, transportation, and storage costs. The lumber used for pallet production is projected to increase well into the future. By the year 2000, pallet production may consume as much as 9 billion board feet of lumber per year (U.S. Department of Agriculture, Forest Service, 1982).

Substitute products have begun to impact on the demand for hardwood lumber in pallet production (McLintock, 1987). The auto industry began the use of plastic pallets and bins in the mid to late 1980s (Singh, 1986). The U.S. military use of pallets has also been affected by competition from plastics (Anonymous, 1988). New molded flakeboard pallets may pose a threat to solid wood pallets.

Flooring   The use of hardwood lumber for flooring decreased significantly from 1960 to 1980. In 1960, hardwood flooring took about 12.8 percent of the hardwood lumber produced; however, by 1980 this had dropped to 1.2 percent. Interestingly, much of the earlier use of hardwood flooring was in effect mandatory because of mortgage lending policies concerning home construction. As these policies were eased and hardwood flooring was no longer required to qualify the home for a particular mortgage, use of hardwood flooring dropped dramatically and carpeting came into more widespread use.

McKeever and Hatfield (1984) described how drastic the drop in hardwood flooring production has been. From 1950 to 1955, hardwood flooring production climbed from 1 billion board feet to 1.3 billion board feet; however, after 1955 hardwood flooring production declined rapidly. By 1982, production had dropped to below 100 million board feet. Hardwood flooring production rebounded by the late 1980s but volumes remained well below earlier peaks.

The drop in consumption of hardwood lumber for flooring was offset somewhat by the increase in pallet production (Fig. 13-4). The flooring industry apparently suffers little from foreign competition but it continues to be vulnerable to changes in residential construction practices and rates and to substitute products. Lowered demand for

FIGURE 13-4
Indices of pallet and flooring production: 1960 through 1988. (*Luppold, 1988; Nolley, 1989.*)

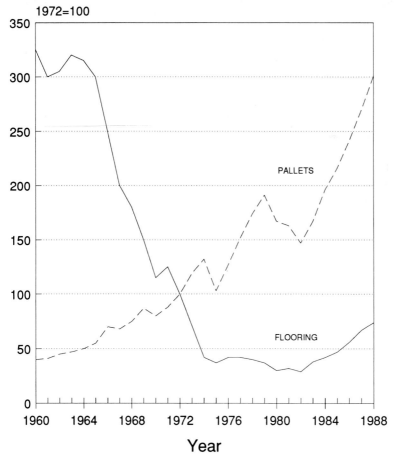

hardwood flooring can result from a decrease in housing starts or because of a reduction in the flooring use per unit. Spelter and Phelps (1984) reported a 90 percent decrease in wood flooring use per single-family home between 1950 and 1976.

Railroad Ties   In the 1960s and 1970s, there was a moderate increase in the use of ties. This was because railroads could not defer track maintenance and started upgrading tracks to improve service and safety. There has also been an increase in the size of railroad ties because a greater portion of ties have gone to main lines, rather than secondary or spur lines. Spur lines use smaller ties because they do not have to bear as heavy a load or bear loads as frequently as main lines. In 1950, there was an average of 38.4 board feet in each crosstie. By 1980, the average tie used by railroads had increased in size to 41.0 board feet (Cardellichio and Binkley, 1984).

In addition, there has been an increase in the proportion of hardwood ties and a decrease in the proportion of softwood ties used. In the 1950s, hardwoods represented only 75 percent of the railroad ties produced. However, by 1980, 90 percent of all crossties produced were made of hardwoods. Price has been noted as the primary reason railroad tie buyers have switched to using more hardwoods; however, hardwoods also offer superior physical and mechanical properties in some instances (Cardellichio and Binkley, 1984).

Some threats to wood ties have emerged from concrete ties. However, nonwood ties have yet to develop into strong substitutes for wood ties.

Marketing of Hardwood Lumber

Channels of Distribution   The hardwood lumber industry, as we have already noted, is a fragmented industry comprised of many small firms. These small firms produce volumes of particular grades and species that are many times in quantities too small for efficient marketing. Most small hardwood lumber mills sell their lumber products through a broker. Many brokers are independent and operate out of an office with little or no storage facilities for lumber. These brokers simply arrange for the sale of the products from smaller sawmills and typically receive a 5 percent commission on the selling price of the lumber as their compensation.

Other intermediaries operate as part of a lumber trading operation at another sawmill. These people purchase lumber from other sawmills and sometimes combine it with their own production to generate sufficient volumes for boxcar- or truckload-size orders.

Brokers may buy and sell from each other to piece together sufficient volumes in a specific grade or species to fill a large order. Smaller independent brokers may, on a regular basis, gather material from very small sawmills and package it together for resale to larger intermediaries which are operating as part of a sawmill's lumber trading service.

Brokers and sawmill trading offices sell to a variety of customers including other channel intermediaries (Fig. 13-5). These additional intermediaries include concentration yards and distribution yards. They differ in that concentration yards purchase lumber from sawmills, brokers, or wholesalers and may grade, sort, dry, or surface the lumber to increase its value. Distribution yards may also process the lumber they purchase but they are oriented more toward the end user. Distribution yards are often located near lumber

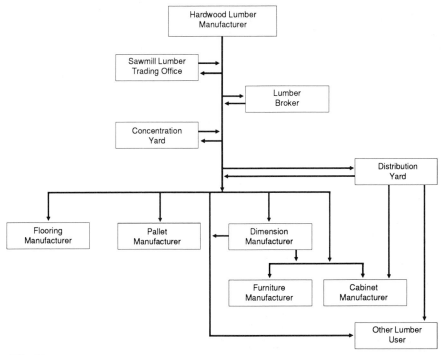

FIGURE 13-5
Typical channels of distribution for eastern hardwood lumber.

users in large cities and provide a variety of lumber species and sizes (one-stop shopping) for small lumber users. Distribution yards typically service small accounts that purchase less than boxcar lots of lumber.

The use of various channel intermediaries tends to vary with company size. Medium-sized companies are often reluctant to use brokers or wholesalers. Channel use also varies with market and business conditions. For example, a company that generally sells its lumber directly to end users may utilize a broker to move its lumber during periods of decreased demand.

At almost any point in the channel of distribution large sales can go directly to large hardwood-lumber-consuming firms. For example, most of the large furniture producers purchase their hardwood lumber directly from an independent broker or the sawmill itself. These large accounts typically buy hardwood lumber rough and green and do their own dry kilning and further processing.

Pricing   Hardwood lumber prices are generally negotiated between buyer and seller from a common starting point for the species, grade, and thickness involved. This starting point is often the average price for the item published in trade publications such as the *Weekly Hardwood Review* and *Hardwood Market Report*. In fact, negotiations often take the form of the published price plus or minus a specified amount to account for local availability, weather, the relationship between buyer and seller, price trends, and other factors.

Lumber prices vary greatly among species. For example, the average price for poplar (No. 1 Common, 4/4, kiln-dried) through the first 10 months of 1989 was $351 per thousand board feet. The same thickness and grade of walnut lumber was $1655 per thousand board feet. Prices and price variability also differ among the various grades of lumber within a species.

Within-species variation of lumber from the various geographic regions (southern, Appalachian, northern) regarding characteristics such as color, growth rate, percentage of heartwood, and the incidence of bacterial infection has resulted in regional preferences. These preferences cause prices for the same grade and species of lumber to differ depending on the region of origin. Consequently, in publications such as *The Weekly Hardwood Review,* separate prices are often reported for lumber from the southern, Appalachian, and northern regions.

Not all species are produced in all regions. However, for species such as red oak that are produced in all the regions, lumber originating in the Appalachian region generally commands the highest price.

While pricing is an important consideration when buying or selling hardwood lumber, some firms are able to consistently obtain a premium over the reported market price by offering superior service and consistent quality. In addition, lumber buyers in several of the major markets for higher grades of lumber have reported that price is not the most important consideration when choosing a lumber supplier. Instead, they indicated that accurate grading (an assurance that they will get what they pay for) is of primary importance (Bush et al., 1991).

Marketing Strategies   Partially as a result of hardwood lumber being bought and sold as a commodity, many firms have adjusted their strategies to the realities of the marketplace. Among the largest hardwood lumber producers, most firms use a low-cost strategy (Bush and Sinclair, 1991). Some firms are attempting to incorporate a differentiation strategy into their low-cost strategy.

It is clear that if everyone gets the same price for a similar basic product then one opportunity to enhance profitability is to reduce costs. Another opportunity is to differentiate the product so that it is no longer considered a basic commodity product. Within the hardwood lumber industry, this differentiation process is done in a number of ways. The most common revolve around the notion of processing hardwood lumber beyond the rough and green state. Such processing includes kiln drying, planing/surfacing, cutting to length, ripping to width, and end trimming, and can even go as far as producing rough dimension parts. A combination of these further processing steps removes the lumber product from the commodity category and can allow for more pricing flexibility.

Northwest Hardwood (a major red alder producer) has carried this process further than most. In its red alder production, virtually all the alder lumber is kiln-dried, surfaced four sides, and precision-end-trimmed. A large number of proprietary guides are used to better match the partially finished lumber product with a required end use. This allows Northwest to charge a higher price on a board foot basis for the lumber product but also allows the customer to achieve a higher yield of finished product per board feet of lumber purchased. In this way, Northwest offers its customers what it believes to be a higher value in its product and hopes to achieve a higher price for it.

Marketing Differences by Industry Segment
Markets differ in the following ways:

Large Companies The very largest hardwood lumber companies are often business units of integrated forest products corporations. They tend to be production cost– and volume-oriented and, in eastern markets, typically sell large amounts of green lumber to high-volume end users such as furniture manufacturers and pallet producers as well as kiln-dried lumber to wholesalers and distribution yards.

Large companies benefit from economies of scale in distribution (freight charges, for example) and offer the customer a competitive price, a wide selection of species, and the ability to ship large volumes on short notice. Product availability is a competitive advantage of companies in this category and, as a result, lumber users may purchase from these firms when their normal suppliers are unable to meet their species or volume needs.

In large companies, the number and turnover of employees may limit the development of long-term relationships with customers. Because of the volumes of lumber the largest companies deal with and the sizes of their sales staffs, they tend to serve many markets for hardwood lumber rather than focusing on a particular market segment. In addition, companies in this category may be more willing to become involved in price-based competition than medium-sized companies.

Typically, western companies in this category sell kiln-dried and planed lumber. While large, they tend to be less volume-oriented than their eastern counterparts and are more likely to work closely with customers. These companies often have the financial strength to develop markets for alternative species or grades and have done this through promotional programs and proprietary grading systems.

Medium-Sized Companies The industry's medium-sized companies generally compete using different methods than those used by the very largest companies. Medium-sized companies, while production cost–conscious, often attempt to differentiate their products through consistent quality, packaging, customer service, and long-term relationships with customers. Medium-sized companies, typically, produce kiln-dried lumber and prefer to sell their own production rather than distribute lumber produced by independent mills. They are often involved in exporting (either through a domestic export company or directly to a foreign agent) and, in domestic markets, may concentrate on serving hardwood lumber distributors.

Because of their lower employee turnover, smaller sales staffs, and the involvement of the companies' owners in selling, medium-sized companies often develop long-term relationships with customers. These relationships help to protect them from competition in their geographic area.

Small Companies The category that includes the smallest companies in the industry is undoubtedly the largest in terms of number of companies and the most diverse. Small companies are often too small to have their own sales force and, consequently, are heavily dependent on lumber brokers and wholesalers to market their products (Luppold, 1987). Dependence on channel intermediaries to perform many of the marketing functions often results in company personnel that are production-oriented and limits relationships with customers.

Alternatively, small companies may develop a close relationship with a local lumber user such as a furniture manufacturer. These users often consume a large portion of the company's production and the arrangement results in the lumber producer becoming a pseudo-captive supplier. There are several advantages to this type of arrangement—sales force requirements are minimal, lumber is moved rapidly and consistently to allow for a constant cash flow, the lumber user may help protect the producer from excessive swings in the lumber market, and the lumber user may help to finance improvements in or additions to the producer's production facilities.

Trends in the Industry

Several trends are evident in the U.S. hardwood lumber industry, including shortened distribution channels, increased specialization of orders, and movement of the inventory carrying function.

A shortening of distribution channels is resulting from a decrease in the use of channel intermediaries and increased direct transactions between lumber producers and lumber users. This trend probably is not a great change for large firms which have their own sales forces. However, for smaller firms that market their products through channel intermediaries, this could represent a major change in the way they do business.

Increased specialization in the products that customers desire is also occurring. Requests for specific lengths or mixes of lengths, specific widths, and specific grade mixes (for example, a certain percentage of each grade in mixed-grade orders) are becoming more common. This trend could continue to the point of certain markets buying dimension parts rather than traditional grade lumber.

Some companies have already positioned themselves to take advantage of this trend by developing flexible sorting and grading systems and by diversifying into dimension operations. For many of the remaining firms, this trend could require investment in new processing equipment and result in increased fixed costs.

The final trend involves the movement of the inventory carrying function back to the producer. Adoption of just-in-time inventory systems and the increased cost of carrying inventory are pushing this trend. Movement of the inventory carrying function back to the producer may increase the competitive advantage of larger hardwood lumber companies that can provide the customer with prompt delivery of a wide variety of species and grades.

Proprietary grading systems are used by only a few producers and such systems are not in widespread use in domestic U.S. markets. However, a few larger companies have adopted complete proprietary systems and other firms use variations of NHLA rules.

Proprietary grading systems can help to move a company's lumber from a commodity with a widely reported market price to a specialty product where price is determined by product value. A company is also afforded a competitive advantage if its proprietary grades fit the needs of a market segment well enough to create barriers to entry into the market and switching costs for the user. For example, companies that wish to enter the market served by producers of proprietary grades face the expense of developing their own grades, changing their production system to produce the grades, and convincing customers of the value of their grades. Proprietary grades can also help a company in developing brand loyalty among its customers. A disadvantage of proprietary grading systems is the additional resources they require to develop, produce, and market. An example of this is the additional staff time required for the customer support that may be required by customers.

WOOD FURNITURE

Historical Overview

American furniture manufacturing began with the earliest colonists, even before 1650, as basically a handicraft.[3] Individual pieces of furniture were handmade from woods available locally by local craftworkers with limited tools and sometimes limited skills. For most of the history of the early colonists and later the first several decades of the early republic, the furniture industry in America suffered from strong competition from European imports. The European producers, because of superior production methods, were very competitive in the American market despite the high ocean transportation costs.

The War of 1812 interrupted trade between America and Europe and also resulted in a 30 percent tariff being imposed on all imported articles. This gave American furniture makers some protection from European imports and allowed them to move from small operations to larger manufacturing plants which could specialize in specific types of furniture. This spurred the development of the strong furniture manufacturing centers of Jamestown, New York, Grand Rapids, Michigan, and High Point, North Carolina.

Jamestown, New York, was probably the first center of high-quality furniture and also became the furniture-making center of the new United States. By the mid-1800s, the eastern forests so vital to the manufacture of furniture at Jamestown had become depleted and the industry there turned increasingly to the manufacture of metal furniture.

The furniture industry then looked west to Grand Rapids, Michigan, where timber was more plentiful at that time. By 1880, Grand Rapids had attained national recognition and its manufacturers were leading in the adoption of technological advances in furniture manufacturing machinery. Grand Rapids also was the early developer of a Furniture Market, which was first held there in 1878. This method of marketing has evolved to modern times and is the predominant method of marketing furniture even today.

Furniture manufacturing did not begin in High Point, North Carolina, until 1888. But the depletion of favorable timber resources in the Grand Rapids area and the abundant raw material resources combined with cheap labor further south resulted in the furniture industry shifting to western North Carolina and southwestern Virginia around the turn of the century. During the early decades at the beginning of this century, these producers concentrated on lower-quality furniture typically sold within the region. However, when cotton prices collapsed in the early 1920s, a recession was triggered in the south which destroyed the local markets for lower-quality North Carolina furniture. In an attempt to

---

[3]Most of this historical overview of wood furniture is taken from Wisdom and Wisdom (1983).

develop alternative markets, these manufacturers exhibited their products in New York. This resulted in considerable criticism of the poor quality of the furniture and prompted North Carolina manufacturers to develop medium-priced reproductions of higher-grade furniture. This was the real beginning of North Carolina's development as a leading furniture-producing state.

Although the industry has remained concentrated in the central Appalachian region, especially for traditional upholstered and case good categories, other hubs of furniture production have also sprung up. California, Texas, and Florida have developed significant furniture-manufacturing industries. And more recently, the geographic focus of the upholstered action furniture industry has been in Mississippi.

Major Market Segments

The furniture industry is comprised of many separate segments. Perhaps the easiest way to look at this industry is through the standard industrial classification codes, which divide it into the following segments: wood household furniture (SIC 2511), upholstered household furniture (SIC 2512), metal household furniture (SIC 2514), wood TV and radio cabinets (SIC 2517), wood household furniture (not elsewhere classified) (SIC 2519), wood office furniture (SIC 2521), and metal office furniture (SIC 2522).

Wood household furniture is far and away the largest segment within the furniture industry, in terms of both numbers of employees and value added by manufacturing. Upholstered household furniture is the second largest segment in terms of numbers of employees; however, metal office furniture has a higher value added by manufacturing. Metal household furniture and wood office furniture are about the same size industries, in terms of both number of employees and value added by manufacturing. The wood TV and radio cabinets segment has been a very small part of the industry and is considerably smaller today than it was back in the 1960s or 1970s.

With the exception of wood TV and radio cabinets and the metal office furniture segments, the furniture industry can be characterized as fragmented, with most of the production being in the hands of a large number of relatively small firms. This is gradually changing; however, at the present time the top firms control only a relatively small segment of the market.

Industry Characteristics and Trends

The household furniture industry has been considered to have many characteristics of a fragmented industry and exhibits some characteristics of the economists' model of pure competition. Some of these characteristics include (U.S. Department of Commerce, International Trade Administration, 1985):

- Many manufacturers, with no firms controlling a large market share.
- Limited recognition of specific furniture manufacturers' brands in the marketplace.
- Products from one manufacturer to another are perceived by consumers as being relatively homogeneous.

- Technology is largely available to all firms such that no one firm can develop a sustainable competitive advantage based on superior technology.
- Within a given quality/price point category, very strong price competition is evident among furniture manufacturers.

Other characteristics of the wood furniture industry in the United States include a pattern of below-average profitability and a low rate of capital investment and productivity growth (U.S. Department of Commerce, International Trade Administration, 1985). Labor productivity fell at an annual rate of 0.2 percent between 1972 and 1981 (U.S. Department of Commerce, International Trade Administration, 1985). By the late 1980s, however, the furniture industry was experiencing slight gains in labor productivity (Herman, 1987).

During the 1980s, the furniture industry saw numerous changes. One of the more remarkable was the pattern of the consolidation of smaller firms into larger, more powerful firms. This resulted as one furniture producer after another acquired smaller furniture producers and also as larger consumer goods concerns such as Interco purchased a variety of furniture manufacturers.

As a result of the increasing size of certain furniture producers, there has been an increase in brand name advertising. For example, Thomasville Furniture Industries embarked on a national television advertising campaign to promote its brand name products.

The furniture industry, especially the wood household furniture industry, has suffered from strong import pressures. The foreign share of the wood household furniture market grew from 6 percent in 1979 to an estimated 13 percent in 1983 to approximately 25 percent by the end of the 1980s (National Dimension Manufacturers Association, 1985; Widman, 1990). Imports of all furniture reached $3.3 billion in 1989, with the leading sources being Taiwan, Italy, Canada, Mexico, and Yugoslavia. Taiwan alone was responsible for about one-third of the total (McKee, 1990a).

Factors Impacting on Furniture Demand

There are numerous factors which can influence household furniture sales. Personal income, especially disposable personal income, has a dramatic impact on furniture sales. Some analysts have reported that the percentage of disposable income going for furniture purchases has been declining over a period of time and the furniture industry banded together in the late 1980s to form the Home Furnishings Council in an effort to alter this trend through market research and a promotion campaign.

Interest rates also have a strong impact on furniture sales. Their impact comes from two directions. The interest rate level determines, to some extent, the level of housing starts. And housing starts obviously influence furniture sales. Further, the majority of furniture is sold on credit. Many states have usury laws limiting the maximum amount of interest which can be charged for consumer loans. For most states, the maximum rate is 18 percent per year. When market interest rates climb above 14 percent, the spread in interest rates between the 14 percent the retail furniture store is charged by the bank and the 18 percent it charges its customers becomes too small for profitable financing. When this happens, retail firms become less inclined to offer attractive financing to consumers and furniture sales suffer.

The movement of the population also has a major impact on furniture sales. A significant proportion of furniture and bedding sales is made each year to persons who have just changed their residence. This new furniture is required to better meet space needs and decorating schemes in the new residence.

The demographics of the U.S. population and the resulting pattern of household formations again influence the sale of household furniture. For a number of years and continuing up to the present time, the household formation rate has exceeded the rate of increase of the total population. This has resulted in an increased demand for housing and furniture above and beyond what would have been expected by the increases in population alone. This has occurred because the average household size has been declining. The Bureau of Census has estimated that the average household declined from 3.14 persons in 1970 to 2.76 persons in 1980 to 2.62 in 1989, and they are predicting a further decline to 2.21 persons in 1995 (McKee, 1990b). A major factor in the declining average size of households has been the dramatic increase in the number of single-person households. As each household unit obtains its own residence, that residence typically is furnished, creating more demand for wood furniture.

As shown in Fig. 13-6, annual household expenditures for furniture vary greatly depending upon the age group of the heads of the household. The 25- to 54-year-old group

FIGURE 13-6
Annual furniture expenditure per household. (*Epperson and Wacker, 1989.*)

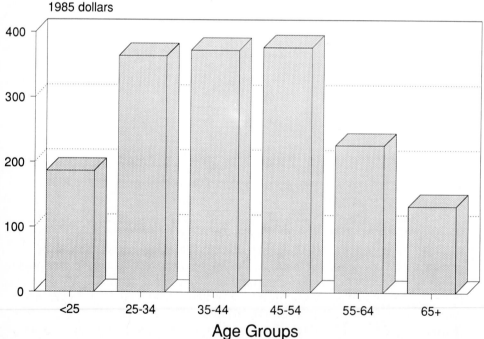

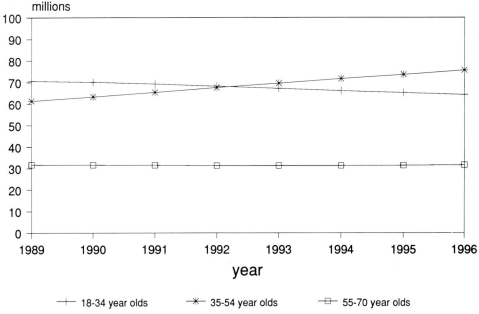

FIGURE 13-7
U.S. population age trends. (*Wall Street Journal, 1989.*)

is shown to have the largest annual expenditures for furniture. If we look at Fig. 13-7, it is easy to see that the 35- to 54-year-old group will be increasing as a proportion of the U.S. population out toward the end of the century. This is obviously the result of the maturing of the baby boomers. This fruitful combination of strong household furniture expenditures in the 35- to 54-year-old group and favorable demographics has resulted in the dramatic increase in projected furniture purchases by the 35- to 54-year-old category as shown in Fig. 13-8.

Clearly, various conflicting trends are evident in the U.S. marketplace for furniture producers. The demographics, in terms of both household formation and age group structure, are favorable for increased furniture sales. However, the continuing importation of foreign-produced furniture, coupled with the smaller percentage of disposable income going toward furniture purchases, is pushing the market for U.S.-produced wood furniture in the opposite direction.

Household Furniture Marketing

Household furniture is mostly sold directly from the manufacturer to the retailer (Fig. 13-9). The Furniture Market or Show serves as the primary vehicle by which manufacturers exhibit their products to the retailers.

Furniture Markets have evolved from the early Market in Grand Rapids, Michigan, to a wide array of Furniture Markets today. These Markets consist of large permanent showrooms where manufacturers display their product lines to retailers and to a much

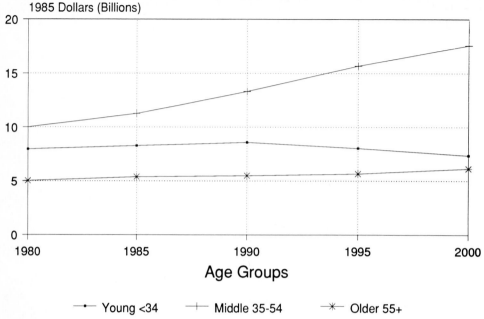

FIGURE 13-8
Furniture purchase trends by age group. (*Epperson and Wacker, 1989.*)

FIGURE 13-9
Channels of distribution for furniture. (*Conners, 1986.*)

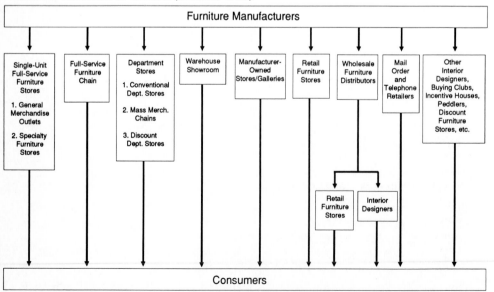

lesser extent wholesalers. Only a small percentage of furniture sales goes through wholesale channels.

Major U.S. Furniture Markets are held in Atlanta, Dallas, Chicago, San Francisco, and High Point. The Southern Furniture Market in High Point, held in April and October each year, is the world's largest. Many other smaller regional markets are also held in such places as Tupelo, Jamestown, Seattle, and Los Angeles. Some of these smaller markets have tended to specialize in various areas. For example, Tupelo has specialized in action furniture, Chicago in high-fashion designer furniture, Atlanta in carpeting, and Dallas in home lighting. Major international Furniture Markets are also held annually in Cologne, Germany, and Milan, Italy.

Although trade shows are widely used to market other products through industrial channels (especially from manufacturer to wholesaler or retailer), the furniture trade show or market has truly developed into the principal marketing tool of many furniture manufacturers. These Furniture Markets can, however, be double-edged swords. They can be advantageous to manufacturers because they allow the manufacturers to display their entire product lines in room-like settings. Such shows also provide an opportunity to meet major customers. Manufacturers can use the reactions at the Furniture Markets to test-market their products and most orders for furniture are placed at the Market or shortly thereafter.

The other side of this sword, though, is that the system of Furniture Markets encourages strong, direct competition, especially on price. It also promotes the pirating of furniture designs from one manufacturer by another because patents on furniture designs are very difficult to obtain or enforce. And last, the wide array of Furniture Markets puts pressure on the manufacturing firms to have something new to generate interest at each Market.

Pricing decisions within the household furniture industry are quite different than those in other segments of the forest products industry which we have previously discussed. Within the household furniture industry, retail furniture dealers/stores typically establish product price lines. Each product line, say for a bedroom suite, has three levels: low, medium, and high. These levels generally correspond to the quality of the furniture and the price. Manufacturers then try to produce goods to sell within these price lines or price point categories.

As shown in Fig. 13-10, a furniture manufacturer will typically select a product line or price point it wishes to target. Furniture is then designed, costs estimated, and samples taken to a Furniture Market. If the response is very favorable, the furniture may be raised in price to the next price point. If the response is poor, the price may be lowered to a lower price point or placed in the factory closeout category, or the product line may simply be canceled.

Generally speaking, prices are set in response to the acceptance of the product. If the price will not cover the costs, a cost reduction effort is typically made. Within a price point category, all products have very similar prices and product differentiation is the most common and effective method of competition between firms.

In an effort to better control their channels of distribution and perhaps avoid some of the cutthroat competition, many larger furniture manufacturers have started a furniture gallery program. There are basically two types of furniture galleries: in-store galleries

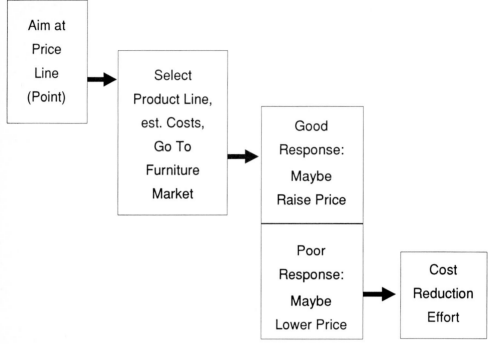

FIGURE 13-10
Pricing household furniture.

and stand-alone gallery stores. For an in-store gallery, a special showroom area within an existing retail furniture store is devoted to the products of a single manufacturer. This store may also sell products from other manufacturers. But within this special showroom area, the retail store sells only the products of a single manufacturer. This manufacturer then provides special decorating, advertising, selling services, and training to the retail store.

Stand-alone gallery stores are retail furniture stores devoted to the product lines of a single manufacturer. These stores can be independently owned or owned by the furniture manufacturing company. Ethan Allen Furniture Galleries and La-Z-Boy Showcase Shoppes are two of the early pioneers in the stand-alone gallery store category.

The gallery concept is growing tremendously within the furniture industry and is another example of the larger firms being able to exercise their power in the marketplace. *Furniture Today* (1989) predicted that the number of galleries will more than double during the early 1990s.

RTA Furniture

Ready-to-assemble furniture is specifically designed and manufactured to be sold in a flat package, which allows consumers to take it home or to the office and assemble it the same day. Shipping furniture unassembled lowers costs by eliminating assembly costs and reducing shipping costs. The savings are then passed on, in part, to the consumer in the form of a lower purchase price. This product has gained rapid acceptance and is now thought to be the fastest-growing segment of the world's furniture market (Pepke, 1988).

In several European countries, RTA represents 20 to 40 percent of furniture sales. In Great Britain, RTA is said to have a 70 to 75 percent market share (Kleeman, 1986). Clearly this large market share in Europe raises the question of whether RTA will become dominant in the United States as well. While this question remains unanswered, U.S. RTA furniture shipments were growing by 15 to 25 percent per year in the mid to late 1980s (Scarangelia, 1987; Sinclair et al., 1990).

RTA Furniture Marketing   Although major RTA producers use furniture markets, the marketing of RTA differs somewhat from that of traditional assembled furniture. This is evident in Fig. 13-11, which shows that for the average RTA producer nearly 62 percent of sales was made through manufacturer's reps to retailers and 23 percent was made through the company's own sales staffs to retailers. Similar to traditional assembled furniture, only a small percentage of sales moves through wholesalers or by mail order directly to customers.

Another difference between RTA furniture and traditional assembled furniture is the types of retail outlets which are used. Table 13-2 shows that only a small percentage of RTA furniture sales move through traditional furniture stores. The dominant retail outlets for RTA furniture are the discount mass merchants, small specialty furniture/lifestyle stores, and home improvement centers.

FIGURE 13-11
Channels of distribution for typical RTA furniture producers by average percentage of sales. (*Sinclair et al., 1990.*)

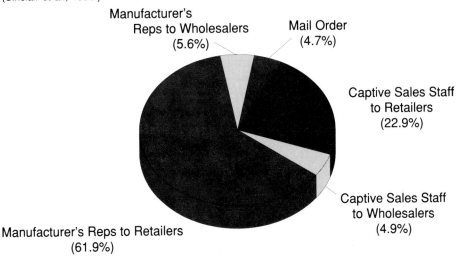

Manufacturer's Reps to Wholesalers (5.6%)

Mail Order (4.7%)

Captive Sales Staff to Retailers (22.9%)

Captive Sales Staff to Wholesalers (4.9%)

Manufacturer's Reps to Retailers (61.9%)

RTA (ready-to-assemble) bedroom furniture.

Positioning for RTA Furniture   A critical aspect of marketing strategy revolves around product positioning. That is, what basis of comparison does a company use to stack its product up against and how does it then promote this comparison in its literature and advertisements? Most RTA producers position their products on value (best quality for a given price), quality, style, or price (Fig. 13-12). Only a very few firms position their products directly against assembled furniture.

Positioning strategies can vary by price point categories for furniture manufacturers. Low-price RTA producers position their products more on price than the other two

TABLE 13-2
AVERAGE PERCENTAGE OF RTA SALES BY RETAIL OUTLET

Retail outlet	% of Sales
Discount mass merchant (such as K mart)	27.9
Small furniture speciality/lifestyle store	26.7
Home improvement center	11.8
Mass merchant/department store	10.4
Traditional furniture store	6.6
Warehouse club	5.2
Catalog showroom	5.1
TV/electronics store	4.4
Mail order/catalog retailers	0.6
Other	1.3
Total	100

Source: Sinclair et al., 1990.

% of respondents positioning on

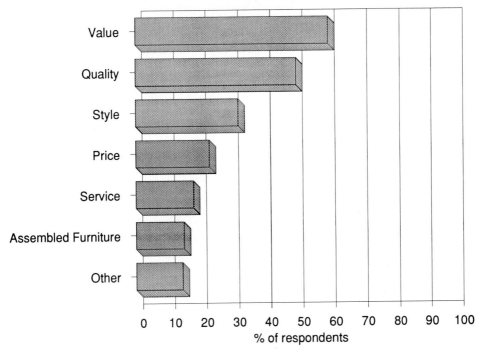

FIGURE 13-12
Percentage of RTA furniture producers positioning their product on various attributes. (*Sinclair et al., 1990.*)

groups. Mid-price-range producers appear to position much more on value than low- or high-point producers. And, not surprisingly, high-end producers appear to position much more on quality than low- or mid-price-point producers. Price point category seems to have little impact on some positioning strategies as very few firms try to compete on the basis of service or directly with assembled furniture. The use of style for positioning also has very little relationship to price point category.

Quality Perceptions and Material Use   Many writers have noted the growing market for high-quality goods and a growing market for high-quality furniture has been predicted for the 1990s (Epperson and Wacker, 1988). A segment of RTA furniture producers is positioning itself for this quality market. However, retailers of household RTA furniture have noted strong price resistance among consumers when RTA prices reach the level of assembled furniture prices.

The increasing growth of RTA furniture in the U.S. marketplace might have an impact on the raw material mix which is produced for the overall furniture industry. Most RTA furniture is produced using a flat panel design and uses substantial amounts of wood composite panels. This use can be clearly seen in Fig. 13-13, which shows the use of industrial particleboard increasing more strongly than any other material in 1989 and 1990. However, it is also interesting that the increasing use of wood veneers and hardwood lumber falls right behind that of industrial particleboard. Only the use of vinyl overlays and OSB shows any declines.

A dichotomy is evident between firms producing high-price-point products versus those producing low-price-point products. High-price-point producers are pursuing a high-quality strategy and believe the use of solid-wood parts and veneers to be important. Low-price-point producers are most often using a low-price strategy and believe the use of solid wood and veneer is not important.

Product Lines   RTA manufacturers produce products for all areas of the home; however, items produced for the living room/den and home office come from a higher percentage of manufacturers. Of specific product items, the most popular ones are entertainment centers, TV stands, bookshelves, coffee/end tables, and modular wall units. In recent years, RTA producers have increasingly targeted the growing office furniture market, both home offices and commercial offices. For some RTA producers, office furniture is the predominant portion of their business.

FIGURE 13-13
Material use predictions for RTA furniture producers for 1989 and 1990. (*Sinclair et al., 1990.*)

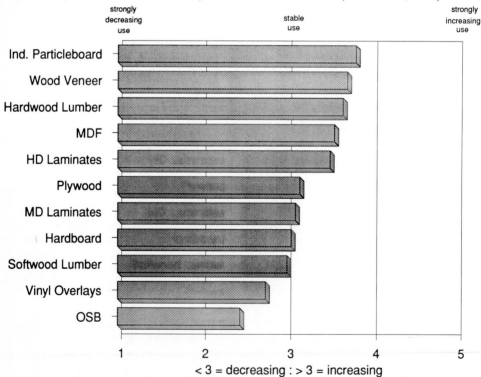

WOOD PALLETS

Among the secondary wood-processing industries, the pallet-manufacturing industry is one of the most recently developed. It began during World War II because of the need to efficiently move large volumes of war materials. Today, the wooden pallet represents one of the most economical systems for unitized load handling.

Even though this industry has been in existence only a short period of time, its growth rate has been phenomenal. As a result of rapidly expanding technology and the need for increased unitized load handling, the pallet industry has grown from producing 57 million pallets in 1959 to over 500 million by the early 1990s.

The pallet industry is the largest consumer of U.S. hardwood lumber. However, the typical pallet-producing firm remains relatively small. McCurdy and Ewers (1987), in their survey of the U.S. pallet industry, noted that the average pallet firm produced 193,000 pallets per year and had an average of 19 employees. The most commonly produced pallet size was a 48 × 40-inch pallet, typically used by the grocery industry. This size accounted for approximately one-third of the pallets produced; however, over 200 different sizes of pallets were reported.

Pallet marketing is predominantly handled by the firms producing the pallets. Only a small portion of pallet production is sold through wholesalers or brokers (McCurdy and Ewers, 1985). Brokers and wholesalers are typically used only when the pallet buyer is too far from the manufacturer to receive adequate service (Brindley, 1984). The industry, for the most part, tends to be located near the industrial and farming centers of the United States and most pallets are sold within 100 miles of the point where they are produced (McCurdy and Ewers, 1987).

There is very strong price competition among pallet producers. Pricing is typically done on a cost-plus basis. Lumber cost has historically been 50 percent or more of the cost of the finished pallet (Sendak, 1973; White and Brindley, 1980). Because of the high lumber cost in pallet production, pallets are often priced according to the number of board feet of lumber in them. Although pallets are sold largely on the basis of price, availability/ service and quality are also mitigating factors.

There are essentially no brand names or even widely accepted standards in the pallet industry. The 48 × 40-in grocery pallet comes closest to a standard. The National Wooden Pallet and Container Association does provide some pallet specifications and also promoted a logo-marked pallet as a way for its member firms to differentiate their pallet products. The logo-marked pallet program was much like the voluntary product grading programs which were successfully established in the softwood lumber and plywood industries. These programs set minimum accepted qualities in the marketplace and provided standards needed for strong market growth.

Perhaps the most exciting development in recent years in the marketing of wooden pallets has been the Pallet Design System (PDS) which was jointly developed by the U.S. Forest Service, the National Wooden Pallet and Container Association, and Virginia Polytechnic Institute and State University. PDS is a microcomputer-based pallet design system which allows a pallet producer to custom-design pallets for their intended uses. This system can result in reduced lumber and fastener costs while at the same time using state-of-the-art engineering. The state-of-the-art engineering is important because it may serve to reduce a pallet manufacturer's potential liability should the pallet fail in service.

The pallet industry is undergoing significant changes. The development of ever more sophisticated automated pallet production equipment has led to larger pallet producers and many producers are integrating lumber and pallet production at a single site. The increasingly expensive equipment for state-of-the-art production, coupled with the potential efficiencies to be gained by vertical integration, are changing the face of this traditionally fragmented industry. There has also been emerging competition for the nailed wooden pallet from plastic pallets, slip sheets, and molded flakeboard pallets.

SUMMARY

The hardwood industries are rapidly becoming more sophisticated. This industry is strongly impacted by international business. Hardwood lumber exports represent as much as 20 percent of the value of hardwood lumber production. On the other hand, imports of hardwood furniture and furniture parts are having a major impact on the industry also. Major lumber-producing firms are moving into the hardwood lumber industry and consolidation is also occurring in the wood furniture industry. Economies of scale and vertical integration within the wooden pallet industry may even serve to reduce the fragmented nature of this industry.

Most products within the hardwood industry are sold with strong price competition being the norm. However, some firms are attempting to strategically differentiate their products from the competition to gain some relief.

BIBLIOGRAPHY

Ackerman, J. C. 1987. "Markets for Solid and Composite Wood Products for Furniture and Cabinets." *Forest Products Journal* 37(10):11–15.

Anonymous. 1988. "Have We Been Had . . . Or Have We Been Had? You Be The Judge." *Pallet Enterprise* 8(1):23–26.

Anonymous. 1987. *Secondary Products: Markets, Competition, and Technological Improvements.* Plenary session presentation at the Forest Products Research Conference. Madison, WI. October 6–8.

Araman, P. A. 1988. "Secondary Products: Markets, Competition, and Technological Improvements." *Proceedings: Forest Products Research Conference 1987: The Role of Utilization Research in Enhancing U.S. Competitiveness in Forest Products.* Madison, WI. October 7. pages 40–49.

Araman, P. A. 1987. "New Patterns of World Trade in Hardwood Timber Products." *Plenary Session Proceedings: Annual Agricultural Outlook Conference, U.S.D.A.* Washington, DC. 11 pages.

Brindley, E. C., Jr. 1984. "Pallet Markets in the United States." *Pallet Enterprise* (September/October):3–5.

Bush, R. J., and S. A. Sinclair. 1991. "A Multivariate Model and Competitive Strategy in the U.S. Hardwood Lumber Industry." *Forest Science* 37(2):481–499.

Bush, R. J., and S. A. Sinclair. 1989. "Labor Productivity in Sawmills of the Eastern and Southeastern United States." *Wood and Fiber Science* 21(2):123–132.

Bush, R. J., S. A. Sinclair, and P. A. Aramon. 1991. "Determinant Product and Supplier Attributes in Domestic Markets for Hardwood Lumber." *Forest Products Journal* 41(2):33–40.

Bush, R. J., S. A. Sinclair, R. M. Shafer, and B. G. Hansen. 1987. "Equipment Needs and Capital Expenditure Budgets for Eastern Sawmills and Pallet Manufacturers." *Forest Products Journal* 37(11/12):55–59.

Cardellichio, P. A., and C. S. Binkley. 1984. "Hardwood Lumber Demand in the United States: 1950 to 1980." *Forest Products Journal* 34(2):15–22.

Conners, S. B. 1986. *A Preliminary Analysis of Furniture Distribution in the U.S.* Presented at the Forest Products Research Society Meeting. Spokane, WA. June.

Corlett, M. L. 1989. *Forest Industries 1988–89 North American Factbook.* Miller Freeman. San Francisco.

Epperson, W. W., and B. E. Wacker. 1988. *1988 Home Furnishings Compendium.* Wheat First Securities. Richmond, VA.

Furniture Today. 1989. "*Furniture Today* Survey: Numbers Will Double by 1993." *Furniture Today* (February 27):1, 24–30.

Greber, B. J., and D. E. White. 1982. "Technical Change and Productivity Growth in the Lumber and Wood Products Industry." *Forest Science* 28(1):135–147.

Hansen, B. G., and P. A. Araman. 1985. "Hardwood Blanks Expand Export Opportunities." *Forest Industries* 112(11):33–35.

Herman, A. S. 1987. "Productivity Gains Continued in Many Industries during 1985." *Monthly Labor Review* 110(4):48–52.

Kleeman, W. B., Jr. 1986. "RTA is Ready to Roar." *Wood and Wood Products* 91(2):48–52.

Losser, S. V. 1986. "Current Trends and Developments in the Wood Component Parts Industry." In *Eastern Hardwoods: The Resource, the Industry, and the Markets.* Forest Products Research Society. Madison, WI. pages 55–61.

Luppold, W. G. 1989. *Shifting Demand for Eastern Hardwood Lumber.* Presented at the Conference on Hardwood Forest Product Opportunities: Creating and Expanding Businesses. Pittsburgh, PA. October 17.

Luppold, W. G. 1988. "Current Trends in the Price and Availability of Pallet Lumber." *Pallet Enterprise* 8(5):30, 32.

Luppold, W. G. 1987. *Material Usage Trends in the Wood Furniture Industry.* Research Paper NE-RP-600. USDA Forest Service, Northeastern Forest Experiment Station. Princeton, WV.

Luppold, W. G. 1984. "An Econometric Study of the U.S. Hardwood Lumber Market." *Forest Science* 30(4):1027–1038.

Luppold, W. G. 1982. *An Econometric Model of the Hardwood Lumber Market.* Research Paper NE-512. USDA Forest Service, Northeastern Forest Experiment Station. Princeton, WV.

Luppold, W. G. 1981. *Demand, Supply and Price of Hardwood Lumber: An Economic Study.* Unpublished Ph.D. dissertation. Virginia Polytechnic Institute and State University, Blacksburg, VA.

Luppold, W. G., and G. P. Dempsey. 1989. "New Estimates of Central and Eastern U.S. Hardwood Lumber Production." *Northern Journal of Applied Forestry* 6(3):120–123.

McCurdy, D. R., and J. T. Ewers. 1987. "Trends in the United States Pallet Industry, 1980 to 1985." *Forest Products Journal* 37(6):46–48.

McCurdy, D. R., and J. T. Ewers. 1985. "The U.S. Pallet Industry." *Pallet Enterprise* 4(6):8–12.

McKee, C. J. L. 1990a. "Furniture Imports Top $3 Billion in 89." *Furniture Today* (April) 14(3):1, 86.

McKee, C. J. L. 1990b. "90's Demographics Will Change Industry Strategies." *Furniture Today* (January) 14(16):1, 17.

McKee, C. J. L. 1989. "U.S. Wood Imports Dip Slightly in 1988." *Furniture Today* (March) 13(28):80.

McKeever, D. B., and C. A. Hatfield. 1984. *Trends in the Production and Consumption of Major Forest Products in the United States.* Resource Bulletin FPL-4. USDA Forest Service, Forest Products Laboratory. Madison, WI. July.

McKeever, D. B., and D. G. Martens. 1983. *Wood Used in U.S. Manufacturing Industries, 1977.* Resource Bulletin FPL-12. USDA Forest Service, Forest Products Laboratory. Madison, WI. December.

McLintock, T. F. 1987. *Research Priorities for Eastern Hardwoods.* Hardwood Research Council. Asheville, NC.

Morrissey, W. C. 1985. *Hardwood Lumber Use in Furniture.* Unpublished manuscript. Raleigh, NC.

National Dimension Manufacturers Association. 1985. *A Comprehensive Diagnostic Analysis of the U.S. Hardwood Dimension Industry.* National Dimension Manufacturers Association. Marietta, GA.

Nolley, J. W. 1989. *Bulletin of Hardwood Market Statistics: Spring 1989.* General Technical Report NE-128. USDA Forest Service, Northeastern Forest Experiment Station, Princeton, WV.

Pepke, E. 1988. *Ready-to-Assemble Furniture.* USDA Forest Service Publication NA-TP-12. St. Paul, MN.

Reynolds, H. W., and C. J. Gatchell. 1982. *New Technology for Low-Grade Hardwood Utilization; System 6.* Research Paper NE-504. USDA Forest Service, Northeastern Forest Experiment Station. Broomall, PA.

Robinson, V. L. 1965. "A Changing Hardwood Market: The Furniture Industry." *Forest Products Journal* 15(7):277–281.

Scarangella, D. 1987. " '86 Was a Good RTA Year; '87 Should Be Even Better." *Furniture Today* 11(16):8–9.

Scarangella, D. 1986. "The RTA Challenge: More Stylish Product, Entry into Furniture Stores are Keys." *Furniture Today* 10(45):22–25.

Sendak, P. E. 1973. "Wood Pallet Manufacturing Costs in Pennsylvania." *Forest Products Journal* 23(9):110–113.

Sinclair, S. A., M. W. Trinka, and W. G. Luppold. 1990. "Ready-to-Assemble Furniture: Marketing and Material Use Trends." *Forest Products Journal* 40(3):35–40.

Singh, S. P. 1986. "Automotive's Swing to Plastic . . . Is It 'Goodbye' to the Wooden Pallet? What Do We Do about It?" *Pallet Enterprise* 6(5):3–5.

Spelter, H., and R. B. Phelps. 1984. "Changes in Postwar U.S. Lumber Consumption Patterns." *Forest Products Journal* 34(2):35–41.

Spelter, H., R. N. Stone, and D. B. McKeever. 1978. *Wood Usage Trends in the Furniture and Fixtures Industry.* Research Note FPL-0239. USDA Forest Service, Forest Products Laboratory. Madison, WI.

Tomasko, B. J. 1988. "Consumers Have Become Largest WMMPA Customer, Tomasko Says." *Southern Lumberman* 249(12):55–58.

Ulrich, A. H. 1987. *U.S. Timber Production, Trade, Consumption, and Price Statistics, 1950–1985.* Miscellaneous Publication No. 1453. USDA Forest Service. Washington, DC.

U.S. Department of Agriculture, Forest Service. 1982. *An Analysis of the Timber Situation in the United States, 1952–2030.* Forest Resource No. 23. U.S. Department of Agriculture. Washington, DC.

U.S. Department of Commerce. 1985. *Lumber Production and Mill Stocks.* Curr. Ind. Reps. Ser. MA-24T. U.S. Department of Commerce. Washington, DC.

U.S. Department of Commerce, Bureau of the Census. 1985. *Millwork, Plywood, and Structural Wood Members, N.E.C. 1982 Census of Manufacturers.* Industry Series No. MC82-I-24B. U.S. Department of Commerce. Washington, DC.

U.S. Department of Commerce, International Trade Administration. 1985. *A Competitive Assessment of the U.S. Wood and Upholstered Furniture Industry.* U.S. Department of Commerce. Washington, DC.

Verity, C. W., C. J. Brown, B. Smart, and C. E. Cobb, Jr. 1988. *1988 U.S. Industrial Outlook.* U.S. Department of Commerce, International Trade Administration. Washington, DC.

Wall Street Journal. 1989. *Wall Street Journal* (March 23):B1.

White, M. S., and E. C. Brindley, Jr. 1980. *Economics of a Small-Scale Pallet Manufacturing Plant Using Low-Grade Southern Hardwood. Utilization of Low-Grade Southern Hardwoods.* Forest Products Research Society. Madison, WI.

Widman, 1990. *Markets 90–94. The Outlook for North American Forest Products.* Widman Management, Ltd./Miller Freeman Publications, Inc. San Francisco, CA.

Wisdom, H. W., and C. D. C. Wisdom. 1983. "Wood Use in the American Furniture Industry." *Journal of Forest History* 3(27):122–125.

PART FIVE

INTERNATIONAL MARKETING AND MARKETING ORGANIZATION

An increasing volume of international trade is in containerized shipments as seen here.

INTERNATIONAL MARKETING

Being global is about people, attitudes, and responses to global trends. Ultimately, we must develop a way of thinking which embraces strategic marketing and customer service as compelling issues.

Arkadi Bykhovsky

Certainly since the recession of the early 1980s few things have captured the attention of U.S. forest products firms like export markets. These markets have been viewed as an opportunity for a firm to diversify its markets and insulate itself from recessions in the domestic marketplace. Using international markets can be both rewarding and frustrating for forest products firms. A variety of options exists for marketing in overseas markets. Exporting can be done through a firm's own sales force selling directly to overseas buyers, through a foreign agent or trading company, through the local buying office of a foreign company, or through an international trading company. These are only a few of the many options available for exporting; many more exist. In addition, joint ventures are becoming commonly used to build manufacturing facilities in the United States, especially for producing wood or paper products for export to Japan. In this case, one partner in the joint venture provides the marketing skills and the other the manufacturing capabilities. For firms with a higher aptitude for risk taking, direct ownership can be another option for participating in international markets. In this case, a firm would own, outright, a subsidiary in a foreign country and would have total responsibility for manufacturing and marketing of the product in the foreign country.[1]

Success in overseas markets requires market segmentation. Language, customs, and income patterns differ from country to country. Further, market and marketing data in most foreign markets are not as plentiful nor as accurate as similar data available in the United States. A major road block to successful international marketing is an unwillingness to learn about and adjust to different cultures and their needs. Regulations also vary widely between countries as do the degrees to which they are enforced. This is highlighted by the following:

In France everything is permitted except that which is prohibited: In Germany everything is prohibited except that which is permitted: In Russia everything is prohibited including that which is permitted; but, in Italy everything is permitted, especially that which is prohibited. (Buckley, 1989)

---

[1]Bilek and Ellefson (1990) provide an excellent overview of the organizational arrangements of the foreign operations of U.S. wood products companies.

While exports from domestic production facilities have clearly gained the most attention, many North American firms have made direct overseas investments in manufacturing facilities. The primary reason overseas investments are made is to better service overseas customers. Most major forest products firms view their ability to compete in world markets as important but not critical to their future (Bilek and Ellefson, 1990). Even though not viewed as critical, most major North American forest products firms have extensive overseas operations. In one study of twelve major U.S. forest products firms, 200 foreign subsidiaries were identified (Bilek and Ellefson, 1990).

DECIDING TO ENTER THE EXPORT MARKET

Participating in international markets is clearly more complicated and risky than participating in domestic ones, but potentially rewarding as well. The decision to participate in export markets requires careful consideration and study. Figure 14-1 presents a seven-part approach as an aid to making this decision and understanding its ramifications (Westman, 1986).

The first step in this process is understanding management's commitment to export marketing. Many U.S. firms have been accused of not being in export markets for the long term. While firms should maintain some flexibility to take advantage of export market dynamics, in general only those firms which are prepared to be in the market on a long-term basis should consider the necessary investments in exporting.

To enter into an export market a firm must analyze its own objectives, strengths, and weaknesses. What are the firm's short- and long-term goals in regard to exporting? Does it have the skilled personnel to effectively compete in export markets? Does it have the financial and timber resources required along with the necessary production facilities? Is financing available and does the firm have sufficient knowledge of export marketing? These and many other factors must be considered in this analysis and the firm must be prepared to bolster itself in its areas of weakness.

At this point in the process, if the firm has decided to enter the export arena, market contacts must be developed and current market information on export markets should be collected. There are a number of ways to develop much of this information. Trade associations, the U.S. government, and various state agencies have export market development programs from which individual firms can secure information. The U.S. Department of Commerce has specialists for forest products which are typically located in the International Trade Administration. Further, the Department of Agriculture also has, within its Foreign Agriculture Service, a number of wood products export specialists and for a nominal price offers a variety of publications which provide up-to-date export market information (Foreign Agricultural Service, 1990). Virtually every state in the union, through its Department of Commerce, Department of Economic Development, and/or State Forestry Organization, has someone working on exports and in many cases forest products exports. Additional sources of information include commercial banks with international departments, local universities, and international development centers. Also, U.S. port authorities, marine insurance agents, freight forwarders, and major export companies can provide a wealth of information concerning export markets.

Once contacts and information are collected, a market analysis is required. In this step

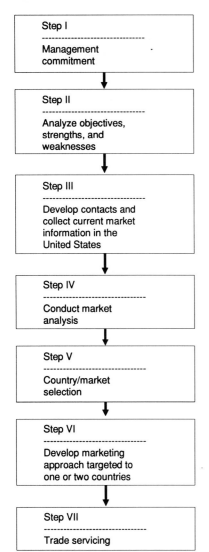

FIGURE 14-1
A seven-step approach to export marketing. (*Adapted from Westman, 1986.*)

U.S. export statistics are studied and compared along with foreign import statistics. Current market developments and trends are identified. And import barriers are identified and understood. Additionally, other factors, including political, economic, geographic, and cultural factors, should be considered at this time.

As noted earlier, export markets are clearly more complicated than domestic markets. Each foreign market should be treated as a separate market. Lumping foreign markets together will certainly increase the probability of failure. Therefore, it is important at this point in the process to select the several countries which seem most promising. In order to do this, the demand potential and trends must be identified. The products required by the countries must be identified along with their specific standards and product specifications. Any special language requirements must be identified as well. The local business practices and most appropriate distribution channels must be understood, along with any special licensing or phytosanitary requirements. Last, shipping costs and other legal considerations should be considered.

Once all this information is developed, the analyst can then choose one or two countries which seem the most promising and then develop a targeted marketing approach aimed at them. At this point it is time to consider the organization of the exporting firm and its relationship to the targeted market. A commitment needs to be considered concerning what percentage of production will be exported. Foreign importers for the specific countries should be contacted and marketing or sales trips scheduled to the country or market. Careful attention must be paid to the grade specifications and measurements required by the foreign market. They will likely be different from domestic needs.

As in domestic markets, it is not enough to simply sell a product. Products must be tailored and modified specifically for the markets in question. Therefore, for long-term success in export markets products must be modified in response to customer changes and demands. Firms must pay close attention to the needs of importers and periodic visits to the market are necessary to maintain good customer relations and also to develop new contacts and enlarge the market. In many foreign markets (especially Japan), follow-up contacts and service after the sale are critical to repeat business. Maintaining a good reputation is important because a poor reputation is even more difficult to overcome in foreign markets than domestically. Periodic reassessments of the marketing approach used are also important.

CHANNELS OF DISTRIBUTION

North American wood products are moved to overseas markets in a variety of different ways (Fig. 14-2). The channels of distribution can be roughly divided into direct and indirect channels. Direct channels mean those instances where the manufacturer deals directly with a foreign buyer and indirect where the domestic manufacturer deals with a series of channel intermediaries prior to the product reaching the overseas market.

Several studies have been done exploring the channels of distribution used by wood products firms. In one study, 208 small wood products firms (less than 250 employees) in the deep south were surveyed. Sixty-one percent of the responding firms said they were active exporters, while 39 percent indicated they were not exporting. Of those that were exporting, 55 percent used direct channels and 45 percent used indirect channels (Brady, 1978). Export channels of Oregon lumber producers have been studied and the results, shown in Table 14-1, clearly indicate the declining importance of indirect channels of distribution and the increasing number of Oregon producers who are using direct channels (i.e, their own sales staffs).

Figure 14-2 provides a simplified look at export channels of distribution for North American wood products. When using indirect channels of distribution, domestic producers typically sell their products to exporters. The exporters perform the function of arranging buyers and sellers to fill orders in the marketplace. For hardwood products, especially, exporters may also prepare the products for further processing with such activities as kiln drying, planing, trimming, regrading, and packaging. The term "exporter" is used in Fig. 14-2 generically and such players in the channels of distribution may also have other names like "export management companies," "export trading companies," or "export merchants."

Instead of having in-house export shipping departments, some exporters and many domestic producers avail themselves of the services of freight forwarders. Table 14-2 provides a list of the common services provided by a freight forwarder. Shipping products overseas requires a level of expertise, not always available to the average North American forest products firm. The freight forwarder stands ready to provide services for manufacturers and exporters. The freight forwarder can book passage on the appropriate steamship lines, can complete the bill of lading, can provide a wide variety of information necessary for completing U.S. customs documents and documents of the destination country, can review letters of credit, and can calculate shipping costs. Further, the freight

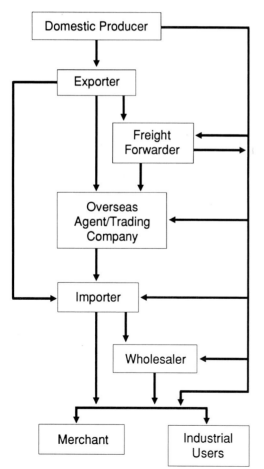

FIGURE 14-2
Simplified export channels of distribution.
(*Adapted from Rich, 1981.*)

TABLE 14-1
EXPORT MARKETING CHANNELS OF OREGON LUMBER PRODUCERS

Marketing channels used	Mid-1960s survey, % of firms using them	Mid-1980s survey, % of firms using them
Export agent/foreign freight forwarder	33	24
Other distributors	38	21
Licensing/franchising	0	2
Other U.S. firms (e.g., joint ventures)	5	9
Own sales staff	14	33
Parent company (if respondent firm is a subsidiary)	10	11
Total	100	100

Source: McMahon and Gottko, 1988.

forwarder can monitor and supervise port deliveries and vessel loadings, making sure that a given load or container is appropriately loaded or delivered.

Several types of ocean shipping vessels are available for exporting wood products. For the exportation of wood chips there are specialized chip-carrying vessels available; however, most operate under long-term contracts. Chips can also be shipped in vessels designed to carry bulk commodities such as grain.

Wood products were traditionally shipped in large conventional break bulk carriers where packages of wood products were simply loaded into the hull of the vessel. A recent development has been the advent of the container ship, which clearly dominates trade

TABLE 14-2
SERVICES OF A FREIGHT FORWARDER

Serves as a booking agent

Completes bill of lading

Provides information on or can handle the following:
 Documents for U.S. export
 Documents for import into destination country
 Available steamship lines
 Types of available vessels

Reviews letters of credit

Calculates shipping or other costs as follows:
 Inland freight charges
 Port charges
 Insurance charges
 Ocean freight charges

Monitors and supervises port deliveries and vessel loadings

Source: Henley, 1981.

between the east coast and northern European ports and is gaining in popularity in Pacific trade also. The containers come in a variety of sizes; however, the sizes closely approximate the large trailer pulled by a tractor-trailer truck. No doubt, you have seen such containers being pulled down your local highways and may not have realized it. These containers come in a variety of lengths including 20, 40, 45, 48, and 53 feet. Most are 8 feet wide × 8 feet 6 inches tall, although some are 9 feet 6 inches tall. Container shipment of lumber (especially kiln-dried lumber) is gaining popularity because it provides increased protection for the product and also helps prevent theft and vandalism.

Once the material is loaded on board a ship and is moving overseas, the next level in the channels of distribution comes into play: the agent. As in domestic markets, the agent does not take title to the material, but rather represents the seller. The agent works for a commission on the sale and is paid only for a successful sale. Agents are a strong force in export distribution channels and for many countries a majority of their wood products imports go through agents. For example, in Europe estimates are that 90 percent of the imported U.S. hardwood lumber goes through agents (Buckley, 1989). For large customers, some exporters will skip the agent. In Japan and some other countries, the role of the agent is taken over by the trading company. Many trading companies in Japan even have their own U.S. buying offices and eliminate the role of the exporter in the channels of distribution. Some North American firms such as MacMillan Bloedel also have their own trading companies in Japan.

The agent or trading company then provides the product to the importer, which may also act as a wholesaler selling directly to merchant retailers and other industrial users in the foreign country. The importer may also sell to wholesalers, who then in turn sell the product to merchant retailers and industrial users. Of interest for wood building products exporters is a growing trend in developed countries toward home-center-type retailing of do-it-yourself products. In some countries in Europe and, to a certain extent, in Japan home center retailing is catching on and some firms may choose to export directly to these new retailers.

Export Documents

Exporting can be a complicated activity which requires a high level of expertise. One area which can cause problems is the wide variety of export documents required (Table 14-3).

One of the more common documents is the ocean bill of lading. This is in essence a receipt for the materials being shipped and also serves as a contract for transportation between the shipper (exporter) and the ocean carrier. In other circumstances, it can also be used as a certificate of ownership. Obviously it is important to have this document correct.

A dock receipt is typically used to transfer accountability for materials shipped between the inland and ocean carriers at the port. Delivery instructions are provided to the ocean shipper and/or inland carrier. These instructions confirm the arrangements made to deliver the cargo to a particular steamship line or even a particular pier at the port. The shipper's declaration is required by the U.S. Department of Commerce and is a key source of the information Commerce uses to compile export statistics. Many times this is prepared by the freight forwarder.

A consular invoice may be required by some countries to control and identify goods

TABLE 14-3
TYPICAL SHIPPING DOCUMENTS REQUIRED IN EXPORTING*

Documentation	Prepared by
Bill of lading—a detailed description of the cargo including destinations; two types are necessary: an inland and an ocean bill of lading	Freight forwarder
Shipper's declaration—for compiling U.S. statistics/enforcing U.S. export controls	Freight forwarder
Dock receipt	Freight forwarder
Banking papers	Banks
Delivery instructions	Freight forwarder
Letters of transmittal	Freight forwarder
Notice of exportation	Freight forwarder
Certificate of origin	Freight forwarder
Consular invoice	Freight forwarder
Export packing list—itemizes product shipped	Freight forwarder
Domestic packing list—itemizes product shipped	Exporter
Insurance certificate	Freight forwarder
Phytosanitary certificate	Manufacturer/exporter
Letter of credit—a promise to pay a specified amount of money upon receipt by the bank at the buyer's request in favor of the seller	Importer

*More detailed information about exporting documentation may be obtained from the international departments of major banks, freight forwarders, or the United States Council of the International Chamber of Commerce, 1212 Avenue of the Americas, New York, NY 10036, or the International Trade Institute, 5055 North Main Street, Dayton, OH 45415.
Source: Westman, 1986.

being shipped to them. It is typically prepared on special forms and can be prepared by the freight forwarder. The commercial invoice is simply a bill for the goods. It is often used to assess the true value of the exported goods to determine customs duties and may also be used in the preparation of consular invoices. A certificate of origin is used to assure the buying country which country produced the goods. Additionally, such items as insurance certificates, phytosanitary certificates, letters of transmittal, and notices of exportation may be required. In recent years, the phytosanitary certificates have assumed greater importance as more and more countries have developed increasingly stringent sanitary requirements for exported and imported goods.

PRICING

The pricing of export material is a more complicated process than the pricing of domestic items. For domestic commodity items, a producer or buyer can turn to one of the many price newsletters, such as those published by Random Lengths or Crow's, and look up the current market price for the item in question. However, it is not nearly that simple in export markets. Random Lengths and the *Weekly Hardwood Review* publish export price guides but they seem to be used more as indications of typical activities in the market rather than as a strict guide to pricing. Also, these price guides usually lag the market by as much as 4 to 6 weeks. Many exporters have noted that the market volume is thin enough in certain grades to enable prices to move very rapidly, either up or down. As a result, export pricing requires careful negotiations between the buyer and the seller. Table 14-4 provides the elements of pricing goods for export using a cost-based approach. The many aspects which add cost to the product clearly call for the careful consideration of all these factors prior to setting or negotiating a final price.

Getting Paid

Even in domestic markets, getting paid for the goods shipped can sometimes be a hassle. This problem can be greatly magnified in overseas dealings. There are a number of ways to help ensure payment and the most appropriate way will vary by country and the creditworthiness of the buyer. When dealing in export markets, the manufacturer/exporter is well advised to seek the advice of a source knowledgeable in this area, such as a commercial banker or other financial institution.

Cash in advance is clearly the most desirable and safest way to receive payment; however, this method is unlikely to be employed with any frequency. At the other end of the risk spectrum is the open account. In this method, the seller simply allows the buyer to have an open account and ships the buyer the product as needed, billing the buyer at the

TABLE 14-4
ELEMENTS OF PRICING GOODS FOR EXPORT

Element	Price basis*					
	F.o.b. (Factory)	F.o.b. (Point of export)	F.o.b. (Vessel)	C. & F. (Named port)	C. & I. (Named port)	C.i.f. (Named port)
Cost of producing wood products and profit	$	$	$	$	$	$
Packaging	$	$	$	$	$	$
Processing/handling	$	$	$	$	$	$
Inland transportation		$	$	$	$	$
Unloading fees at port		$	$	$	$	$
Forwarders' charges			$	$	$	$
Consular fees			$	$	$	$
Ocean freight				$		$
Marine insurance	—	—	—	—	$	$
Total	$	$	$	$	$	$

*F.o.b. = free on board; C. & F. = cost and freight; C. & I. = cost and insurance; C.i.f. = cost, insurance, and freight.
Source: Westman, 1986.

appropriate time. Open accounts are typically used by exporters only for long-term customers of established creditworthiness.

For most shipments of wood products to the United Kingdom and the continent of Europe, payment is made through a sight draft (Rich, 1981). When using this method, the exporter or its bank sends the appropriate negotiable documents, including invoices, ocean bills of lading, etc., to the buyer's bank overseas. The buyer's bank in turn then releases the required documents to the buyer upon payment for the goods. The buyer's bank then typically releases the funds to the appropriate U.S. bank (Rich, 1981). There are other variations on the draft which allow the buyer to delay payment for a specific period of time or require payment by a specified time. A sight draft is also known as "cash for documents."

A safer way to handle payments in exporting is through an irrevocable letter of credit. When using a letter of credit the buyer typically places the required funds in escrow prior to the shipment of goods. The bank holding the escrowed funds then issues a letter of credit. This letter of credit then obligates the bank to make payment when the bank receives the required documents (Fig. 14-3). The letter of credit is the predominant way of securing payment in forest products trade to much of the Pacific Rim and in some European countries as well (Rich, 1981; Hansen, 1989). Revocable letters of credit may also be issued but they can be amended or canceled by the issuing bank and thus offer little protection to the seller (Anonymous, 1990a).

PACKAGING AND PRESENTATION

In export markets, packaging and presentation are much more important than in domestic U.S. markets, especially for wood products. Lumber, plywood, and other products must be consolidated into consistently sized units and must be securely packaged. Carefully prepared units of the same length of lumber are the norm. This is especially important when loading containers. Even-length lumber allows more efficient use of the container's volume. Packaging usually involves strapping the unit and frequently includes a moisture-resistant wrapping. The wrapping serves not only to protect the unit but also to display the exporting company's logo, and may include information about the unit. Brand names and color coding (especially colored ends of boards or plywood edges) have become important in recent years as some exporters have developed strong brand recognition in specific overseas markets.

GRADES AND PRODUCT STANDARDS

Product specifications in the export marketplace can, at times, be confusing. There are standard sets of rules available. For example, an early set of standard rules for grading export southern pine lumber was the 1923 *Gulf Coast Classification Rules*. These were revised in 1982 by the Southern Pine Inspection Bureau. However, many European customers for southern pine still request material graded according to the 1923 rules.

For Douglas fir, hemlock, sitka spruce, and western red cedar, the Pacific Lumber Inspection Bureau has a set of grading rules entitled the *Export R List*. These rules were originally copyrighted in 1919 and revised in 1971. For both the western species and

Irrevocable Documentary Letter of Credit

Issuing Bank:

The Bank of The Republic of China, Ltd.
Kaohsiung, Taiwan ROC

L/C Number:

91BOR12345-6789

Date of Issue:

May 28, 1991

Date and place of expiry:
August 31, 1991
at counters of negotiating bank

Applicant:

Taiwan Forest Products Importing Co. Ltd.
89 Chung Shan W. Road
Kaohsiung, Taiwan ROC

Beneficiary:

U.S. Forest Exporting Co, Inc.
1234 Main Street
Columbus, Ohio 43216 USA

Advising Bank:

The Huntington National Bank
Columbus, Ohio 43287

Amount:

USDLRS 100,000.00
(One Hundred Thousand U.S. Dollars)

Credit Available with: __Any Bank__   **By: Negotiation** X   **Payment** __   **Acceptance** __   **Deferred Payment** __

We hereby issue our irrevocable documentary letter of credit in your favor payable by presentation of your drafts drawn ___at sight___ on ___us___ when accompanied by the following documents:

- Signed Commercial Invoice in 3 copies.
- Full Set of Clean On Board Ocean Bills of Lading made out to the order of The Bank of The Republic of China, Ltd., dated not later than August 15, 1991 and marked freight collect and notify applicant.
- Packing List in 3 copies.
- Certificate of U.S.A. origin.

Covering: White Pine Lumber, FOB USA port
Partial shipments: permitted   **Transhipment:** prohibited

Documents to be presented within ___15___ days after the date of the transport document but within the validity of this credit.

Instructions to Advising Bank: Please advise beneficiary without adding your confirmation

Instructions to Negotiating Bank: In reimbursement of drawings, please claim on our account with The Huntington National Bank, Columbus, Ohio

We hereby engage with the drawers, endorsers and bona fide holders of drafts that documents presented under and in compliance with the terms of this credit will be duly honored if presented as stated above on or before the expiration date. This credit is subject to the Uniform Customs and Practice for Documentary Credits, International Chamber of Commerce Publication No. 400 (1983 Revision).

FIGURE 14-3
Example of an irrevocable letter of credit.

southern pine, the rules are recognized widely in the market for the upper grades such as clears in Douglas fir and saps in the southern pine. These grades are largely clear and there is little room for interpretation of the rules. However, when getting into the prime and merchantable grades for southern pine and similar categories for the western species, the marketplace makes a dramatic shift toward considerably more use of proprietary sets of rules. Many times in these rules the specifications for a given grade are largely determined through negotiations between the supplier and the customer.

Many of the specific export categories of North American lumber are sold in larger dimensions than is typical for domestic markets and the material is resawn into specialty products such as molding and millwork items in the destination country. Many North American lumber trade associations have been working hard to increase the overseas market acceptance of American Lumber Standards Committee and Canadian Lumber Standards Committee grading and dimensions. Some success in this area has been achieved in Japan and Great Britain. Additionally, a considerable amount of this material is exported to the Caribbean basin. However, the Caribbean customers do prefer an upgraded product from the standard North American grades in that they prefer much less wane and a fuller cut, and strongly prefer longer lengths. Most overseas customers for North American grades of lumber dislike wane.

In general, overseas buyers of North American softwood lumber products prefer their material to be graded at a higher standard than is typical in the domestic markets. They prefer a fuller cut since the product is many times used for resawing by overseas manufacturers. They also prefer a clearer piece of lumber with fewer and smaller knots and less stain. This has caused some problems for new U.S. and Canadian high-tech sawmills designed with the maximum lumber yield from the log in mind. Such an orientation ensures that the lumber produced will have significant amounts of wane and other characteristics which are less acceptable to export customers. As a result, several large lumber producers have kept old large log sawmills in operation and have converted these mills to sawing for the export marketplace. The differences in the quality of material demanded and the sizes make it difficult in a modern high-speed production mill to produce the upper export grades while at the same time efficiently producing North American grades for domestic markets.

In hardwoods, the basis for export-grade material remains the National Hardwood Lumber Association rules. However, as in softwood lumber, these rules typically serve only as a starting point and most hardwood lumber producers in the export marketplace produce a so-called special or proprietary export grade. This grade is typically an upgraded FAS (Foreign Agricultural Service) grade with less wane and better trimming.

Within North America most softwood and hardwood lumber is sold based on the board foot or in panels sold on the basis of a square foot, while in most international markets the units are different. Even the U.S. Department of Commerce is now reporting export volumes based in cubic meters. Conversion between typical volumes of North American production, such as thousand board feet, to a more standard international measure, such as cubic meters, can be tricky and more than one conversion factor may be available. The buyer and seller in international markets are cautioned to be sure they understand the units being used. Tables are available which give the conversions of various basic forest products such as logs, lumber, and panels from standard board foot and square foot

TABLE 14-5
COMPARISON OF VARIOUS LOG RULES FOR 16-FOOT LOGS

Diameter (inches)	Board feet			Cubic volume	
	Doyle	Scribner	International 1/4 inch	Cubic feet	Cubic meters
8	16	32	40	5.59	0.158
9	25	42	50	7.07	0.200
10	36	54	65	8.73	0.247
11	49	64	80	10.56	0.299
12	64	79	95	12.57	0.356
13	81	97	115	14.75	0.418
14	100	114	135	17.10	0.484
15	121	142	160	19.63	0.556
16	144	159	180	22.34	0.633
17	169	185	205	25.22	0.714
18	196	213	230	28.27	0.800
19	225	240	260	31.50	0.892
20	256	280	290	34.91	0.989
21	289	304	320	38.48	1.090
22	324	334	355	42.24	1.196
23	361	377	390	46.16	1.307
24	400	404	425	50.26	1.423
25	441	459	460	54.54	1.544
26	484	500	500	58.99	1.671
27	529	548	540	63.62	1.802
28	576	582	585	68.42	1.938
29	625	609	630	73.39	2.078
30	676	657	675	78.54	2.224

Source: Binek, 1973.

measures to cubic meters. Examples of these tables are provided in Tables 14-5 through 14-7.

For many wood products, overseas customers desire a different grade of product which is measured in a different way than in domestic markets. To be successful longer term in export markets, the supplier must learn and understand at a minimum the grades and units of measurements its customers will want.

MAJOR INTERNATIONAL MARKETS

Historically, Canada has exported a much higher percentage of its wood products production than has the United States. Exports remain a mainstay of the Canadian industry. On the other hand, while exports have received a great deal of press in the United States, exports from the United States comprise a much smaller proportion of its total wood products production. U.S. lumber exports, for example, in the late 1980s and early 1990s are projected to range between 3.3 and 3.5 billion board feet while from a smaller

TABLE 14-6
VOLUME CONVERSION FACTORS FOR LUMBER

Cubic feet	× 0.03703	= yd^3
	× 0.02832	= m^3
	× 0.01200	= Mfbm*
	× 0.0200	= load
	× 0.00606	= Pstd†
	× 12	= board feet
Cubic meter	× 35.314	= ft^3
	× 1.3079	= yd^3
	× 0.424	= Mfbm
	× 0.706	= load
	× 0.214	= Pstd
Load	× 50	= ft^3
	× 1.416	= m^3
	× 0.600	= Mfbm
	× 0.303	= Pstd
Mfbm	× 83.33	= ft^3
	× 2.358	= m^3
	× 1.667	= load
	× 0.505	= Pstd
Pstd	× 165	= ft^3
	× 4.670	= m^3
	× 3.300	= load
	× 1.980	= Mfbm

Volume equivalents for lumber

	Mfbm	ft^3	m^3	Load	Pstd
Mfbm	1.000	83.33	2.358	1.667	0.505
ft^3	1.200	100.00	2.832	2.00	0.606
m^3	0.424	35.31	1.000	0.706	0.214
load	0.600	50.00	1.416	1.000	0.303
Pstd	1980.000	165.00	4.670	3.300	1.000

*1000 board feet.
†Petrograd standard.
Source: Binek, 1973.

industry Canadian lumber exports overseas are projected to range from 3.9 to 4.3 billion board feet (include the significant volumes sold by Canada to the United States, and the importance of export markets for Canada becomes clear) (Widman, 1990).

As noted previously, the world market for lumber and other forest products varies significantly on a country-by-country basis. Therefore, to better understand international markets, we will spend a bit of time on the major importers of North American forest products.

TABLE 14-7
CONVERSION OF Mfbm* AND m^3 TO ft^2 AND m^2 FOR PANEL PRODUCTS

Thickness		Mfbm = 2.36 m^3		1 m^3 = 35.31 ft^3	
inch	mm	ft^2	m^2	ft^2	m^3
	2.00	12,696	1180	5380	500.00
1/10	2.54	10,000	929	4237	393.90
	3.00	8468	787	3588	333.33
1/8	3.18	7985	742	3385	314.47
1/7	3.63	6995	650	2965	275.48
5/32	3.97	6396	594	2711	251.89
	4.00	6348	590	2691	250.00
1/6	4.23	6003	558	2545	236.41
3/16	4.76	5336	496	2261	210.08
	5.00	5078	472	2153	200.00
	6.00	4232	393	1794	166.67
1/4	6.35	4000	372	1695	157.48
	8.00	3174	295	1345	125.00
	9.00	2821	262	1196	111.11
3/8	9.53	2664	248	1129	104.93
	10.00	2539	236	1076	100.01
	12.00	2116	197	897	83.33
1/2	12.70	2000	186	848	78.74
	13.00	1953	182	828	76.92
	14.00	1813	169	769	71.43
5/8	15.88	1599	149	678	62.97
	16.00	1592	148	673	62.50
11/16	17.40	1459	136	619	57.47
	18.00	1410	131	598	55.56
3/4	19.05	1333	124	565	52.49
	20.00	1269	118	538	50.00
7/8	22.22	1142	106	484	45.00
	23.00	1104	103	486	43.48
	24.00	1058	98	448	41.67
	25.00	1016	94	430	40.00
4/4	25.40	1000	93	424	39.37

*Mfbm = 1000 board feet.
Source: Binek, 1973.

Japan

Japan is the largest importer of North American wood products. Figure 14-4 shows the overwhelming size of the Japanese market compared to other U.S. export markets. Traditionally, Japanese lumber needs were mostly serviced from the British Columbian coast and U.S. Pacific northwest coast. However, increasing volumes now go to Japan from the British Columbian interior and other interior regions. As can be seen in Fig. 14-5, during the late 1980s roughly 41 percent of U.S. lumber exports went to Japan. Canada and the United States supplied nearly 70 percent of the Japanese lumber import

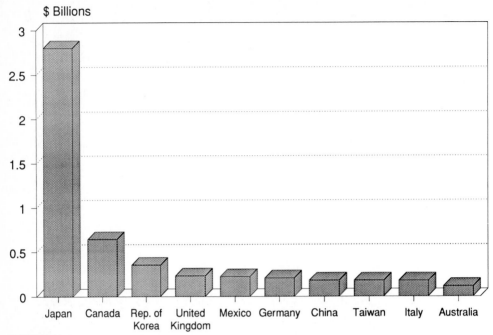

FIGURE 14-4
Value of U.S. exports of wood products by country, 1989. (*Foreign Agricultural Service, 1990.*)

needs during the late 1980s. Our major competitors in meeting Japanese lumber needs have been southeast Asia, Chile, the USSR, and New Zealand.

A majority of the softwood lumber imports into Japan are used for housing. The Japanese housing market has a strong wood housing component and also a repair and remodeling sector much like that of the North American housing market, although the repair and remodeling sector in Japan is much smaller and has not experienced the growth rates of its counterpart in North America.

The Japanese housing market demands high quality and has traditionally relied on post-and-beam construction. This requires a special grade of product known as the "baby square." Baby squares are preferentially sawn from hemlock and are either $3\%_{16} \times 3\%_{16}$ inch or $4\frac{1}{8} \times 4\frac{1}{8}$ inch square. Problems in securing the required skilled labor necessary for the construction of a traditional Japanese home have resulted in an increase in North American–style light-frame platform construction. This trend has been furthered by the promotional efforts by various trade associations and U.S. and Canadian governmental agencies. The trend toward North American–style housing construction will open the market for larger shipments of standard North American lumber grades to Japan. Additionally, the number of factory-built homes has increased, resulting again from the decrease in skilled construction tradespeople, causing an increased demand for North American grades of softwood construction lumber. Changes in the Japanese building codes in the late 1980s, which included allowing three-story wood houses, should also help demand.

In an interesting development, some west coast U.S. exporters are packaging entire 2 × 4 home kits into containers for shipment to Japan. The kits are sold directly to Japanese builders.

The Japanese sawmill industry continues to be a significant provider of lumber into its domestic market. However, this industry is dependent upon imported logs for roughly 40 percent of its raw material. As more and more countries place log export restrictions on their resources, it is likely that the Japanese sawmill industry will provide a smaller and smaller portion of the Japanese lumber requirements.

Asia, with Japan leading the way, also imports large quantities of other U.S. wood products. For example, in 1989 Asia took 32 percent of all U.S. hardwood exports while

FIGURE 14-5
U.S. lumber exports by destination in the late 1980s. (*Adapted from Widman, 1990.*)

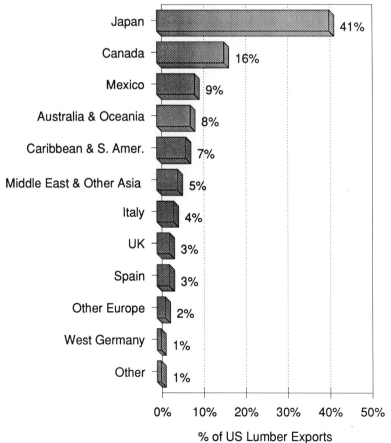

Europe took 31 percent and Canada 24 percent (Widman, 1990). The major species sold to Japan include alder, oak (red and white), ash, and yellow poplar. Japan is also the largest (by far) importer of U.S. wood chips, softwood veneer, and softwood logs.

Canada

Canada is a major forest products trading partner for the United States. Canada supplied about 30 percent of the U.S. softwood lumber needs and over 50 percent of the U.S. newsprint demands in the late 1980s. Forty to sixty percent of Canada's OSB/waferboard production is regularly exported to the United States. Canada is the world's dominant newsprint producer and accounted for nearly 60 percent of global exports in 1988 (Widman, 1990).

On the other hand, Canada, during the late 1980s, was the second largest market for U.S. exports of softwood lumber, the third or fourth largest market for U.S. softwood plywood exports (depending on the specific year) and the largest single-country market for U.S. hardwood lumber (on a volume basis)[2] and furniture. Overall, Canada ranked number two behind Japan on the value of U.S. wood exports received during the second half of the 1980s (Foreign Agricultural Service, 1990).

United Kingdom

The United Kingdom has become an increasingly important offshore market for North American lumber producers. It is a major offshore market for Canadian producers and during the late 1980s was the second largest overseas market for North American lumber exporters, the largest market for U.S. plywood exports and the fourth largest market for U.S. hardwood lumber exports (Foreign Agricultural Service, 1990; Widman, 1990). Unlike for Japan, much of the construction lumber shipped into the United Kingdom is $1\frac{7}{8}$-inch green carcassing (framing) grade. This may change in the future as the United Kingdom is expected to make a major shift to kiln-dried material. Uncertainty abounds in the United Kingdom and the rest of the European Community as they work together toward a united economic zone. The exact regulations and building code requirements which may go into effect and their impact on the resulting importation of lumber products are not known.

The power of negative publicity became strongly evident in the United Kingdom in the mid-1980s. Timber frame housing had captured approximately 25 percent of the market in Great Britain. However, a television program was aired that exposed some unfavorable and unfortunate construction practices. This negative publicity resulted in a drop in the market share of timber frame housing from roughly 25 percent to 6 percent. Since that time, strong efforts by the Canadian, U.S., and Scandinavian timber trade organizations have helped timber regain a portion of the British housing market.

Canada, in the late 1980s, surpassed Sweden as the largest supplier of softwood lumber to the United Kingdom. Other major suppliers include the USSR, Finland, and

---

[2]Much of the U.S. hardwood exports to Canada are re-exported outside of North America.

Portugal. A greater move toward kiln-dried material may benefit some U.S. exporters equipped to provide this product.

Taiwan and Republic of Korea

In 1989, Korea was the third largest importer of U.S. forest products and Taiwan the eighth (Foreign Agricultural Service, 1990). Both countries have strongly growing economics and very limited timber resources. Strong domestic demand and growing export-oriented industries have combined to increase demand. During the late 1980s, Korea and Taiwan were among the top five or six largest importers of U.S. softwood logs, hardwood veneer, hardwood logs, and pulp chips. Korea, by the late 1980s, was the largest buyer of U.S. wastepaper (Widman, 1990). Taiwan, due to its strong furniture industry, for most of the late 1980s ranked behind only Canada and Japan in the volume of U.S. hardwood lumber purchased (Foreign Agricultural Service, 1990).

Continental Europe

The European countries are able to supply only about half of their softwood lumber needs (Widman, 1990). As shown in Figure 14-4, the United States already exports significant quantities of lumber to Italy, Spain, and West Germany. Widman and Associates (1990) have indicated that the traditional suppliers to continental Europe will have great difficulty increasing production. The USSR is becoming more concerned with supplying its own domestic demand as its focus begins to center more on improved standards of living at home. Other traditional export producers, including Sweden and Finland, also have increasing domestic demand and timber supplies have relatively fixed production levels. All in all, this could spell opportunities for North American producers.

In the hardwood area, the backlash against tropical hardwoods and the pressure exerted by European consumers on companies that use tropical hardwoods is intensifying. This is clearly forcing many European importers to look at alternative North American species as substitutes. Germany was already the largest importer of U.S. hardwood veneer in 1989, with Italy also being a significant importer (Foreign Agricultural Service, 1990). Belgium-Luxembourg, Italy, Germany, Spain, France, and the Netherlands were all major importers of U.S. hardwood lumber in the late 1980s (Foreign Agricultural Service, 1990).

Europe also imports significant amounts of U.S.-produced softwood plywood. In the late 1980s the leading importers were Belgium-Luxembourg, the Netherlands, Denmark, Germany, Italy, and Sweden (Foreign Agricultural Service, 1990).

Australia/Oceania

Australia was, during the late 1980s, a strong market for U.S. and Canadian exports of softwood lumber. Much of the material exported was of construction-grade (sometimes also called "scantling-grade") material, with some volumes of large sizes for re-manufacturing. Over the long term, Australia has committed itself to becoming self-sufficient in forest products. This has resulted in new plantations of radiata pine and eucalyptus. If this is successful, it could result in a significant slowdown in Australia's importation of construction-grade lumber. North American producers may still have an advantage in the market in their ability to provide high-grade, relatively clear material.

Caribbean and Latin America

The Caribbean and Latin American markets, including Mexico, have been strong markets for North America lumber exports. The Caribbean area receives a large volume of lower-grade Canadian production and southern pine construction lumber from the U.S. southeast. Because of transportation advantages, the U.S. southern pine industry will continue to be a dominant supplier for this market. Much of the southern pine into the region is sold with a fuller cut and undressed.

Middle East/North Africa

North American shipments into the Middle East and North African regions have been, at times, quite volatile. Lumber shipments into these regions have varied as a result of government decision making, which many times has been dependent upon the price of oil. There is a continuing need for housing in this region, providing an opportunity for lumber exports, but this consideration must be coupled with an awareness of the political unrest so common in the area.

SUMMARY

International marketing can be an excellent way to diversify a firm's customer base and to enhance its profitability. However, it requires expertise, a commitment to stay the course, and a carefully laid strategy. Business practices, customs, and product grades can vary country by country, requiring careful analysis and expertise for success in these markets. While we have discussed international markets with a commodity viewpoint because of the availability of data on commodities, don't be deceived. Much trade is in specialty if not custom products. Product specifications and customer demands typically exceed those in domestic markets.

North America is a major world player in forest products and will likely remain so. Most major North American firms and trade associations have export market development efforts. Export markets are not a flash in the pan. They will be with us for the foreseeable future and will likely grow in importance.

BIBLIOGRAPHY

Alabama International Trade Center. 1984. "Export Opportunities for Alabama Lumber." *Export, International News for Alabama Business* (January–March) 4(1). Published by Alabama International Trade Center, P.O. Box 1996, University of Alabama, Tuscaloosa, AL.

Anderson, R. G. 1984. *Italian Report: A Study of the Italian Market for American Timber Products, Market Research and Economic Services*. Findings of a market study conducted by the

American Plywood Association (APA) with assistance from Centro Studi Industria Leggeras, Scrl (CSIL), Milan. American Plywood Association. Tacoma, WA.

Anonymous. 1990a. *The Maryland Guide to International Trade.* The Maryland International Division Office of International Trade and the Maryland Chamber of Commerce. Annapolis, MD.

Anonymous. 1990b. "Packaged Homes Gaining Popularity in Japan." *Random Lengths Export Newsletter* (September 20):1.

Bethel, J. S. 1983. *World Trade in Forest Products.* University of Washington Press. Seattle, WA.

Bilek, E. M., and P. V. Ellefson. 1990. "Organizational Arrangements of Foreign Operations of U.S. Wood-Based Firms." *Forest Products Journal* 40(1):4–10.

Binek, A. 1973. *Forest Products in Terms of Metric Units* (Published by author). Westmount, Que.

Brady, D. L. 1978. *An Analysis of Factors Affecting Methods of Exporting Used by Small Manufacturing Firms.* Unpublished Ph.D. dissertation. University of Alabama. Tuscaloosa, AL.

Buckley, M. 1989. "Cultural Considerations and Distribution Systems: The European Community." In *Hardwood Exports: Building a Business Strategy.* Lexington, KY. pages 39–49.

Evans, C. 1990. *A Guide to Exporting Solid Wood Products.* 2nd edition. USDA Agriculture Handbook No. 662. U.S. Department of Agriculture. Washington, DC.

Bykhovsky, A. 1991. "Meeting the Global Challenge." *Timber Trades Journal.* (April) 27(357):20–21.

Foreign Agricultural Service. 1990. *Wood Products: International Trade and Foreign Markets.* Publication WP2-90. USDA, Foreign Agricultural Service, Forest Products Division. Washington, DC.

Gottko, J., and R. McMahon. 1989a. "Oregon Lumber Producers: Attitudes and Practices of Exporters and Non Exporters in the 1980's." *Forest Products Journal* 39(6):59–63.

Gottko, J., and R. McMahon. 1989b. "Oregon Lumber Producers: Evaluating Exporting in the 1980's." *Forest Products Journal* 39(11/12):29–32.

Guldin, R. W. 1984. "Exporting Southern Softwood Lumber: How Small Firms Can Begin." *Forest Products Journal* 34(3):37–42.

Guldin, R. W. 1983. *Sales and Distribution Channels for Exporting Southern Forest Products.* USDA Forest Service Report FSRP-SO-192. Southern Forest Experiment Station. New Orleans, LA.

Hammett, A. L., and K. T. McNamara. 1991. "Shifts in Southern Wood Products Exports: 1980 to 1988." *Forest Products Journal* 41(2):68–72.

Hansen, J. 1989. "Getting Paid Internationally." In *Hardwood Exports: Building a Business Strategy.* Lexington, KY. pages 74–79.

Henley, J. 1981. "What a Forwarder Does." In *Proceedings of Exporting Southern Forest Products.* September 14–16. Houston, TX.

Jaakko Poyroy International OY. 1983. *Financial Performance of Major Forest Product Companies in Western Europe and North America.* P.O. Box 16, Kaupintie 3, 00441 Helsinki, Finland.

Jacobsen, J., and H. W. Wisdom. 1988. "The Influence of Socioeconomic Factors on the Demand for Oak Lumber in Europe." *Forest Products* Journal 38(9):64–66.

Japan Forest Technical Association (editors). 1983. *Wood Demand and Supply in Japan.* Nippon Mokvzai Bichiku Kiko Publishers.

Jélvez, A., K. A. Blatner, R. L. Govett, and P. H. Steinhagen. 1989. "Chile's Evolving Forest Products Industry: Part 1. Its Role in International Markets." *Forest Products Journal* 39(10):63–67.

Luppold, W. G. 1983. *Analysis of European Demand for Oak Lumber.* USDA Forest Service Research Paper NEFES/84-14. Northeastern Forest Experiment Station. Broomall, PA.

Luppold, W. G. 1982. *Econometric Model of the Hardwood Lumber Market*. Northeastern Forest Experiment Station. USDA Forest Service Report #FSRP-NE-512; NEFES/83-38. Broomall, PA.

Luppold, W. G., and R. E. Thomas. 1991. "A Revised Examination of Hardwood Lumber Exports to the Pacific Rim." *Forest Products Journal* 41(4):45–48.

McLennan, Kay L. 1984. Office of Transportation: "Assistance Available for Agricultural Exporters." *Foreign Agricultural Magazine*. Superintendent of Documents, GPO. Washington, DC 20402.

McMahon, R., and J. Gottko. 1988. "Oregon Lumber Producers: Export Marketing Trends Between the Mid-1960's and Mid-1980's." *Forest Products Journal* 38(4):44–46.

National Lumber Exporters Association. 1984. *The Development of Overseas Markets for Hardwood Dimension*. 805 Sterick Building, Memphis, TN 38703. October. 96 pages.

Pollini, C., G. Leonelli, and D. L. Sirois. 1991. "The Forest-Based Industry of Italy: Problems and Prospects." *Forest Products Journal* 41(3):50–54.

Random Lengths Publications. A biweekly *Report on World Markets for Forest Products*, Jessie Taylor, editor. *Random Lengths Export*. Random Lengths Publications, Inc. Eugene, OR.

Rich, S. U. 1981. "Distribution Channels in the Wood Products Export Market." *Forest Products Journal* 31(2):9–10.

Schumann, D. R., and R. T. Monahan. 1978. *The Lumberman's Guide to Exporting*. Northeastern Area State and Private Forestry. USDA Forest Service. January. 20 pages.

Thomas Publishing Company. 1983. *Inbound Traffic Guide* (October/November) 3(4). New York.

Western Wood Products Association. 1982. *Special Report, European Lumber Grades and Trade*. Report of a European market survey trip by H. A. Roberts and B. J. Hill. November. Portland, OR.

U.S. Department of Agriculture. Office of Transportation. 1985. *Exporting Oak Logs and Lumber to the European Community, 1983 and 1984*. Final Draft. January.

U.S. Department of Commerce. 1984. *A Competitive Assessment of the U.S. Solid Wood Products Industry*. International Trade Administration, Basic Industries Sector, Office of Forest Products and Domestic Construction. Washington, DC.

Westman, William W. 1986. *A Guide to Exporting Solid Wood Products*. USDA Agriculture Handbook No. 662. Washington, DC.

Widman. 1990. *Markets 90–94*. Widman Management Ltd. Vancouver, B.C.

Widman. 1989. *Markets 89–93*. Widman Management Ltd. Vancouver, B.C.

Wisdom, H. W., J. E. Granskog, and K. E. Blatner. 1986. *Caribbean Markets for U.S. Wood Products*. USDA Forest Service Research Paper SO-225. Washington, DC.

CHAPTER 15

MARKETING ORGANIZATION

Essential to the implementation of marketing strategy is a properly designed marketing organization structure.

Rich, 1978b

As Rich notes in the above quotation, marketing strategy is difficult, if not impossible, to implement without a properly designed marketing organization. Many different forms of marketing organizations exist. Marketing organization within a forest products firm might be best designed around the major marketing functions of product management, market management, and sales force management (Rich 1978b). On a broader scale, a marketing organization should have two major objectives:

1 To facilitate the flow of information
2 To achieve coordination of efforts

Before we continue further into the details of various forms of organizational structures, it is helpful to back up for a moment and look at some important aspects of organizations. Organizations have at least two critical aspects, their structure and their organizational behavior. The structure involves the organizational framework within which people work. This will be the aspect of organizations which we will focus on in this chapter. However, equally, if not more, important is organizational behavior or how people actually act within the structure. In many instances, organizational behavior can operate in a fashion that will make marketing systems work in spite of structural deficiencies (Bonoma, 1985).

TYPES OF POSITIONS

There are two broad categories of positions which fit within a forest products organization. These categories are line or operating positions and staff positions. Line or operating positions are generally concerned with the production, distribution, and sales of products. In short, people in these positions are responsible for the day-to-day work of producing, shipping, and selling the various goods which forest products firms manufacture.

On the other hand, a staff position is one which generally serves a support role for line positions. These support roles can be in such functional areas as finance, personnel, accounting, legal, advertising, and marketing research.

AUTHORITY

Different types and levels of authority are also associated with these types of positions. Line authority, typically associated with line or operating positions, has traditionally included the authority to hire, sign, direct, alter the pay of, and at times fire employees. Recent changes in corporate policy and the legal code have served to limit some of this traditional line authority within stricter guidelines.

Staff authority can serve two roles: advisory and functional. In an advisory role, a staff position provides expert advice in areas such as advertising, marketing, accounting, or legal matters. Usually individuals with this level of authority have no direct ability to force the acceptance of their advice. However, it is important to remember that their power can be considerable and is measured by their ability to influence decisions within the company.

In some instances, certain staff positions can exercise functional control or the ability to give direct instructions to other employees who are not under their direct control. For example, a regional sales manager may receive instructions from people in such staff positions as finance, accounting, personnel, marketing research, or other staff areas. The sales manager may be required to take action based upon these instructions.

ORGANIZATIONAL TYPES

Now that we understand the types of positions within a company and the various levels of authority associated with them, it is time to move on to how these various areas might be integrated into an organizational structure. It is important to remember that most companies have organizational structures which have evolved over time and may, in fact, be a mixture of various prototypes. However, there are several widely recognized prototypical organizational structures. These can be grouped into two basic categories: those which serve the needs of centralized companies and those which serve the needs of companies that operate on a decentralized basis.

Centralized Structure

A functional organizational structure is one which is used by firms operating under a centralized management scheme. All functional areas report to a chief executive or, in the case of Fig. 15-1, to the vice president of marketing.

An advantage of this system is that it does allow for some specialization in functional roles and does permit a chief executive to be personally involved in the management of the business. It is a simple system with clear roles and levels of responsibility. Information and coordination needs are typically low and it can work well when a narrow product line is sold in a relatively limited market structure.

One weakness of this system is that most of the tough decisions tend to be bucked up to the top executive. Additionally, as products and/or markets tend to proliferate, the simplicity of the form can lose its effectiveness.

Although the strength of the functional organizational structure lies in its simplicity, much can be learned about the relative importance of various functions by their

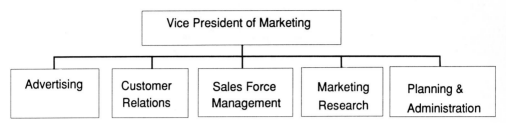

FIGURE 15-1
Typical function-based organization.

placement within the organization. This can be of benefit to individuals as they are making career choices. For example, in Fig. 15-2 the functional organization is split into a marketing function and a sales function, giving roughly equal weight (based on the organizational chart) to marketing and sales. Each of the two main functions of sales and marketing report directly to a vice president of marketing. Contrast this with the organization shown in Fig. 15-3, which shows the marketing manager and the field sales manager reporting to a general sales manager, who then reports to a vice president. Clearly, in this latter organization sales probably has a dominant role over marketing.

Decentralized Structures

Company growth and increased diversity in product lines and markets can serve to put considerable strain on a functional organization. One of the more important changes in the management of U.S. corporations during the post–World War II period was a shift from a functional organization to one of the more decentralized organizational structures (Rich, 1978a).

FIGURE 15-2
Function-based organization separating the sales function. (*Adapted from Buell, 1984.*)

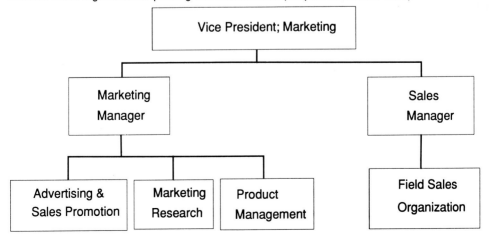

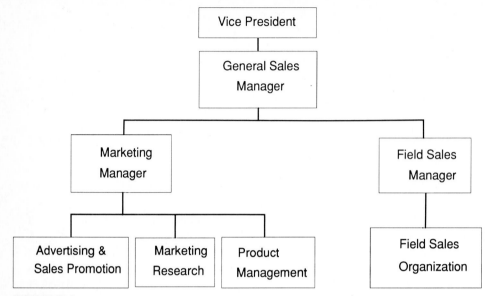

FIGURE 15-3
Function-based organization showing marketing reporting to sales function. (*Adapted from Buell, 1984.*)

In taking a decentralized approach to structure, the organization pushes the decisions and responsibilities further and further down within the organizational structure. Some common structures which have evolved are market-based organizations, geographic-based organizations, and product-based organizations.

A typical market-based organization is shown in Fig. 15-4. This structure can be very useful and helpful in forest products companies because their markets are diverse and

FIGURE 15-4
Typical market-based organization.

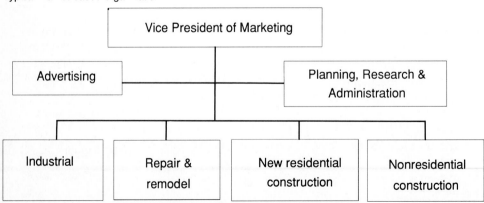

radically different channels of distribution are needed to reach the different market segments. This is especially true when moving from wood building products to paper products. Market-based systems can be very effective when the product lines sold to a given market are quite similar.

Another decentralized structure used by many forest products companies has been a geographic-based structure. This has worked well for some companies with widely scattered manufacturing facilities producing different products from the locally available timber resources. In these situations, it was logical to organize the manufacturing and marketing structures around these geographic bases, with their own production facilities producing products from specific tree species for specific markets. Weyerhaeuser, for example, was organized on a geographic basis for many years until the early 1990s when it switched to a product-based organization.

Over time, many companies have moved toward a product-based organizational structure. The product-based system has been shown to be typically easier to apply and has worked well for a number of firms. A typical product-based organizational structure is shown in Fig. 15-5. This shows a separate manager for each of the major product areas within the firm. It has been noted that the product-based organizational system is used by 90 percent of the consumer package goods companies and by over 50 percent of the industrial goods companies in the United States (Buell, 1984).

This system has the advantage of allowing each important product to receive the full attention of one product manager, who has direct access to higher-level corporate management. Some of the disadvantages of this particular system revolve around the fact that many times the product manager has no line authority over the functional departments that manufacture, distribute, or even sell the specific product.

A matrix organizational structure is one which has not been adopted to any great extent by forest products companies; however, it deserves mention here for completeness. It is a complex organizational structure, as shown in Fig. 15-6, and can be difficult to manage well. One reason for this management difficulty is that many people have more than one immediate supervisor. This can cause contradictory orders in this structure and as a result it has not worked particularly well for many industrial companies. The matrix organization has worked well in some professional organizations, particularly consulting firms, where project teams are developed across functional areas within the corporation (Buell, 1984).

Operating Divisions   As companies grow larger and more complex, many times product-, geographic-, or market-based systems are no longer adequate. At this time, firms usually divide themselves into separate operating units or divisions. Often, these divisions are considered stand-alone profit centers within the company. As shown in Fig. 15-7, the management within a profit center is responsible for both its costs and its revenues. Upper-level company management then evaluates the profit center management team based upon its profit or contribution to total company revenues. The profit center managers are given strong authority and responsibility in this management structure.

Some divisional structures contain only manufacturing and marketing functions and must depend upon the company's main office for other functional areas such as finance,

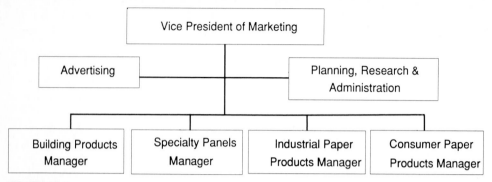

FIGURE 15-5
Typical product-based organization.

personnel, accounting, legal, and other areas. On the other hand, the divisionalization may be entirely complete, with each division having its own functional staff groups and leaving the main company office with only a skeleton staff. This type of organizational unit is many times called a "strategic business unit" (Fig. 15-7). One study showed that sixteen of the twenty top U.S. forest products companies had adopted a strategic business unit concept as the key to their organizational structure (Rich, 1979a). The strategic business units within the companies studied were defined primarily based upon the product line and the markets served. The manufacturing facilities and the necessary supporting raw material supply and gathering operations may also be part of the strategic

FIGURE 15-6
Typical matrix organization.

PRODUCTS

FUNCTIONS		BUILDING PRODUCTS	CONSUMER PAPER	INDUSTRIAL PAPER	SPECIALTY WOOD PRODUCTS
	LEGAL				
	FINANCE				
	GOVERNMENT AFFAIRS				
	HUMAN RESOURCES				
	MARKETING RESEARCH				

Divisional Form	Key Characteristics
Revenue Center (i.e., Marketing)	Responsible for revenues. Expenses will be incurred; however, profits are not measured
Cost Center (i.e., Legal)	Responsible for costs
Profit Center	Responsible for <u>costs</u> and <u>revenues</u> - management measures <u>profits</u>
Investment Center	Profit center that also considers <u>assets</u> required to achieve products
Strategic Business Unit	Similar to investment center but SBU managers have considerably more authority and responsibility

FIGURE 15-7
Various divisional structure forms.

business unit or there may be separate timberland units which support the other various business units.

The managers of a strategic business unit have considerable authority and responsibility both to achieve profits and to manage the assets required to achieve those profits. A strategic business unit's scope, resources, products, and markets must be such as to allow the establishment of a realistic framework for forecasting strategic planning and goal setting at the strategic business unit level (Rich, 1979a).

An Actual Company   All of this can seem somewhat distant; however, let's look at Fig. 15-8, which shows the organizational structure of a major forest products firm's building products unit. The building products unit is headed by an executive vice president similar in stature and prestige in the company to vice presidents in charge of the paper products division and others. The distribution part of the building products division is organized on a geographic basis, with five regional vice presidents in charge of the wholesale distribution centers. The marketing, advertising, and national accounts systems are organized under a vice president in the distribution division. Under this vice president there are a number of functional areas such as product development, industrial products, national accounts, marketing services, and advertising.

The various distribution centers operate as autonomous units, free to purchase the products they sell from company facilities or production facilities of other firms. This is evident on the right-hand side of Fig. 15-8, which also shows a wood products sales

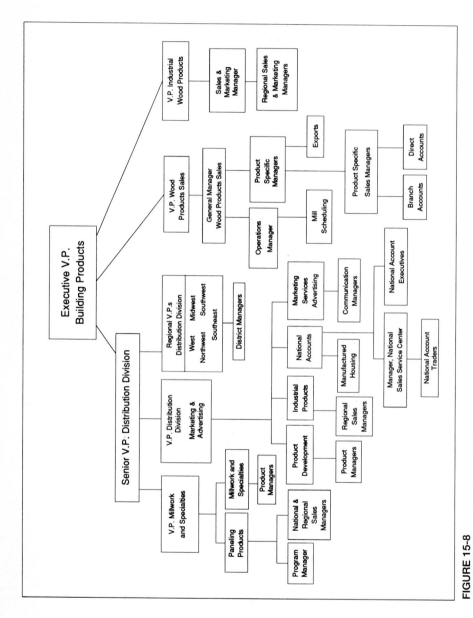

FIGURE 15-8
A major forest products firm's building products marketing organization.

organization. This sales organization functions as the sales arm of the building materials production facilities. Each specific product area, such as softwood lumber or structural panels, is headed by a product-oriented manager who then reports to the general manager for wood products sales who reports to a vice president of wood products sales. The vice president of wood products sales, like the senior vice president of the distribution division, reports to the executive vice president in charge of the entire strategic business unit.

The millwork and specialties group and the industrial wood products group operate under their own vice presidents. This shows the emphasis the company is placing on these product areas.

SUMMARY

Understanding a marketing organization can be useful in career planning and in everyday management activities in a firm. Don't forget to study both the formal structure and the organizational behavior of a firm, as each is important to understanding how a system functions. Centralized structures are widely used for smaller firms, but most larger forest products companies have adopted decentralized marketing/management structures. The largest of the large firms are mostly organized into strategic business units, with each unit having its own internal structure.

BIBLIOGRAPHY

Barnes, H. M. 1980. "Education in Wood Science & Technology: An Update 1978–79." *Wood and Fiber Science* 11(4):252–260.

Bonoma, T. V. 1985. *The Marketing Edge.* The Free Press. New York.

Buell, V. P. 1984. *Marketing Management.* McGraw-Hill Book Company. New York.

Phelps, R. B. 1980. *Timber in the United States Economy.* General Technical Report WO-21. U.S. Department of Agriculture, Forest Service. Washington, DC.

Rich, S. U. 1979a. "Making the Strategic Business Unit Concept Work." *Forest Products Journal* 29(4):8.

Rich, S. U. 1979b. "Setting up the Strategic Business Unit." *Forest Products Journal* 29(2):10.

Rich, S. U. 1978a. "Product Division and the Strategic Business Unit in Organization Structure." *Forest Products Journal* 28(12):9.

Rich, S. U. 1978b. "Designing a Marketing Organization." *Forest Products Journal* 28(10):36.

PART SIX

APPENDIXES

APPENDIX A: Case Studies

HOW TO GET THE MOST OUT OF YOUR CASE STUDY ASSIGNMENTS[1]

Cases are sometimes viewed with a shuffle of the feet and shrugged shoulders by students early in a term. However, it has been my experience that by the end of the course and certainly after the course has concluded, most students tell me that the cases were the most interesting part of the course. Since many forestry and forest products students have never had a case assignment, let me briefly describe what is involved.

A case, as used in our current context, is a description of a marketing-management issue or issues which have actually been faced by forest products executives or managers, together with the facts and opinions upon which executive decisions had to depend. Simply stated, then, the use of the case method calls for discussion of real-life situations that have been faced by forest products executives. As in real life, expect to make decisions with incomplete information.

In this assignment you may encounter three general types of cases. One type can be classified as an *evaluation* case. This describes what a company has done, and the principal purpose of the case discussion is to evaluate the soundness of what the company's management actually did. A second type is commonly known as a *problem* case, that is, a situation where management faces a specific problem, such as increasing or decreasing a price. A problem-type case calls for the consideration of alternative actions and finally for a decision. Obviously, problem or decision-type cases are structured so that the executives in the situation—and you—are asked and required to make a decision. This will be the most common case for this book. Third, there is the *general appraisal* type of case in which information is unstructured. In this type, it is up to you to

---

[1]Adapted from "The Case Method of Learning" by S. U. Rich.

"So then I says to him, I says, let's get into
the Lumber business..."

determine how things are going, and to evaluate whether there are problems calling for action, and, if so, what action to take. It should be noted finally that some cases may contain elements of all three of the above categories.

Since each case, like each forest products marketing-management situation, is in some respects unique, there is no one standard pattern for solving all the cases or problems in life you will encounter. However, the following procedure is presented as a guide in developing a method of approach to case preparation, particularly for the problem-type cases:

1 Read through the case quickly to get a general feel for what is going on. Then go back and read the case very carefully to develop a complete mastery of the case facts. Where figures are presented, use them to make comparisons, or do some calculations to determine the relationships between the figures which may help you better understand problems in the case.

2 Define the main issue (or issues) in the case. Write it (or them) out in precise terms.

3 Determine the main goals/objectives of the company in the case and write them down.

4 Break the main issue (or issues) down into subsidiary issues. Distinguish between problems calling for immediate solution by the company, or by a particular executive in the company, and more long-range issues, the solutions to which the company can decide later.

5 Gather the important facts and figures in the case which have a bearing on the main issue or the subsidiary issues. Discard the facts which may be irrelevant to the problem at hand.

6 Examine all the alternative courses of action which might be followed in resolving the main issue and subissues. List the pros and cons of each alternative.

7 Decide which alternative you recommend in each instance, and develop arguments to support your recommendation. Check to see if your recommendations are consistent with the company's goals/objectives. Don't ignore potential people problems.

8 Lay out a program of action for the company to follow in implementing your recommendations. Provide a timetable, indicating what action should be taken immediately and what might be deferred until later.

It is up to you individually to come to your own conclusions concerning what should be done, if anything. It is important to emphasize that the process of arriving at your own personal position on each case is an important aspect of the case assignment. If you clearly take a position in your own mind, then you will be quite personally involved in the discussions and intensely interested. On the other hand, if you simply read a case and remain neutral about it or take the opinions of your classmates as your own, then you are unlikely to be affected and thus will not learn a great deal from the preparation of the case or a class discussion.

JAMAICAN HARDWOODS LIMITED

Development of a Marketing Plan to Introduce Tropical Hardwood Boards to the U.S. Home Center Marketplace

Mr. Robert Wynter who, after 10 years with Jamaican Hardwoods Ltd., recently completed his Master's in Forest Products Marketing, had just been promoted to Vice President of Marketing. The President, Samuel Robee, tossed a consultant's report (see attached) on his desk his first day as V.P. and said, "Bob, use this as background and give me an action oriented plan to sell our tropical hardwood boards into the U.S. do-it-yourself market."

Bob Wynter was not sure getting into the highly competitive U.S. market was a good idea. Jamaican Hardwoods Limited was a furniture producer with a small domestic market and excess production capacity. Their government mandate was to provide jobs and operate for the long-term good of the Jamaican economy. They imported hardwoods from Central and South America and produced Caribbean-style furniture. After years of trying to sell furniture to the U.S., they had given up and were now looking for a new way to use their excess manufacturing capacity.

Bob stared out his window and began to turn his attention to reading the report from Vega Marketing Research. So many decisions must be made to effectively plan to introduce tropical hardwood boards to the U.S. market. What species? Channels of distribution? What's the target customer group? Do we sell to final customers through retail or mail order channels? How do we price our product? Might there be a better product for our idle facilities? Do we target market? Etc. By now, Bob's head was swimming so he decided to list all the key elements of a good marketing plan and under each element he noted the alternatives Jamaican Hardwoods had. This took quite some time, but Bob knew his plan for the President better be good or his new job could be in jeopardy.

A Market Research Report To
Jamaican Hardwoods Limited

THE HOME CENTER MARKET FOR HARDWOOD BOARDS

Prepared by:

Vega Marketing Research
1868 Mad Dog Road
Blacksburg, VA 24060

Background Information

The transition of local lumberyards to home centers was driven by many market forces. Local retail lumber dealers were required to service very diverse segments in the retail building products market. One segment consisted of high volume builders who had increased demands for products, service and credit. However, the different needs of smaller builders still had to be served. At the same time, the growing repair and remodeling market had to be dealt with. This market consisted of professionals and the expanding do-it-yourself (DIY) market, which had new demands for information and sales assistance. These widely divergent needs resulted in the development of new types of retailers such as dealer-builders, construction supply dealers, full service dealers, the cash and carry dealer and finally home centers and warehouse home centers.

On the merchandising side, product lines were exploding and the small corner display area was nowhere near the merchandising area required. By the 1990's, the average home center stocked 20,000 to 25,000 different items in 25,000+ square feet of indoor space while warehouse home centers averaged 30,000+ different items in 70,000 to 140,000 square feet of indoor space. Figure 1 shows home center sales by product line.

Current Distribution Systems

Census data appear to show that independent wholesalers are still the dominant means by which lumber and building materials are transferred from producer to retailer. This is illustrated in Table 1, which shows that independent wholesalers accounted for over three-quarters of total sales in the last census.

Vega - 2

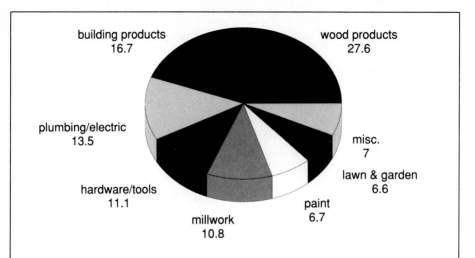

FIGURE 1
Distribution of sales by product line for home centers/warehouse home centers.

Table 1.
Sales of lumber, plywood, and millwork through independent and manufacturer-owned intermediaries (U.S. Bureau of the Census).

	Total U.S. Sales					
	1963	1967	1972	1977	1982	1987
	-------------- % --------------					
Independent Wholesalers	85	80	80	75	84	87
Manufacturers' sales branches and offices	15	20	20	25	16	13

Distributors, both independent and captive, are now called upon not only to handle the physical distribution of products and materials, but also to provide additional marketing and inventory control services. Retailers are calling on wholesale distributors (and manufacturers) more frequently for help with in-store selling, product demonstrations, and training of store personnel. Increasingly, distributors are called upon to develop their own individualized merchandising programs for home center retailers as well as administering national manufacturers' programs at a local level.

The emergence of strong networks of manufacturer-owned wholesale distribution centers was a major force in the marketplace in the 1970's and 1980's. The strongest networks appear to have been developed by Georgia-Pacific and Weyerhaeuser, although many other firms also have captive distribution centers. Most distribution networks work from a strong geographical base, and many function as full line distributors for other manufacturers' products as well as their own. The 1990's have seen many major firms backing away from their distribution network and concentrating more on manufacturing.

Impact of the Repair and Remodeling Market

The increasing importance of the repair and remodeling market had a large impact on the distribution of building products. The potential in this market is better understood when one reviews recent statistics. In 1988, sales to do-it-yourself home owners reached $65 billion, double what they were a decade earlier. Contractors purchased $35 billion of materials for repair and remodeling in 1988, up from only $15 billion in 1980. Sales to both groups are expected to show similar growth in the 1990's.

To capitalize on this growing market, one-stop home improvement centers material-ized, carrying thousands of items for adding to, repairing, renovating and maintaining a home. These centers were geared toward the DIY home owner and repair and remodeling contractor, who not only wanted to find a large variety of items available in small quantities, but also expected to find instruction manuals and material return service. Total home center sales have grown from $24.3 billion in 1975 to well over $100 billion in 1990.

Most of these home centers are public companies which drew on department stores and the financial community for their management personnel. Names like Wickes, Lowe's, Hechinger, Scotty's, Home Depot and Payless Cashways became household words in many areas.

Total sales in the repair and remodeling market reached well over $100 billion in 1990 and continued increases are expected to push sales over $400 billion by 2000. Sales to DIY customers are predicted to reach $168 billion by 2000. The repair and remodeling market is strong enough to be called ''recession-proof.''

Hardwood specialties should enjoy a portion of this large market, and in fact empirical evidence gained from many informal conversations with various hardwood specialty firms and home center retailers indicates a strongly growing demand for high value hardwood products at the retail level. One Vice President of Marketing indicates that as much as 7 to 10 percent of the hardwood lumber production may go to this market in the next several years. This new demand appears to be growing with the DIY movement which began in the late 1970's. It could prove to be a significant outlet for a wide variety of hardwood products such as boards, flooring, moulding, edge glued panels, plywood panels and other products in the 1990's.

Table 2.
The top 10 do-it-yourself retail building products sales leaders arranged in order of sales.

	1989 sales (millions)
Home Depot	2,740
Lowe's Companies	2,650
Builder's Square	2,000
Payless Cashways	1,900
Hechinger	1,200
Grossmans	1,100
Home Club	1,000
Wickes Lumber	1,000
84 Lumber	800
Wickes Companies	675

In addition to the boost in demand from the DIY movement, the increasing consumer demand for quality is believed to be spurring hardwood specialty sales also. Two income families with more disposable income are engaged in numerous remodeling products requiring the use of hardwoods.

Results of a Preliminary Market Research Study

A survey was conducted of hardwood specialty buyers representing 1500 typical home centers and the Top 100 home centers (by sales). The survey objectives were to characterize the home center market for hardwood specialty products and to estimate its size. The percentage of respondents who stocked different products varied as shown below:

Percent of respondents who stocked the specified products

Hardwood Product Type	Typical Firms	Top 100 Home Centers
Boards	66%	81%
Panels	77	74
Moulding	59	79
Flooring	68	64

Vega's research team extrapolated the data to estimate the home center market for hardwood specialty products by using the stocking percentages above to estimate the total number of home centers carrying a product, then multiplying the number of home centers by the average sales per home center. This gave a rough market size estimate of approximately $1.4 billion for hardwood boards, panels, moulding and flooring.

Survey respondents believed demand for hardwood moulding, boards and plywood would increase, while future demand for hardwood flooring was reported as stable. Sales in the DIY customer segment of the hardwood board market are predicted to grow by 144 percent in the next five years. We estimate the home center market for hardwood boards is approximately $511 million. The most important customer project reported by typical firms was cabinets while shelving was the most important project to Top 100 home centers. The Southern region with the smallest percentage of firms stocking boards has a large potential for growth.

Market Segmentation: Urban vs. Rural

Preliminary data indicate differences between urban and rural locations in the hardwood board market. Home center stores located in rural areas seem less likely to carry hardwood boards. The major reasons given were the lack of demand and the high price of the products. However, larger demand was believed to be present at lower prices.

On the other hand, a higher portion of home center stores located in urban or high-populated areas carried hardwood boards. The products in these locations are being sold with an emphasis on quality. Price, as compared to quality, did not seem to be limiting demand for hardwood boards in urban locations. However, preliminary data indicate that lower price is the single most desirable element of hardwood board substitutes.

The Product Line

Although quality is important, there seems to be a variety of quality types sold. For Top 100 home centers, boards were mostly S4S (surfaced on 4 sides) and clear on both faces. The most common board quality type for the typical home centers was surfaced on two sides with one or two clear faces. These boards are most likely FAS grade boards planed on

Vega - 8

two sides. The final products for both home center segments bring a premium price; consequently, managers indicated that consumers demand a product relatively clear and free of knots and cracks.

Oak was the most important wood species in terms of sales. Walnut, maple, ash and cherry, respectively, followed oak in terms of sales importance. Yellow poplar ranked below all the above species. Other important species which were mentioned by respondents included birch, mahogany, and some tropical hardwoods. Species importance appeared to vary depending on the geographic location of the retail stores, however, more data is needed for a definite pattern.

Although indications are still unclear, the most common board widths reported were 6, 8, 10 and 12 inches with 6- and 8-inch boards most important to sales. Common board lengths were 6, 8, 10 and 12 feet.

The major complaint with hardwood boards according to retailers was inconsistent product quality. Apparently they are having a problem purchasing a consistent product.

Purchasing

The majority of home center stores who stock hardwood boards did not indicate significant difficulties in purchasing other than quality concerns. A variety of local or regional producers were supplying the needs of the typical home center stores contacted. Average lead time for hardwood board shipments was approximately two to four weeks. Lead times were too long for some retail operators.

Minimum order size required by manufacturers varied depending on region and geographical location. Minimum order sizes for hardwood boards ranged from one board foot to 1000 board feet with 500 board feet and 1000 board feet reported as the most common minimum order sizes. Some retail firms expressed a need to buy smaller amounts, say a dozen boards of a given size.

Top 100 home centers were more frequently supplied by large established suppliers who provide a standardized product with display rack and point-of-purchase literature. Only a few established suppliers are in the marketplace. ChoiceWood and Canfield Forest Products are the largest and they deal mostly (if not totally) with domestic U.S. species.

The major concern expressed with merchandising hardwood boards was displaying the product. The majority of typical retailers, especially rural retailers, appear to display their product without a set strategy. Rural retailers were likely to stock random length, random width material which is difficult to display in racks.

Co/op advertisements by big urban home centers and word-of-mouth communication from satisfied customers of small rural home centers were rated as the most effective advertising strategies. Information provided to customers by sales personnel about the products was noted as an important aspect of in-store promotional strategy. Clearly, sales personnel need an understanding of hardwood boards and mouldings.

The Consumer

Hardwood board customers are split in two distinct categories, the DIY customer and the professional-remodeler/new construction. Typical hardwood board purchases ranged from five to 100 board feet for DIY customers up to several hundred board feet for professionals. The major uses of hardwood boards were cabinets, furniture and shelving.

Vega - 10

Retailers' perceptions of their hardwood board customers indicate that at least 75 percent of them are male. In terms of sales, white-collar, salary-income customers were ranked as the most important with blue-collar, hourly-income customers ranked next in importance. Retailers believed customers between 30 and 49 years old were most important in terms of sales.

Manufacturers

Current data indicate the presence of a variety of mostly small manufacturers of hardwood boards for retail sale. Most manufacturers seem to supply local or regional markets and no one producer dominates the market. However, ChoiceWood is becoming the clear market leader. Canfield Forest Products has also introduced a major hardwood board program to the home center market. Most major producers are, however, concentrating on developing major accounts at Top 100 home centers.

New Competition for Home Centers

New competition which could limit home center sales in this product line is emerging in the form of hardwood retail stores and mail order suppliers. The specialty retail store for hardwoods is an idea which is catching on and gaining momentum. Most of the current stores are small but plans are being made by several U.S. hardwood producers to open their own stores. Mail order retailing is big business and now includes hardwood lumber and veneers in the merchandising mix of many catalogs aimed at craftsmen.

Summary

We at Vega were delighted to serve Jamaican Hardwoods through our analysis of the home center market for hardwood boards. Urban home centers are marketing a higher quality and more sophisticated product mix while rural home centers generally have a less developed product mix and lower product quality. Growing national suppliers in this market could present a problem in capturing Top 100 home center accounts, however, the small firm category does present an opportunity for a very aggressive firm. This firm would need a well developed product mix and sophisticated merchandising strategy to succeed.

CALIFORNIA REDWOOD PRODUCTS[1]

Analysis of Industrial, Consumer and Commodity Markets

Jim Bush has owned and operated California Redwood Products for the last 40 years. The mill produces redwood lumber from logs harvested from its own land in Northern California. While Jim has never had a big profit he also has never suffered a loss. Two months ago, in January 1986, he received an offer of forty million dollars from a large wood products firm for his equipment, land and mill. The prospective purchaser recently lost a large tract of land to the government, for expansion of the Redwood National Park, and is primarily interested in Jim's land because it has a large quantity of high quality, old growth redwood trees. The purchaser would most likely close down the manufacturing facility. During the past forty years, most of the easily accessible (and therefore inexpensive) stands of redwood trees have been harvested. The large remaining acreage of virgin Redwood trees requires a total of ten million dollars to build all the necessary access roads.

The mill is old and in need of substantial improvements. One engineering firm estimated that it would cost thirty million dollars to turn the production facility into a modern, high volume, lumber manufacturing plant. The engineering report advised that due to the age of the existing equipment it would be inadvisable to replace it one section at a time.

Most of the fifty-five non-union mill workers have been with the company for a long time. They have always allowed Jim to adjust salaries to ensure the firm's profitability as demonstrated by their acceptance of pay cuts during a recent four year long recession. While the equipment in the mill is mostly obsolete, the workers have managed to maintain operations by their ability to fix almost anything that moves. They are extremely worried about the closure of the mill. Organizers from the International Woodworkers of America are exacerbating the workers' concern while lobbying for a union ratification vote.

The softwood lumber industry has recently recovered from the worst recession since the depression of the 1930's. This recession has resulted in over fifty permanent closures of lumber mills in Oregon and Northern California from 1980 to 1985. Redwood and Cedar prices, normally recession proof, dropped substantially since the halcyon days of 1978–9.

There is guarded optimism over future demand and price levels for all softwood lumber. Since demand for redwood and cedar products is expected to exceed dwindling U.S. supplies of redwood logs, there is a consensus that prices will be forced higher for all redwood products. However, there are two potential, competitive threats to increasing demand for redwood lumber: 1) increasing production and use of treated Southern Yellow Pine for exterior decks and 2) increasing imports of Canadian Red Cedar. Western Red Cedar imports to the United States from British Columbia, Canada have increased from 704 million board feet (MMBF) in 1983 to 867 MMBF in 1985. This growth of imports is expected to continue since there is no shortage of Cedar logs in British Columbia.

Jim had always hoped that one of his two children would take over the business. Bob, his son, after completing a graduate degree in civil engineering in 1975 has been employed by the Oregon Department of Transportation to design structural support systems for wood and steel bridges. Alice, his daughter, is four years younger than Bob and completed an MBA three years ago, after divorcing her childhood sweetheart. She presently works in the personnel office of a large forest products firm in Oregon. Her main function, since starting with the company, has been to oversee extensive layoffs resulting from mill closures and/or modernization programs.

Both children are enthusiastic about their own ideas. They both feel that there is no future in

[1]This case was written by David H. Cohen, Assistant Professor of Forest Products Marketing, University of British Columbia, Vancouver, B.C.

continuing to produce low-grade, commodity lumber products. Bob wants his father to sell the land and invest part of the proceeds into producing engineered wood products for structural use in non-residential construction. He wants to create a company that will serve an industrial market. Alice wants her father to use the land as collateral to borrow money to build a new manufacturing facility that would produce high quality redwood boards for the Home Center market. She wants to turn the company into a consumer-oriented marketing company. Jim is unsure whether to follow the advice of either.

BOB'S PROPOSAL

Bob has always been convinced that the mill cannot be profitable without major changes. He has refused to become involved with his Dad's mill because he feels that production inefficiencies, resulting from patchwork modifications of an outdated facility, can no longer be offset by inexpensive timber supply and cheap labor. He also feels that a small company, such as California Redwoods, cannot compete with the larger firms who dominate the market for lumber commodities. He wants his father to sell the business in northern California and set up a facility in southern Oregon (where he knows many of the structural engineers) to manufacture wood structural members for building construction.

He has gathered a lot of material about the growing market for engineered wood products (especially wood I-beams and trusses) in non-residential construction. A wood I-beam is a support member constructed of two strong pieces of wood (2″ by 2″, 2″ by 3″ or 2″ by 4″ in lengths of 8 feet or longer) which are called flanges joined by a ³/₈″ to ⁵/₈″ thick piece of plywood or oriented strandboard which is called the web (see Exhibit 1A). The depths of I-beams range from 7″ to 30″ depending on structural requirements. A wood truss is a support member composed of two pieces of lumber or laminated veneer lumber (with sizes similar to wood I-beams) joined by metal tubes (see Exhibit 1B). The design requirements (i.e., the size of the lumber and metal tubes) are determined by the loads the beam must support.

The solid lumber must have very few defects, such as knots or spiral grain, and often are tested mechanically (machine stress rated—MSR), before use, to ensure that engineering design requirements are above minimum specifications. The cost of installing a machine to stress rate lumber is over one million dollars and slows the production flow of a lumber mill. Presently, the only mills using MSR technology in the Pacific Northwest have contracted to sell most of their production to Trus Joist Corporation, the largest producer of wood structural members in the United States. The only alternative to solid wood for flanges is laminated veneer lumber. Trus Joist Corporation is the only producer of this product in the Pacific northwest. Laminated veneer lumber is a wood composite made up of high quality wood veneers with each layer of veneer having parallel grain direction (as compared to plywood where grain directions are perpendicular between adjacent veneers). There are many west coast mills producing both plywood of sufficient quality to be used as webbing in the manufacture of wood structural members and veneer suitable for production of laminated veneer lumber.

Several industry experts describe the non-residential construction market as being more stable than the residential market as shown in Exhibit 2. There is much talk in the forest industry of the shrinkage of the traditional ''bread'n'butter'' residential market for wood products as the ''baby boomers'' mature and the demand for new housing shrinks. Non-residential construction is mentioned frequently as a potential market for increasing sales of wood products. Trus Joist has not been very successful in breaking into the non-residential market with engineered wood products and the majority of their sales has been to the residential market. However, there has been some industry success in increasing the use of wood trusses in roof framing as shown in

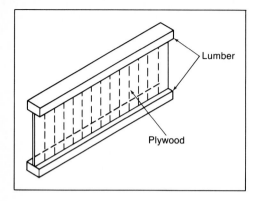

EXHIBIT 1A
Diagram of a wood I-beam.

Exhibit 3. Exhibit 4 shows how important roofs are relative to all the surface area of non-residential buildings. Exhibit 5 shows the acceptance of wood for framing in non-residential buildings, especially in buildings of three stories or fewer.

There are only eleven firms producing wood trusses in the United States. Trus Joist Corporation, the largest of these firms, is based in Idaho and had a profit of almost eight million dollars on sales of 121 million dollars in 1984. Of these eleven firms, only three are located in the Pacific Northwest. Because the design properties of wood I-beams and trusses are proprietary, each company provides engineering support for its product line. This support includes product literature, installation guides, structural analysis and engineering design services. Bob feels that this market has the potential for another firm equal in size to Trus Joist, especially if they concentrate on the non-residential market.

The market for wood trusses is industrial in nature and is different from the present customer groups that California Redwood serves. Compared to consumer markets, industrial markets require increased personal selling and service as well as substantial technical support. Construction firms use wood trusses only if they are specified in the plans provided by architects and structural engineers. When wood trusses are specified, they are designated by brand name about 70% of the time. In the remaining 30% of the cases they are specified solely by the strength properties required by the specific building design. This particular market requires missionary sales to convince architects and structural engineers not only to try a relatively new product in

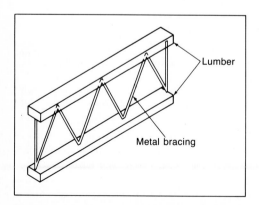

EXHIBIT 1B
Diagram of a wood truss.

EXHIBIT 2
VALUE OF NEW U.S. CONSTRUCTION
(In 1977 U.S. Billions of Dollars)

Year	Residential	Non-residential
1964	65	48
1968	60	59
1972	87	51
1976	64	41
1980	61	51
1984	83	55

Source: Bureau of the Census, *Current Construction Report,* 1980 and 1985.

innovative ways but also to specify products from a new manufacturing company. Bob would have to convince potential customers that his product is superior, in some manner, to his more established competitors. These competitors include not only other firms producing wood trusses and I-beams, but also the steel and concrete industries.

Wood structural members have several advantages over concrete or steel, especially in light industrial or commercial buildings such as small factories or strip malls. Compared to concrete or steel, they have a lower strength to weight ratio, are visually more pleasing, are more easily formed in curved shapes, and are less expensive to ship. In addition, the long distance of steel mills from the Pacific Northwest creates a substantial price advantage for wood members over steel in the western U.S. region. However, the majority of structural engineers and architects who design non-residential buildings are often totally unaware of the possibilities of using engineered wood products for structural support members.

Bob feels that, because of his structural engineering background, he could convince many of the local engineers to change to wood trusses from their present preference for steel, concrete and laminated beams. He also feels that he could convince firms already using wood trusses to specify his company's products, since he already knows many of the structural engineers—Bob feels that the new company's strongest asset would be his own personable manner, his contacts and his engineering knowledge. Most wood trusses are purchased directly from the manufacturer and few

EXHIBIT 3
ROOF FRAMING IN NON-RESIDENTIAL CONSTRUCTION BY PROJECT SIZE AND MATERIAL

Project size	Metal	Concrete	Wood truss	Lam. beams	Other
			%		
Small	50.3	3.2	29.4	16.0	1.0
Medium	61.5	2.0	20.4	15.2	0.9
Large	72.1	4.9	6.2	16.8	0.0
Average	60.1	3.4	19.8	16.0	0.7

Note: Small = < 10,000 sq. ft., medium = 10,000 to 50,000 sq. ft., large = > 50,000 sq. ft.
Source: U.S.D.A. Forest Service, Bulletin FPL 15, Jan. 1985.

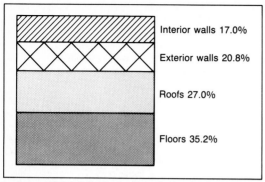

EXHIBIT 4
Building components (1982) for surface area of nonresidential buildings.

Source: U.S.D.A. Forest Service. Forest Products Lab Bulletin FPL 15, Jan. 1985

wholesalers carry them in stock. This requires any manufacturer of engineered wood products either to carry a large inventory or to produce to order. There are costs associated with both options; either high inventory costs or flexibility in production volumes along with delays in delivery times.

Bob envisions a company that would produce none of the lumber, plywood or metal components of the final product but would solely design, manufacture and sell the finished products (i.e., I-beams and trusses). He would like Jim to help in staffing and running the new facility and Alice to help with the marketing and management while he concentrates on designing the products and sales. Bob does not have exact costs on the installation of a manufacturing facility, he has a rough estimate of the costs required to set up this facility and get it operating until it is self sufficient (see Exhibit 6). These costs include the installation of the necessary equipment to produce laminated veneer lumber. Annual sales potential would be ten million dollars, based on existing prices for similar products from Trus Joist. Annual costs are estimated at 4.8 million dollars. This includes purchases and inventory of component materials and labor costs for 23 mill employees and 7 management and sales personnel. It does not include interest or taxes.

Bob wants to locate the first facility in southern Oregon for three reasons: 1) there are

EXHIBIT 5
NONRESIDENTIAL BUILDINGS, TYPE OF SUPERSTRUCTURE FRAMING
BY NUMBER OF STORIES AND FRAMING MATERIAL

No. of stories	Framing type			
	Wood	Metal	Masonry	Concrete
			%	
1	27	23	39	11
2	40	27	25	8
3	26	32	24	18
4+	0	47	10	43

Note: For every $1,000 of construction average wood use was 26 board feet of lumber, 21 sq. feet of plywood, 0.5 sq. feet of non-veneered panels.
Source: Spelter et al., *Forest Products Journal* 37(1):7–12.

EXHIBIT 6
ESTIMATED COST OF SETTING UP AN ENGINEERED WOOD PRODUCTS FACILITY
(All Figures in Thousands of 1986 Dollars)

Cost of land	4,000	
Site preparation	1,000	
Services	500	
Land and development		5,500
Building foundations and yard	750	
Superstructure	1,250	
Building		2,000
Specialized equipment	9,000	
Inventory of components	500	
Equipment		9,500
Operating expenses for 1 year of construction	250	
Operating capital for first 6 months of operation	250	
Operating capital		500
Total cost of set-up		17,500

sufficient quantities of Douglas Fir and Hemlock trees which are the most suitable west coast species for structural use (redwood is not suitable for structural products), 2) he is familiar with the local engineers which would facilitate sales and 3) the facility would be within reasonable shipping distance of a large potential market (Washington, Oregon and northern California). He envisions adding manufacturing facilities once this facility was successful.

Bob wants his father to accept the 40 million dollars and use some of the funds to design and build a manufacturing facility in southern Oregon to produce engineered wood members. He feels that he could not only design and run this facility, but also convince engineers and architects to buy the final products. He feels that the only opportunity for success in the highly competitive and increasingly concentrated wood products industry is to produce specialized, value-added industrial products for a market not being served by the industry giants. Bob wants to carry on the family tradition of producing wood products but wants to do so in a modern and forward thinking manner that serves an untapped industrial market.

ALICE'S PROPOSAL

Alice wants her father to keep the land, borrow money to construct a new mill facility on the existing site and build roads to access the remaining redwood stands. She wants to produce high quality, kiln dried, redwood boards for the fast-growing Home Center market. She has studied the market (spending her own money to purchase privately produced studies on the Home Center Market) and is convinced that by entering this market quickly, California Redwoods can establish itself as the leading supplier of specialty redwood boards in this value-added market niche.

She feels the market potential in this consumer segment is demonstrated by the phenomenal growth of repair and remodeling (R & R) expenditures shown in Exhibit 7. This growth is partially a result of the age of the U.S. house base (Exhibit 8) and has contributed to the high sales growth of Home Centers (Exhibit 9) since 1975. Home Center sales growth directly affects the market for redwood products since lumber accounted for over 23% of all Home Center sales in

EXHIBIT 7
REPAIR AND REMODELLING EXPENDITURES
(In 1982 U.S. Billions of Dollars)

1971	16
1972	18
1973	19
1974	21
1975	25
1976	29
1977	31
1978	38
1979	42
1980	46
1981	46
1982	45
1983	49
1984	70
1985	80

Source: Home Center Report, Home Center Research Bureau, 1986.

EXHIBIT 8
Age of houses in the United States, 1980.

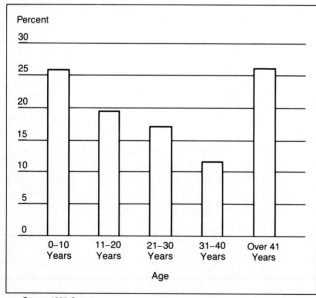

Source: 1980 Census

EXHIBIT 9
HOME CENTER SALES
(In 1982 U.S. Billions of Dollars)

1975	24.3
1977	29.3
1979	37.4
1981	44.8
1983	50.2
1985	62.2

Source: Home Center Report, Home Center Research Bureau, 1986.

1983 and redwood lumber was stocked by an estimated 83% of stores. Growth in Home Center sales is expected to continue until the end of the decade, when experts foresee sales reaching one hundred billion dollars annually. During the recession of 1981–84 this market continued to grow and is now considered almost recession proof.

Home Centers purchase their lumber products from a wide variety of sources. The largest three sources, along with the proportion of dollar value purchased from each, are: 1) Building materials distributors—33%, 2) Direct from manufacturer—27% and 3) Manufacturers distribution centers—18%. Because Home Centers purchase over 25% of their lumber direct from manufacturers, Alice feels that she could sell directly to the larger Home Center chains and increase the profit for her firm by capturing those profits now earned by the wholesalers. She wants to target the largest 5% of Home Center chains (all with at least 5 stores) that accounted for over 50% of total sales in 1984. These firms are easily identified, relatively few in number (for Head Offices), and accessible. They have retail stores throughout the United States and purchase decisions are often made at Head Office. She envisions a focused promotional campaign that would include the use of a brand name, magazine advertising (in magazines such as *Fine Woodworking*) and direct mail to convince the end users to ask for her product from the Home Centers. She wants California Redwoods to develop a new direction as a consumer-oriented, marketing organization that specializes in high quality products aimed at market niches.

Alice feels that California Redwoods could establish a market niche as a producer of high quality, high priced redwood lumber for the Do-It-Yourself builder who shops at Home Centers. To date California Redwoods has produced green, redwood lumber. Half of the production has been remanufactured, by other firms, into lattice, fencing, paneling, decking, trim and a host of other products. She wants to concentrate on producing a finished product consisting of kiln dried architectural grades of redwood lumber targeted for predominantly interior use as decorative items, panelling, molding and trim work. Exhibit 10 shows the existing grades and uses for redwood lumber.

Customers of Home Centers tend to be more concerned with quality and convenience than purely with price. Often they are hobbyists, small professional building contractors or Do-It-Yourself home repair people. It is a market that often will pay a premium price for a premium product. This premium product is often defined by quality, appearance and broad selection of sizes (lengths and widths). Home Center consumers tend to purchase in small quantities. They purchase boards by the running foot, not by thousands of board feet. This market is presently ignored by many of the larger wood products firms, whose traditional market of operations has been in large volume sales of commodity products used in residential construction.

The larger Home Center chains have considerable power over their suppliers. The attractiveness of this market has resulted in fierce competition among small and medium sized manufacturers of wood products as well as building supply wholesalers. They are trying to capture as much of this market as possible through volume discounts, breadth of product line and increasing the service attributes of their product lines with brochures, design plans and pre-packaged kits.

Alice's plan would require a substantial investment in the manufacturing facility, road construction and a marketing campaign designed to enter and control this market niche. She estimates that it would cost one million dollars to clear the existing building of old equipment, repair the superstructure and improve the land for storage of logs and lumber. Manufacturing equipment would cost nine million dollars. It would cost five million dollars over two years to provide the initial road access to the high quality redwood trees. She estimates a start-up budget of 500 thousand dollars with an additional one million dollars required to develop and implement a marketing plan for the first two years. Annual operating expenses would be approximately nine and a half million dollars and includes an annual marketing and sales budget of one million dollars.

The total investment for this mill would be less than the engineer's estimate to revamp the existing mill since annual production volumes would be substantially smaller. In addition, Alice

EXHIBIT 10
REDWOOD GRADES

ARCHITECTURAL GRADES:

CLEAR ALL HEART
Kiln dried heartwood with one side clear of any defects and the other side with only minor defects. Can be smooth surfaced or textured.

USES:
Used when appearance is critical such as paneling, cabinetry, trim and molding.

CLEAR
Same as Clear All Heart but with some sapwood and additional minor defects allowed.

USES:
Similar to Clear All Heart but also exterior above ground use.

B GRADE:
Contains a larger proportion of sapwood and more defects are allowed.

USES:
Similar to the above two. Used extensively for exterior decking.

GARDEN GRADES:

CONSTRUCTION HEART
Contains knots and other defects but no sapwood.

USES:
Decks, posts, fences, exterior stairs and garden uses.

CONSTRUCTION COMMON
Similar to Construction Heart but with some sapwood.

USES:
Above ground exterior use that does not require high degrees of decay resistance.

MERCHANTABLE HEART
Allows larger knots than Construction Common.

USES:
Fences, retaining walls and other exterior, decay resistant uses.

MERCHANTABLE
Unseasoned economy grade with sapwood.

USES:
Temporary exterior uses.

Source: Crow's Digest.

does not envision a high-speed manufacturing line producing commodities, as the engineering report investigated.

She expects three grades of boards to be produced from the mill. The poorest grade would be unseasoned (i.e., not kiln dried) lumber, two inches thick, in varying lengths and widths. Approximately ten million board feet of lumber would be produced from the lowest quality raw material and would sell for an average of $250 per thousand board feet (MBF). Approximately five million board feet of kiln dried, standard quality boards would be produced and would sell for approximately $360 per MBF. The best raw material would be made into high quality kiln dried boards for the Home Center market. These boards have no existing industry designation and are not presently available to consumers. She estimates ten million board feet could be produced and sold at an average price of $1,000 per MBF.

Alice feels that her skills from her MBA and work experience are sufficient to plan and implement this new marketing direction while her father's experience in manufacturing could help set up and run the manufacturing facility. Experienced mill workers from the existing facility would help ensure a well trained and motivated work force. Alice wants to carry on the family tradition but do so in a dynamic manner which would specialize in dominating a profitable, consumer-oriented market niche.

JIM'S DILEMMA

Jim was happy to log his timber, operate his mill, drink beer with his workers and dream of his children taking over the business some day. The past five years have been a rude awakening for him. Logging was getting complicated and expensive (the wood was disappearing); he had to cut wages at the mill to avoid having to borrow money to operate; the union was making a run at his long term employees; and now a big company wanted to give him more money than he ever dreamed of to give up his business. His doctor has recommended that he reduce his work load because of a weak heart but he isn't ready to completely retire yet. His children's recent interest has rekindled his life long hope that his children carry on the family tradition in the wood products industry.

Bob wants to start this new-fangled truss company, and not even in the home town. What would his workers do? Everything costs so much money and most of the people he knows haven't heard of wood trusses. But Bob had an answer to everything.

Alice wanted to keep the company going but change everything. She wanted to borrow millions of dollars, just to keep operating. He had never borrowed over a million dollars in forty years of operations, despite his excellent credit rating; his workers could always fix the equipment to keep it operating without spending big bucks. But Alice had a whole lot of charts and graphs, just like Bob, and at least his workers could keep their jobs.

Jim wishes that things could be like they were six or seven years ago and that his kids wanted to run things just the way he did. What Bob says about industrial markets makes sense but no more so than what Alice says about a consumer-oriented market niche. He is worried that if either scheme failed his good reputation within the industry would be damaged.

He is especially concerned about distribution and sales, since, except for one terrible trial with a sales agent, he has always sold his mill's production to large companies which then distributed the lumber. By developing long term relationships with the buyers of several major companies he has always been able to get a good price for his production. He wants to know how Bob plans to distribute his industrial product and how Alice plans to distribute her consumer product. He has asked each of his children to tell him what skills are needed to succeed in each of these two markets and why they feel that their proposal can succeed, given the skill requirements. He also has asked them to tell him what is wrong with continuing to operate as he has always done, produce basic redwood stock and put it on the commodity market to be sold at the going rate.

With only two weeks left until he has to accept or reject the offer, Jim needs to make a decision soon.

LUMBER SOUTH, INC.[1]

Strategy and Planning for a Medium Size Integrated Firm

Andy Anson, Finance Manager and one of five partners in Lumber South, Inc., was returning home on Delta Flight 106 after visiting some commercial bankers in Chicago. As he stared out the plane window, he knew that his firm must make some changes or stay a weak business. His four other partners were all arguing for different changes that would alter the business. Andy felt the pressure of being a referee while trying to hold the partnership together.

DESCRIPTION OF COMPANY

Lumber South, Inc., is a privately held lumber and plywood producer located in the southeastern United States. Its annual sales grew from $66 million to $100 million in the last 5 years. It is a primary manufacturer of 3 basic product lines: southern pine dimension lumber, hardwood lumber and southern pine plywood. It owns very little of its own timberlands and purchases over 90% of its wood requirements on the open market.

The company operates in the Southeast and mid-Atlantic with mills located throughout the region. It produces no secondary products. The softwood lumber and plywood are sold mostly to brokers/wholesalers which in turn sell to retailers or remanufacturing facilities. The hardwood lumber is sold to brokers/distribution yards which in turn sell to furniture manufacturers, cabinet shops, flooring mills and retailers. The company cut its sales staff in the last recession and targeted larger wholesalers/distributors to reduce selling costs. They still target large wholesalers. Exhibit 1 shows the last 5 year's sales by product type and region.

The company divides its sales regions into 3 geographic areas. The southern region is composed of Florida, Alabama, Georgia and Mississippi. The southeastern region is made up of South Carolina, North Carolina and Tennessee. The mid-Atlantic is made up of Virginia, West Virginia and Pennsylvania.

While sales have increased through the years, profits have remained stagnant. Exhibit 2 shows the lack of profit improvement despite increases in sales. When prices for their products increase, there is a corresponding increase in the cost of their raw materials and manufacturing which squeezes out potential increases in profit. Lumber South owns very little timberlands as a result of a decision made years ago to concentrate its limited capital on manufacturing facilities and buy open market timber. Timberland prices have now risen so high that management believes it is not financially possible to purchase a private timberland base for Lumber South in the foreseeable future.

Softwood manufacturing facilities are a combination of older, relatively inefficient dimension mills, more modern "state-of-the-art" stud mills and an outdated plywood plant. Hardwood lumber is produced at 12 smallish (4 to 7 MMBF) labor intensive mills.

The last recession drained the capital reserves and increased the firm's debt load at a time

[1]This case is co-authored by Dr. David Cohen, Assistant Professor of Forest Products Marketing, University of British Columbia, Vancouver, B.C.

EXHIBIT 1
LAST 5 YEARS' SALES FOR LUMBER SOUTH, INC. (IN $ MILLIONS)

	5th year ago	4th year ago	3rd year ago	2nd year ago	Last year
SYP plywood					
South	9.3	9.5	16.7	16.9	17.9
Southeast	5.9	6.3	9.2	9.1	7.3
Mid-Atlantic	3.1	2.8	4.5	4.4	3.9
Total	18.3	18.6	30.4	30.4	29.1
SYP lumber					
South	14.4	15.3	17.5	18.0	19.3
Southeast	6.0	6.4	7.3	7.5	8.0
Mid-Atlantic	3.6	3.8	4.4	4.5	4.8
Total	24.1	25.5	29.2	30.0	32.2
Hardwood lumber					
South	9.6	10.1	14.2	14.6	15.7
Southeast	9.6	10.0	14.3	14.6	15.7
Mid-Atlantic	4.8	5.1	9.0	7.3	7.8
Total	24.0	25.2	37.5	36.5	39.2
TOTAL	66.3	69.0	97.1	96.9	100.4

when markets and manufacturing technologies were changing. Lumber South was to some extent trapped into selling its products in large lots to large wholesalers/distribution yards. The lack of capital for mill modernization and limited marketing staff has kept them producing mostly a commodity grade product line.

PRINCIPAL PERSONNEL

The company's management team consists of the 5 principal owners of this closely held corporation. Each one is active within the company and controls one aspect of operation. The principal owners and their responsibilities are as follows:

EXHIBIT 2
PROFIT FOR LUMBER SOUTH, INC., LAST 5 YEARS (IN $ MILLIONS)

	5th year ago	4th year ago	3rd year ago	2nd year ago	Last year
Net profit					
SYP plywood	0.2	(4.0)	1.6	1.4	0.9
SYP lumber	0.5	0.2	0.4	0.2	0.4
Hardwood lumber	0.5	0.4	0.6	0.7	0.9
Total	1.2	(−3.40)	2.60	2.30	2.20
Profit as a % of sales	1.81%	(−4.93%)	2.68%	2.38%	2.19%

John Barlow—Sales Manager
Tom Wilkins—Manager, Hardwood Lumber
Bob Jenkins—Manager, Softwood Lumber
Ken Smith—Manager, Plywood
Andy Anson—Manager, Finance

Ken Smith joined the company 40 months ago as an equal partner in return for the addition of his privately owned plywood plant to the company facilities. At the same time, they sold off their older, smaller plywood plant to reduce their debt and finance charges resulting from the recession. Ken's plant, while "state-of-the-art" when it was built in the early 1980's, is no longer considered a modern plywood facility.

MANAGEMENT ISSUES

Andy knows that all the partners want to improve profit margins and increase sales. Status quo operations are no longer acceptable to any of them because of the low profit margins. He found a Chicago bank willing to consider a strong proposal for additional capital for expansion or modernization. But, now the partners must agree on what direction the company should take. Each one believes that their area of responsibility is the one to expand. The bank will insist on a well thought out proposal, not the kind of "by guess and by golly" advice Andy is receiving from his partners.

John Barlow—Sales Manager

John wants to improve the profit picture by increasing the sales staff and selling the existing products for more money. He feels that by de-emphasizing large distributors and selling to channels closer to the end user the company's sales and profit figures can increase dramatically without the need to increase production. He is not sure whether to increase the geographic region of sales. The Northeast and the export markets are two possibilities for expanding the market. John also wants to explore the possibility of producing more specialty products with their higher margins.

John proposes to add 5 additional sales people. Two each for softwood and hardwood lumber and 1 for plywood. These people would be given the task of developing specialty product markets and markets closer to the end user. After one year if this proves successful, he wants to add an export market development manager. John estimates start-up expenses for these additional staff at $500,000 over a two year period.

Tom Wilkins—Manager, Hardwood Lumber

Tom feels that the increasing percentage of the company's sales in hardwood lumber supports his desire to expand the production of hardwood lumber. He would like to acquire several more smallish hardwood mills for the company and shift the company into primarily a hardwood lumber producer. He is not sure whether to produce more finished hardwood products. Presently the company sells only green, rough sawn lumber. Kiln dried hardwood lumber currently sells at a substantial premium to green lumber; however, many lumber users are adding kilns and this premium is likely to decline. The export market for high grade, kiln dried hardwoods is booming but, with no kilns and no export marketing expertise, Lumber South currently does no direct exporting.

EXHIBIT 3
FIVE YEAR PRODUCER PRICE INDEXES* OF LUMBER SOUTH'S PRODUCT LINE

	5th year ago	4th year ago	3rd year ago	2nd year ago	Last year
Southern pine plywood	144.3	162.6	153.6	152.6	155.2
SYP lumber	285.9	319.9	319.8	300.7	299.9
Hardwood lumber	262.4	283.7	319.7	307.2	310.2

*1967 or the base year price index equals 100.

Tom keeps reminding Andy that the relative price of hardwood lumber is higher than plywood or SYP (southern yellow pine) lumber and that its price has increased faster over the last 5 years (Exhibit 3). Tom believes the larger number of small hardwood mills makes it more flexible for management. No single mill can dominate the small firm as a mega-buck plywood mill can. Also, with limited timber supply the small mills compete in more limited geographic regions and can avoid some stumpage price competition. However, Tom also notes that Lumber South must spend more to upgrade and modernize these mills than it has in the past. He estimates $200,000 to $500,000 per mill may be needed over the next 3 to 4 years.

Bob Jenkins—Manager, Softwood Lumber

Bob, on the other hand, feels that the future lies in softwood lumber. He wants to expand the production of SYP lumber and maybe enter some secondary processing. He has seen the explosive growth of the CCA (chromated copper arsenate) treated lumber industry and thinks that this may be the expansion opportunity for the company. However, Lumber South would be entering the CCA treated market late compared to other firms and rumors exist about excess treating capacity. The potential legal liability of using preservative chemicals is also unclear. In other markets, he is getting some competition from Canadian MSR (machine stress rated) graded SPF (spruce-pine-fir) in his higher No. 1 grades. Bob knows that the highly competitive pine lumber commodity markets are forcing his firm to take some action. Relative prices have declined for the past 4 years (Exhibit 3) while industry production has increased over 13.5% (Exhibit 4). He has proposed to modernize his mills and upgrade them to better produce more non-commodity

EXHIBIT 4
TOTAL U.S. PRODUCTION OF LUMBER SOUTH'S PRODUCT LINE LAST FIVE YEARS

	5th year ago	4th year ago	3rd year ago	2nd year ago	Last year
Softwood plywood (Billion ft^2/^{3}/$_8$" basis)	15.8	19.5	19.9	20.2	22.1
SYP lumber (Billion Bf)	8.8	10.3	10.7	10.2	11.7
Hardwood lumber (Billion Bf)	8.5	9.9	8.9	10.0	10.9

products. Other firms are producing export flitches and boards or special truss grade material with good success. One smaller firm has specialized in producing so-called home center studs for the DIY (do-it-yourself) market. These studs have no wane and are very straight for home projects.

Bob believes that the SYP lumber market will continue to expand especially in CCA treated products and in engineered wood products. SYP production has grown steadily and he wants to become a larger player in this market. He believes that timber shortages on the west coast will only enhance opportunities in SYP lumber.

Ken Smith—Manager, Plywood

Ken feels that the future lies with plywood. The market has already seen the absorption of waferboard and OSB (oriented strandboard) and he believes the customer has demonstrated a preference for quality plywood. Once the older mills in the Pacific Northwest close down he sees a bright future for southern pine plywood. He wants to increase capacity in plywood and especially plywood specialties in preparation for good sales and profits in the near future. He is not sure what to do about the rising cost of pine peeler logs, and knows that a substantial capital investment will be required to allow the plywood mill's manufacturing costs to become competitive again.

Estimates are $2 to $4 million to bring the plywood plant into a competitive state and considerably more to get state-of-the-art production. The lower cost OSB producers have held down relative plywood prices as seen in Exhibit 3. However, some structural plywood producers have switched to decorative panels and other pine plywood mills have experimented with producing high quality yellow poplar panels. Numerous options are available which could allow the plywood mill to be more profitable.

SUMMARY

Andy kept having the various arguments go through his head. He just did not know where expansion should occur, but he knew they only had enough money to try one or the other and not everything. He did know, however, that they must do something.

When he hit the office, he knew the other four guys would descend on him with their plans and arguments. They would lobby him for his support as the management team tried to make a decision. He needed to make some decisions before the plane landed.

LEWISTON-COPELAND COMPANY[1]

Determination of Strategy and Planning at the Divisional Level in a Forest Products Company

"We would like to see your division come up with a statement of objectives and strategy which is consistent with that of the corporation as a whole," said Mr. Crawford. "In addition, you should lay out both the short-term and the longer-range plans for your division in as much detail as you can within the limited time available to you." Thus, Mr. Charles Crawford, president of

---

[1]This case is reproduced with the permission of its author, Dr. Stuart U. Rich, Professor of Marketing and Director, Emeritus, Forest Industries Management Center, College of Business Administration, University of Oregon, Eugene, Oregon.

Lewiston-Copeland Company, summarized his instructions to Mr. James Boyd, vice president in charge of the Copeland Paper Division.

Mr. Boyd, along with most of the other top executives of the company, had been attending a three-day long-range planning session held at a resort motel about 25 miles outside of Evergreen City, headquarters of the Lewiston-Copeland Company, and a community of about 150,000 population located in the Pacific Northwest. This three-day session was the culmination of numerous meetings and discussions extending over a two-month period during which the company was trying to embark on a formal system of long-range planning. During the three-day planning session, overall corporate objectives had been agreed upon and general strategy guidelines had been set for the company, although none of the specifics of planning had been spelled out.

Toward the end of the planning session, the suggestion was made that rather than launching all of the company's ten divisions on any type of formal long-range planning, one division should be taken first as a ''pilot division'' and should go through a ''complete planning performance.'' This division was to draw up a statement of divisional objectives and strategy, decide on the appropriate time frame for its plans, consider the roles of division planners versus corporate planners, and draw up some actual plans and programs.

After a relatively short discussion on the matter, the Copeland Paper Division was chosen to be the pilot division. The top corporate executives, whose experience had been mainly in timber, lumber, and plywood, felt that the Paper Division was a good one to start with because the complexity of its products and markets was not as great as in other divisions, and because Mr. Boyd had had long-range planning experience at another firm before joining Lewiston-Copeland several years earlier. He also had a reputation as a self-starter, and displayed a zest for achievement in whatever job he undertook. Although some of his division executives were a little dubious about this honor bestowed upon their division, Mr. Boyd tackled the job with accustomed enthusiasm, determined to make his division a showcase of long-range planning during the three-month period allotted him.

COMPANY BACKGROUND

The Lewiston-Copeland Company was a large manufacturer of forest products with manufacturing, converting, and distribution facilities located throughout the United States as well as in Canada and abroad. Its sales were divided up among lumber, plywood, particleboard and other panel products, logs and timber, paperboard, containers and cartons, pulp, paper, and real estate and housing. The company was organized along product divisional lines, with ten divisions coming under four different business groups. The Copeland Paper Division, along with the Pulp Division, and the Paperboard and Container Division, came under the Fibre Products Group.

All of the divisions were separate profit centers, and some of the product divisions were the raw material sources for other product divisions. For instance, the Pulp Division furnished pulp to the Paperboard and the Paper Divisions as well as selling to outside customers. For its own raw material needs, the Pulp Division bought pulpwood from the Timber Division and wood chips from the Lumber and Plywood Divisions.

The company typically had millsite integration. For example, near one of its large tree farms there might be a manufacturing complex consisting of a lumber and plywood mill and a pulp mill. The latter used logs not suited for lumber and plywood manufacture as well as waste wood in the form of chips from the lumber and plywood mills. At the same site there might also be a paperboard mill and a paper mill using the output of the pulp mill.

The company also had two nonintegrated paper mills, that is, paper mills located by themselves, not near a pulp mill or a source of timber supply. Such paper mills, which were relatively old compared with the integrated mills, bought pulp from distant company-owned pulp mills as well as from outside sources.

COPELAND PAPER DIVISION

The manufacturing facilities of the Copeland Paper Division consisted of some 25 different paper machines, located in both integrated and nonintegrated mills in the Pacific Northwest, the South, and the Northeast. The company had both on-machine and off-machine coaters used in the manufacture of the higher-quality grades of printing papers as well as some types of technical papers. An on-machine coater was an inherent part of the paper machine, and coated only the paper produced on that particular machine, whereas a more flexible off-machine coater stood by itself and could be used to coat paper produced by any paper machine, either at the mill where the coater was located, or shipped in from another mill. On-machine coating was a less expensive method of manufacture than off-machine coating.

The average cost of a new paper machine of the type found in the Copeland mills was $100,000,000. A new coater cost $24,000,000. Plant capacity could sometimes be increased at less cost through rebuilding and speeding up old machines, with expenditures of up to $15,000,000. Typically, however, capacity increases in the paper industry came in sizeable increments, and depreciation on machinery and equipment, along with other fixed expenses, represented about 30 percent of total product costs for the types of papers made by Lewiston-Copeland. Raw material costs accounted for 50 percent, and variable labor costs the remaining 20 percent.

The Division's product line consisted of several hundred different types or ''grades'' of paper, ranging from standard uncoated offset printing paper to silicone-coated papers with water-repellent and non-adhesive characteristics designed for special industrial uses. As a first step toward simplifying the business of his division, Mr. Boyd had recently had the product line classified into 32 product groups, which were then combined into 8 product families, based on their marketing and manufacturing affinity. Among these 8 product families were Premium Printing Papers, usually with fancy coatings; Commodity Printing Papers, such as uncoated offset paper; Converting Papers, such as envelope and tablet stock; Communications Papers, including electrographic and other specialized office copy papers; and Technical Papers such as pressure-sensitive papers.

COMPANY OBJECTIVES AND STRATEGY

The basic corporate objective of the Lewiston-Copeland Company, which had been agreed upon during the recent planning meetings, was to ''sustain an average growth rate in earnings per share of 15 percent per year over at least the next 5 years.'' Having set forth this objective, management then analyzed the major strengths of the company, and the nature of the businesses in which the company was engaged, in order to determine an appropriate strategy to achieve its objective.

The company's major strength was its several million acres of timberlands, much of which had been acquired at a cost far below current market value. The timber was a renewable asset, was readily acccessible, and could be harvested and sold as logs or could be manufactured into forest products. The company also had extensive and well integrated manufacturing facilities, a broad product line, and a sound financial condition with a debt-to-equity ratio lower than that of its major competitors.

Management felt that the company was basically in two major businesses: commodities and specialties. Commodities, such as dimension lumber (''two-by-fours'' and the like), were manu-factured to common specifications, with little or no product differentiation, and were sold on the basis of price and service. The commodity product line had a low rate of technological obsoles-cence, a low value added by manufacture, and a low profit margin on sales. Markets were large, worldwide, and well-established. However, markets were also slow-growing, very competitive, and prices might be highly unstable over short periods of time.

Specialties, such as particular types of overlaid paneling, had a high value added by manufac-ture and often carried a high profit margin, although market size was usually small. There was a real or perceived product differentiation; in fact, some specialties were manufactured to individ-ual customer specifications. Product life was sometimes short due to rapidly changing technology, and the timing of market entry was often quite important. The successful management of a specialty business required a quite different set of skills than did the management of a commodity business. This was true whether the skills involved marketing, manufacturing, engineering, research and development, logistics, or resource allocation. The relative importance of these various functions also differed in the two types of businesses.

The basic strategy guidelines of the company which were developed at the recent corporate planning meetings were stated as follows:

1 Build a solid commodity business by improving and increasing the utilization of our raw material base and delivering products and required services to market at the lowest unit cost; and

2 Use this commodity business as a foundation for supporting investments in higher-risk, higher-return opportunities in present and new markets, by new technologies, and in new businesses.

DETERMINATION OF DIVISION STRATEGY

One of the major strategic issues facing Mr. Boyd was the roles to be played by the Copeland Division's nonintegrated mills and by its integrated mills. In the former were old, slow-running paper machines, suited to the manufacture of specialty papers, which included most of the communications and technical grades. These slow-running machines were also well suited for the many trial runs required in the development of new types of specialties. In the integrated mills were large, modern machines designed for high-volume production of commodity papers, including most of the printing grades.

Complicating the nonintegrated versus integrated mill situation was the fact that the more successful specialty papers often developed into commodity papers. This was because their high profit margin and the expanding market demand attracted competitors, and new types of papers were very difficult to patent. As usage developed, their manufacture became more standardized. In order to keep costs down and keep prices competitive, the Division then had to shift the manufacture of these paper grades from the old, slow-running machines in the nonintegrated mills to the modern, fast-running machines in its integrated mills. Technical difficulties sometimes had to be solved in making this shift. After such a shift, the nonintegrated mills often had the problem of filling the gaps in their machine utilization in order to keep the mills profitable.

Another problem which faced all nonintegrated mills, including those of the Copeland Division, was the fact that their raw material costs were about 15 to 20 percent higher than those at integrated mills. This was because in an integrated facility, pulp was fed in slush form from an adjacent pulp mill directly to the paper machines. To supply a nonintegrated mill, pulp had to be shipped in dry form from a distant pulp mill and then reconstituted into a slurry on the paper machines. Because of their location away from any pulpwood supply, it was not feasible to build a pulp mill adjacent to the Copeland Division's nonintegrated paper mills.

About three-fourths of the Division's sales and four-fifths of its profits were from commodity grades made in its integrated mills, with the remaining percentages coming from the specialty grades in the nonintegrated mills. Within these two categories, however, there was a wide variation in profit margins, particularly in the specialty papers. Certain grades of copy papers and release papers were the most profitable of the entire product line. Some of the older specialty papers had shown a loss for several years, but were kept as long as they covered their own direct costs and made some contribution to common fixed costs. The most profitable grades were often papers which had recently moved out of the specialty and into the commodity class, and enjoyed a strong market demand.

In working up a statement of strategy relating to the product lines and mills in his division, Mr. Boyd had decided that there were four major strategic alternatives from which to choose. The first was to operate the Division's existing businesses without investing capital over and above that required for routine maintenance and replacement. If the company ceased being cost competitive in particular grades of specialty papers it should withdraw from those particular products and the markets they served.

The second alternative was for the company to shut down and/or divest itself of all nonintegrated facilities. It should emphasize the commodity markets and should retain only those specialties which could be produced in the integrated mills.

Alternative Three was to provide adequate capital to modernize and speed up some of the machines in the nonintegrated mills in order to be more cost competitive in the fast-growing fields of copy and release papers. Expenditures would also be made to improve the capability of the integrated mills to produce specialties, including copy and release papers.

The fourth alternative was to close one of the company's nonintegrated mills, to drop all but the most profitable specialties such as copy and release papers, and to increase the emphasis on commodity papers.

SHORT- AND LONG-RANGE PLANNING

In determining the appropriate time frame for his division's planning, Mr. Boyd reviewed the type of planning which had gone on in the past. In the Copeland Division, as in other company divisions, planning had been limited largely to an annual sales forecast, production forecast, and financial operating budget, which were drawn up during the fourth quarter of each year for the following year. At the corporate level, although no regular long-term planning had been carried on, special studies were sometimes made extending as far as 15 years ahead. For instance, if the company were considering the acquisition of timberlands, tied in with the construction of production facilities in a new region, a 15-year cash-flow analysis was made, and Copeland or other divisions affected might be called upon to contribute data.

Mr. Boyd knew that there were many activities in his division which could not be fitted into a one-year planning cycle, but which did not require a 10- to 15-year look into the future. These included new product development, establishing positions in new markets, and adding new production capacity. As an example of the latter, the typical time required for a new paper machine to be constructed and brought to full production capacity, with operating problems all ironed out, was three years.

Regardless of which of the four strategy alternatives under consideration were to be adopted, Mr. Boyd knew that plans would have to be drawn up, built around an appropriate time period for their achievement. The Division's annual operating budgets would also have to be made to mesh with any longer-term plans that were adopted.

DEVELOPING PLANS AND PROGRAMS

Before drawing up any long-range plans for his division, Mr. Boyd felt that it would be necessary to distinguish between those planning activities which had to be done at the corporate level and those which were more appropriately left to the division. Obviously the determination of overall company objectives, goals, and strategies was a corporate headquarters responsibility. Major capital investments in new regions or in new lines of business would continue to be planned at the corporate level, although Mr. Boyd felt that the divisions should play a larger role in such decisions than merely furnishing data inputs when requested.

The main planning areas in which corporate versus divisional responsibility had not yet been spelled out were as follows:

1 Identification and projection of economic, social, and competitive trends affecting the Lewiston-Copeland Company;

2 Identification and study of merger or acquisition candidates;

3 Anticipating and securing future resource requirements, and allocating resources to divisions;

4 Improving the utilization of raw materials, including raw materials not currently being converted at their highest potential economic return;

5 Identification and study of specific new product and market opportunities and of the application of new technology to products and processes;

6 Proposal of courses of action to be taken on obsolete, marginal, and unprofitable businesses, facilities, and products;

7 Setting of specific targets for such performance criteria as earnings and market share;

8 Determination of the particular economic and competitive assumptions upon which specific forecasts would be based;

9 Determination of the general format in which plans were to be drawn up, and the manner in which they were to be presented, approved, and implemented.

Mr. Boyd realized that this list was by no means complete, and he was trying to think of other areas which should be included. Some of the planning areas would apply equally to all Lewiston-Copeland divisions, whereas others would apply particularly or uniquely to divisions in the Fibre Products Group, or perhaps just to the Paper Division.

The role which the Copeland Division market managers should play in planning had yet to be decided. There were eight market managers, one for each of the eight product families. These men, who had all worked up through sales, now occupied these recently-created staff positions. In the case of specialty products, several of the market managers still retained some of their old customer accounts. Mr. Boyd considered these men as sales-oriented, but "not well trained in abstract thinking." Their main function was to help to develop the markets in their respective product areas. They studied demand trends and present and future customer needs and tried to determine how the Division could better satisfy these needs. They helped the line personnel to balance customer demand with machine capabilities. They also provided "top-down" sales forecasts, which were combined with the "bottom-up" forecasts of the field salesmen and sales managers to arrive at a composite sales forecast which became part of the Division's annual operating budget.

Some of the Division executives, particularly those in manufacturing, felt that the creation of the new market manager positions had made the Division too sales-oriented. One of these executives, who had been with the company for many years, remarked, "It's nice to look at the market and see what it wants, but in a capital-heavy industry like ours it may be better to look at the paper machines first and see what they can make, and then ask ourselves how we can sell it. This is particularly true in our big new integrated facilities, where there are machine limitations, pulp mill limitations, and fibre species limitations."

CONCLUSION

As a start toward carrying out the directions given him at the recent planning meetings, Mr. Boyd decided to call his division management staff together and discuss with them what the requirements for success were in the commodity paper business and in the specialty paper business. From such a discussion he hoped there would emerge a consensus as to what division objectives were, and what the appropriate strategy should be to accomplish those objectives. The main planning areas noted earlier would have to be examined, priorities set, plans developed, and programs launched.

To enlist the full cooperation of his management staff in what seemed to them an awesome undertaking, Mr. Boyd knew he would have to show them how planning would help make their daily operating jobs easier. Some of the planning activities he would have to do himself and others he could delegate to his subordinates. One of his most important jobs, he felt, would be to provide everyone with a clear sense of direction as they worked their way through the planning process.

McLAUGHLIN FURNITURE INC.[1]
Pricing Decisions in a Small Furniture Manufacturing Operation

Fred Montague has recently been hired to run the newly created McLaughlin Furniture. McLaughlin Furniture Incorporated is a subsidiary of a much larger furniture firm, John Allen Furniture, Inc. The Board of Directors at John Allen believed that they needed a position in the fast growing RTA furniture marketplace to stay in line with many of their competitors. Not wishing to risk the fine name of John Allen Furniture on an RTA furniture line, they formed McLaughlin Furniture and hired Fred Montague as the Chief Executive Officer.

Fred has worked as the production manager for other RTA production operations over a 15 year career in the furniture industry. However, this is his first opportunity to have a free hand in operating a stand alone company. The Board of Directors of John Allen Furniture, Inc., has given Fred clear profit and loss responsibilities for McLaughlin Furniture. Several members of the Board of Directors of John Allen Furniture, Inc., believe that a manager of a subsidiary should be given total responsibility for that operation. Of course, that means Fred Montague will get rewarded handsomely for success and probably fired for failure. In order to sweeten the success portion, the Board of Directors has agreed to provide Fred with a sliding bonus based on the pretax profits of McLaughlin Furniture Inc. The bonus starts at 2.5% of pretax profits and increases $1/2$% each year until it reaches 5% at which point it is capped.

---

[1]This case is co-authored by Dr. Edward Pepke, USDA Forest Service, Northeastern Area State and Private Forestry, St. Paul, MN.

BACKGROUND INFORMATION

RTA Furniture

Originally, RTA furniture was known as KD for knock-down and unfortunately the products produced by many manufacturers came apart almost as easily as they were put together. This generated considerable consumer bias against KD furniture. However, substantial improvements in construction methods and hardware have increased product quality to a much higher level today.

RTA furniture has had a long history in Europe and today controls a substantial share of the furniture market in both Great Britain and Europe. The original introductions of RTA furniture into the U.S. were poorly constructed compared to today's products. These early products were typically vinyl-wrapped particleboard television stands, book cases and microwave carts. They were purchased as short-term, *disposable furniture,* but the value was still reasonably good for the low price paid. The industry soon realized that long-term survival clearly meant increasing quality, design and function of their products.

RTA furniture is now the world's fastest growing segment in the furniture market. RTA furniture is shipped unassembled to eliminate assembly costs, to facilitate transporting and to reduce shipping costs. A portion of these savings and the assembly are then passed on to the consumer. The unassembled furniture is packaged in boxes to minimize freight damage. RTA furniture has come a long way from the previous poor quality knock-down product to become the consumer friendly furniture of the 1990's. This has occurred because of material, hardware and design innovations. Estimates place U.S. RTA furniture sales at between $2–4 billion.

RTA furniture is typically constructed from particleboard covered with colored or printed vinyl or melamine laminants. And, in some higher quality RTA furniture, wood veneers are used. Office furniture, entertainment centers, electronic and computer furniture and storage units are the most popular pieces of RTA furniture purchased. At one end of the market is the intensely competitive low-priced RTA furniture with its good sales record, and at the other end is the high-priced RTA furniture with its limited affordability and smaller, more exclusive market.

RTA Markets and Marketing

Most RTA furniture is sold from the manufacturer direct to the retailer either by the manufacturer's own sales staff or through a manufacturer's representative. The primary retailers which carry RTA furniture are the mass merchants such as Sears and JC Penney along with the discount mass merchants such as Target, Wal-Mart, Roses and K-Mart. Additionally, there are well known specialty furniture retail chains which carry one or more lines of RTA furniture. Such retailers are Storehouse, Workbench, Room and Board and Conrans. Perhaps the most publicized specialty furniture retailer carrying RTA is the Swedish-based Ikea. Ikea is one of the fastest growing furniture retailers in the U.S. and imports most of its products from overseas; however, they have recently begun sourcing some of their product line from North American manufacturers.

Most RTA furniture must be priced significantly below a competing product in the traditional assembled furniture line. This lower price is to compensate the consumer for the assembly function and also to overcome the price resistance still lingering from the poor quality perceptions of knock-down furniture. As prices of RTA furniture approach those of assembled furniture for comparable lines, sales drop noticeably.

It has been estimated that the top five U.S. RTA furniture manufacturers control about 50% of the U.S. market for RTA furniture. These firms have mostly had their plant expansion and growth based upon a low priced particleboard (covered with vinyl or melamine) product and have recently moved into the higher end of the RTA product lines.

McLAUGHLIN FURNITURE INC.

Figure 1 shows the new product which Fred Montague is expected to manufacture at McLaughlin Furniture. This product is an open book case with four adjustable shelves. Fred and the Board of Directors of John Allen Furniture jointly chose the initial product due to its simplicity and broad based demand. Book cases are one of the hottest selling RTA furniture items and they are sold by a wide variety of retail outlets.

Fred decided to produce a medium quality product using particleboard covered with oak veneer. Within this product category, a comparably sized book case of ⅝″ melamine coated particleboard retails for approximately $75. While, on the other hand, a similarly sized book case made of solid oak might retail for $350 at the high end. Fred believed he could successfully manufacture a medium priced book case between these two extremes. Fred Montague's expertise has always been in manufacturing; therefore, he chose not to get heavily into the decisions regarding marketing and distribution of his product line. Consequently, Fred chose to use the manufacturer's representative as his tool for distributing the new book case product. As is typical in the furniture industry, the manufacturer's representative works as an independent sales person selling the company's products with products of other companies to develop a complete product line. The manufacturer's representative is then paid a commission based on 5% of the manufacturer's selling price of the product.

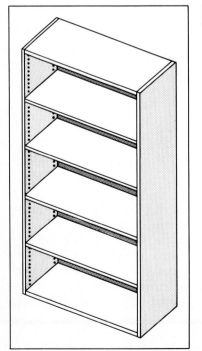

FIGURE 1
McLaughlin Furniture's book case unit.

TABLE 1
LABOR FORCE AND LABOR COSTS

Position	Annual salary or wages
General manager, Fred Montague	$ 45,000
Sales manager	35,000
Office manager and secretary-bookkeeper	25,000
Production foreman	25,000
Panel saw operator	20,000
Panel saw tailer	20,000
Edgebander operator	20,000
Edgebander tailer	20,000
Boring machine operator	20,000
Finishing person	20,000
Packaging and shipping person	15,000
Packaging and shipping person	15,000
Forklift driver and odd-job person	15,000
Total with 25% fringe benefits added	$368,750

TABLE 2
INITIAL CAPITAL REQUIREMENTS

Machinery (panel saw, edgebander, boring machine and finishing equipment)		$ 95,000
Working Capital		
12 months' salaries and wages	$368,750	
Contingency money	100,000	
Receivables	100,000	
Inventory for 26,100 units (100 units/working day)		
Particleboard: oak-veneered,		
5/8-inch thick, 45-pound,		
4-foot by 8-foot (1/unit × $23/board)	600,300	
Fasteners and hardware		
(36/unit × $0.10/piece)	93,960	
Edgebanding (80 feet/unit × $0.018/lineal foot)	37,584	
Catalyzed varnish (128 square feet/unit for 2 coats ×		
$0.0257812/square foot)	86,130	
Shipping boxes (1 box/unit × $2.10/box)	54,810	
Pallets for shipping ($5/pallet × 2/day)	2,610	
Total Working Capital		1,444,144
Forklift and 20 factory trucks		12,000
Miscellaneous tools ($2,000) plus factory supplies ($5,000)		7,000
Office furniture and equipment ($5,000) plus supplies ($1,200)		6,200
TOTAL		$1,564,344

TABLE 3
STATEMENT OF PROJECTED ANNUAL OPERATIONS

Production capacity—unit sales planned		Totals
		24,000 units
Financial plan		
Cost of goods sold		
Materials		
Veneered particleboard	$554,300	
Fasteners and hardware	86,760	
Edgebanding	34,704	
Finishing	79,530	
Shipping boxes and pallets	52,538	
Total materials		$ 807,832
Other		
Labor	$190,000	
Benefits	47,500	
Equipment depreciation	20,800	
Utilities and supplies	27,500	
Total other		$ 285,800
Total cost of goods sold		$1,093,632
General and administrative expense		
Salaries	$105,000	
Benefits	26,250	
Freight	46,200	
Travel	19,800	
Rent	48,000	
Office supplies	1,200	
Depreciation	900	
Miscellaneous expense	10,500	
Total general and administrative expense		$ 257,850

Fred has carefully planned out all the production operations and expenses of the new start-up firm, McLaughlin Furniture Inc. Table 1 shows Fred's estimated labor costs for all employees. Table 2 provides the list of capital requirements Fred estimates for the start-up of McLaughlin Furniture Inc. Table 3 shows Fred's estimated projections for production volumes and costs for 12 months of operation. Fred has had considerable experience at planning and operating furniture operations and the Board of Directors has a great deal of confidence in his abilities.

PRICING DECISION

With the product line, distribution plan and operational costs under control, Fred was left with what was still a very major decision. That is, what price should he set for the book case? The Board of Directors of John Allen Furniture, Inc., realized that pricing was not a strong area of expertise for Fred and while they publicly wished him to make his own decisions in this regard, many Board members have taken Fred aside privately and provided him with advice.

Sam Jones, a successful wholesaler, has urged Fred to keep his pricing system simple and just use a mark-up system. Sam recommended a 20% mark-up above costs. He told Fred that any operation that generated a return of 20% above costs would be in excellent shape.

John Neely, a financial analyst, has encouraged Fred to consider a target return pricing system. John has told Fred to estimate his production for the year and then determine what price he must set to achieve a 20% return on the money which John Allen Furniture has invested in their new subsidiary, McLaughlin Furniture. John reasoned with Fred that a 20% return on equity would signal that McLaughlin Furniture was operating under excellent management. However, John cautioned Fred about using pricing to generate short term profits at the expense of the long-run health of the firm.

Another Board member pulled Fred aside one day and described to him the notion of breakeven analysis and suggested that might be the technique Fred should use in setting prices. By this time, Fred was thoroughly confused and borrowed a book from the library which discussed pricing techniques. Fred read about mark-up and target return pricing methods; however, he also ran into perceived value pricing. The writer of the book argued that the proper price for the product should reflect its true value in the marketplace. Fred did not disagree with this notion but was unsure how he might determine this so-called true value. Fred also discovered that consumers many times use price as a cue to a product's quality. This was an interesting phenomenon to Fred and he began to wonder how high could he set the price and would the ultimate consumers' perception of quality rise along with the price.

Unfortunately, about this time Fred's bubble popped and he began to worry about the competition. If he set his price too high, would he encourage the competition to begin producing book cases like McLaughlin's because of the high profit margins? Yet a very low price to discourage new competitors would reduce the amount of profits Fred got to keep in the short-run but it might boost Fred's bonus in later years by discouraging competitors. Fred didn't know what to do so he kept reading.

Fred began to better understand the role of discounts in pricing products. He made the decision to offer his customers a 2% discount if they would pay their bill within 30 days. Additionally, he began to consider the possibility of quantity discounts for very large orders, but then became concerned over how much of his sales should go to a given firm. For example, a major order to K-Mart could take his plant's entire production for a total year. He wondered if this might give K-Mart too much power over his firm and began to question whether he should set a policy on how much of their production might be sold to any one customer.

Fred, of course, had every reason to maximize profits but mostly because he got to keep a part of them. He believed $350 to represent the extreme top price for a product like his book shelf unit. However, at that price he was afraid his orders would be too limited to operate the plant efficiently. Once production got going he believed they could produce 2000 book shelf units per month. But, if he utilized his plant capacity more efficiently, by adding a night shift, he could produce more than 2000 units per month. Fred's problem was at what price will profits be the highest?

Since furniture retailers as a rule of thumb generally have a 100% mark-up on furniture, Fred knew that the $350 retail price was a selling price of $175 for McLaughlin Furniture. After considering the manufacturers rep's commission and the 2% trade discount this would represent total sales of $3,906,000 annually or a pre-tax profit of $2,587,268. Of course, Fred knew that he could never sell his entire production of book cases at $175 (i.e., the same price as solid oak).

Since Fred had the most optimistic figures already, he decided to calculate his breakeven price to prepare for questions from John Neely and Sam Jones at the upcoming Board meeting next week. Fred also began to call manufacturers reps and visit retail stores to get a better feel for the demand for book cases at various price levels. He came up with the following rough estimates:

Potential sales	at	This wholesale price
>500,000		$ 20.00
250,000		40.00
80,000		60.00
30,000		80.00
15,000		100.00
10,000		150.00
5,000		175.00

Fred now began to prepare his pricing plan for the upcoming Board meeting. He had to make sense of the sales demand estimates and arrive at a pricing method which would maximize profits and keep the Board members satisfied.

JOHN LACY WALNUT COMPANY

Product Line Expansion in the Hardwood Sawmill Industry

INTRODUCTION

Joseph Smith was sitting at his desk in the President's Office of the John Lacy Walnut Company wondering how, at thirty-three years old, he had become the President of his own company. After graduating from college several years ago with a degree in Forest Products, he had worked for a succession of sawmills mostly in the deep south. Rising rapidly he had managed a number of these mills and two years ago found himself employed by a major forest products corporation as the Manager of the John Lacy Walnut Company, a wholly owned subsidiary. The parent corporation decided to close this subsidiary and lay off all of its employees. Joe believed that the subsidiary could be operated as an independent business and be profitable. He made a pitch to corporate management and was able to purchase the subsidiary at considerable personal financial risk.

A key factor in securing the financing for this venture was the Board of Directors Joe was able to pull together for John Lacy Walnut. This Board included a number of powerful and creative people from a variety of backgrounds. A banker, a corporate lawyer, an almost legendary marketing consultant, a member of the Board of Directors of a multi-national paper firm and other experienced business people were on his Board. The new Board provided Joe a source of advice and counsel. They viewed their service as almost fun, being back in a small entrepreneurial start-up situation.

One of their first recommendations was to retain the name, John Lacy Walnut Company, because this firm had been in operation for more than 90 years located in the heart of the midwestern walnut belt. One of the firm's strengths was its reputation as an established, reliable supplier of high quality walnut lumber. Actually, given the smallness of the sawmill (about 4.5 mmbf per year on a one shift basis), the name could be its most valuable asset. Walnut sawmills are a special bunch, with log prices 2 to 5 times that of oak and with log supplies limited they don't tend to be big. John Lacy, even though small by sawmill standards, was one of the biggest walnut producers in the country.

CURRENT PRODUCT LINE

The current product line's profitability was heavily dependent upon premium prices for the higher grades of walnut lumber which had been averaging $1500 to $2000 mbf for FAS kiln dried 4/4 walnut. No. 1 Common had been around $900 to $1000 mbf. These prices had been increased somewhat by the firm's willingness to produce specialty sizes of lumber such as shorts for plaque and trophy manufacturers, and thick specialty pieces for local hardwood specialty stores. However, as with any sawmill, there was only a limited amount of high grade material out of any one log, and lower grade walnut prices dropped drastically with kiln dried No. 2 Common being priced at only 15% to 20% of kiln dried FAS. Therefore, Joe Smith has spent a considerable amount of his time trying to decide how to best market the lower grades of walnut lumber which are necessarily produced. Even with the best logs in his area and a good grade sawyer, Joe's mill typically averaged 15% FAS, 25% No. 1 Common, 29% No. 2 Common and 31% of the lumber below No. 2 Common. A number of potential markets have been explored including pallets, hardwood dimension and flooring which could use the No. 2 Common and especially the below No. 2 Common which almost nobody wanted. The large percentage of lumber in the below No. 2 Common grades was a real problem for Joe and most other walnut mills. Sometimes it was tough to give it away. Joe had just enough kiln capacity to dry the below No. 2 Common, but almost never did because of the poor prices.

PALLETS

The pallet market has the advantage of being able to use large quantities of low grade, unseasoned lumber. Furthermore, pallet production has grown tremendously since the 1960's (see Exhibit 1) and strong demand for pallet stock has pushed up prices. Predictions are for prices to continue to rise considerably faster than the rate of inflation.

The very idea of walnut pallets seemed strange to Joe; however, exceptionally strong demand and increasing prices had him interested. Some investment in cut-up equipment would be needed to convert the lower grades of walnut boards to pallet shook,[1] but with careful coordination between the sawmill and a pallet shook operation Joe believed he could get a 85% to 90% yield from No. 2 Common and below material into pallet shook. Joe contacted the National Wooden Pallet and Container Association and discovered he could secure a list of members in his area as well as product specifications for the shook material.

The prices for pallet shook ($185 to $215/mbf) were not yet high enough to cover Joe's log cost, but then low grade material rarely does. Still pallet shook sales were better than no sales and he reasoned that selling all they could produce would be easy. This would allow them to concentrate on the high end of the market where they traditionally were strong.

HARDWOOD DIMENSION

One of Joe's company Board members suggested that he consider hardwood dimension production as an outlet for his lower quality lumber. A call to the National Dimension Manufacturers Association produced considerable information.

Dimension manufacturers generally operate a cut-up operation to convert lumber into dimension parts. These parts can be rough blanks, semi-machined parts which are carried further in the

---

[1]Pallet shook is 1″ nominally thick boards 4″, 6″ or 8″ wide used on the pallet top and bottom.

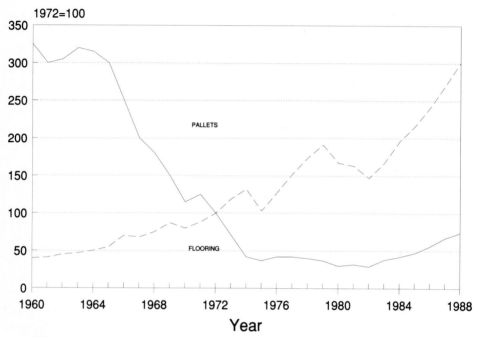

EXHIBIT 1
Pallet and flooring production indexes.

manufacturing process by molding, sanding, trimming, etc., and fully machined parts which are completely ready for assembly into furniture, cabinets, picture frames, etc.

About 50% of the dimension industry's sales are to furniture manufacturers. The remaining half goes to many industries, with the most important being cabinet manufacturers. The U.S. Census Bureau combines hardwood dimension and flooring in SIC 2426 and this SIC category's value of shipments declined in real dollars an average of 3% per year during the last decade. During this same time, 110 or 12% of the dimension and flooring producers went out of business.

Much of this poor business climate can be attributed to strong imports of wood furniture which has limited growth and profits for U.S. furniture producers. In addition, some U.S. industries have begun to import dimension parts for assembly in the U.S. The level of woodworking technology for producing dimension parts is roughly equivalent worldwide, with no nation having a clear advantage. However, a recent survey by the National Dimension Manufacturers Association showed some reasons for optimism. Forty-four percent and 71% of domestic furniture and cabinet producers, respectively, believed their purchases of dimension parts would increase.

Pricing for dimension parts has always been a tricky business. First you have to guess the processing yield, then estimate labor, production and overhead costs, and finally transportation expenses. Joe thought, what a tough process! With most parts being custom produced every job was different and there was no reliable market price sheet to turn to, so he phoned a couple of friends in the business. They confided with him a commonly used rule-of-thumb: multiply the market price of the grade of lumber you use by 4 to get the price for the same board footage of finished parts (i.e., if you buy No. 1 Common at $1,000 mbf then price the finished parts at (4 × $1000) $4000 mbf of parts.

Tough foreign competition, curious pricing methods, a declining market and a significant level of investment; all of these factors were buzzing in Joe's head as he remembered his Board member's advice, "Look into the dimension business. Walnut parts; why everybody needs some to fill out their product line and you can turn that low grade lumber to gold! No one else knows walnut better than John Lacy Walnut. Once our top-of-the line furniture customers for lumber discover we can help them out with dimension parts too, the orders will just roll in!" Joe's thoughts were jumping to hardwood flooring.

WALNUT FLOORING

In his quest for new markets and products to use the lower grades of walnut lumber, Joe came upon the idea of walnut strip flooring. Most strip flooring production uses lower grades of lumber such as No. 2 common and below. Joe reasoned this might be a good match for his lower grades of walnut lumber.

He checked with the National Oak Flooring Manufacturers Association and found that the demand for wood flooring has been improving after reaching a low in 1980 as shown in Exhibit 1. He also found that hardwood flooring production peaked in 1955 at 1.268 billion board feet and by the early 1980's production had dropped below 100 million board feet. Most of the current flooring production was in oak with a small percentage of hard maple and even a smaller percentage of imported exotic woods. After some quick checks, Joe found walnut flooring was available in very limited quantities, but almost always as a special order. Walnut lumber production was also very limited.

A bit of an upswing in demand after years of decline for hardwood flooring had Joe wondering if perhaps wood flooring was making a comeback. With this hope, Joe began to see that while his supplies of lower grade walnut lumber were plentiful they were clearly limited. His mind raced ahead thinking of pricing strategies for walnut flooring. "We only want a small part of the wood flooring business. We'll be specialists in a commodity field, entitled to specialist's profits." Oak flooring was priced at $500 to $1300/mbf fob mill depending on its grade. Joe knew he should price walnut at a premium to the oak. But, how much of a premium?

PRODUCTION AND FINANCIAL CONCERNS

After the image of being a walnut flooring tycoon cleared, reality slammed into Joe. Everything he owned was already tied up in buying the sawmill and his supply of credit at local financial institutions was almost exhausted. All three of the options available would require substantial investment of capital. But, the pallet shook option required much less capital and had a proven, growing market. Joe thought he might talk his banker into this one because of what he thought would be a quicker payback and lower marketing expense.

Joe did some back-of-the-envelope calculations combined with some phone calls to develop the following major capital requirements:

Pallet shook:		
40′ × 30′ building @ $45/ft²	=	$ 54,000
Edger and cut-off saws	=	80,000
Miscellaneous	=	10,000
		$144,000
Dimension parts:		
40′ × 60′ building @ $45/ft²	=	$108,000
Gang rip, cut-off, salvage saws and planer	=	250,000
Miscellaneous	=	10,000
		$376,000
Flooring:		
40′ × 60′ building @ $45/ft²	=	$108,000
Gang rip and cut-off saws, strip flooring machine	=	200,000
Miscellaneous	=	15,000
		$323,000

All three production systems used a basic gang rip, cut-off saw configuration to initiate the lumber breakdown process. However, the pallet shook operation with close coordination with the sawmill could do without the gang rip and use a less expensive edging machine. Joe was told either configuration would process about 10 mbf per 8-hour shift. This would be about 2.5 million board feet of lumber per year on a 1 shift basis. Although for the flooring machine to run an 8-hour shift, the gang rip cut-off saw operation would probably need to be double shifted. Except for this double shift, the production personnel requirements of the 3 systems were very nearly the same. The major differences in the operations were obvious in two areas; space and dry kilns. Pallet shook is processed and sold green, therefore, no additional kiln capacity is needed and outdoor storage of raw materials and finished shook allows a much smaller building.

Of course, 1000 board feet of lumber input to any of Joe's options does not generate 1000 board feet of output. Joe consulted local engineering firms and arrived at a consensus option in processing yields as shown in Exhibit 2. He was very pleased with the very high yields of a pallet shook operation even with a very low grade input.

Joe prepared Exhibit 3 to show his Board of Directors the opportunities available. For example, the 1395 mbf of below No. 2 Common lumber was 31% of the mill's production yet

EXHIBIT 2
LUMBER PROCESSING YIELDS

Grade	Yield of dimension parts	Yield of flooring	Yield of pallet shook
FAS	74%[1]		
No. 1 Common	60%		
No. 2 Common	47%	60%[2]	90%
Below No. 2 Common	35%	50%[2]	85%

[1]Interpret as follows: With a typical order, 1000 board feet of FAS walnut will yield 740 board feet of finished dimension parts.
[2]Includes character marked flooring (i.e., not clear).

EXHIBIT 3
PRODUCTION OF LUMBER AND REVENUES BY GRADE

Grade	Production	Revenues	% Revenues
FAS	675 mbf	$1,181,250	42%
No. 1 Common	1125 mbf	1,068,750	38%
No. 2 Common	1305 mbf	399,656	15%
Below No. 2 Common	1395 mbf	139,500	5%
	4500 mbf	2,789,156	100%

only contributed 5% of the mill's revenue. Joe knew his Directors would push hard to increase that 5% contribution to revenues.

Joe presented his numbers to John Lacy's last Board meeting with a spreadsheet analysis of his debt situation. One member noted the possibility of forming a partnership with a pallet, dimension or flooring firm to supply them with walnut lumber and the John Lacy name. This would be expedient and would solve the money problem, but Joe would give up a lot of control and this was troublesome. An alternative would be to use the other firm's services on a contract basis to produce the products on a custom basis. This could be expensive but would only require the price of the product to cover the costs and would require no further risky leveraging of the company. Joe saw the logic in this idea but was concerned about losing control of the John Lacy Walnut name in one instance and potentially losing control of manufacturing quality in the other.

PRODUCT POLICY ISSUES

A quality image and the reputation of John Lacy Walnut had been carefully developed over a number of years. Joe didn't want to damage this and hurt their profitable high-grade sales. Any of the three new product line extensions would be a new venture for John Lacy and would carry a certain degree of risk to the company image if handled poorly.

MARKETING

Joe believed sales of pallet shook posed no tough marketing problems. He believed his regular sales staff could handle it with no additional people.

Dimension parts could be marketed to furniture and cabinet producers. These folks were known to Joe's sales staff, but given the custom nature of dimension parts a special sales person just for the dimension would likely be needed. Still, it was a market they were somewhat familiar with.

Flooring posed a problem of different proportions. Joe realized that he would probably be producing walnut flooring in relatively small amounts; therefore, he needed to take this into consideration as he was developing his target market. His inclination was to focus on a particular market and do a better job servicing that market than any other supplier. The question was, which market? There were large home center chains in his area selling hardwood flooring. There were a number of specialty hardwood retailers. There were a number of custom floor distributors which had a large clientele and sold typically to contractors and architects, etc. And, last but perhaps not least, was a trend that Joe, and particularly Joe's wife, had noticed among their neighbors and elsewhere. This trend was a movement back to a "country look" with a warmth of exposed wood particularly in the do-it-yourself repair and remodeling markets. Thumbing through magazines recently he had noticed what he believed to be small firms advertising products similar to his in various homeowner magazines such as *Handyman, Country Living, Fine Woodworking,* and others. In the back of his mind, Joe wondered if perhaps positioning his product and targeting it toward a mail order audience might be a good strategy.

Using the traditional flooring distribution channels (i.e., flooring distributors, home centers, etc.) had the potential to give Joe larger initial sales volumes, but the profit margins would be much smaller than the margins on direct or mail order sales. On the other hand, conventional wisdom stated that a significant direct/mail order sales campaign would generate sufficient animosity in the traditional distribution channels to close those channels to Joe in the future should his direct/mail order campaign flop.

Clearly, flooring would require a minimum of one special sales/marketing person on the staff. Joe wondered if expensive advertising would be required.

SUMMARY

The phone rang and Joe's train-of-thought was interrupted. It was his Board member from the paper industry calling to recheck the details of that night's Board meeting. What was Joe to do? He had to explain his strategy for the low grade problem tonight to his Board of Directors. He knew it better be good or they would "eat him alive."

TRUS JOIST CORPORATION[1]

Developing New Uses and Markets for an Established Product

Mr. Mike Kalish, salesman for the MICRO=LAM® Division of Trus Joist Corporation, had just received another moderately-sized order for the product MICRO=LAM laminated veneer lumber; however, the order held particular interest to him. The unique feature of the order was that the material MICRO=LAM was to be used as a truck trailer bedding material. This represented the second largest order ever processed for that function.

Earlier, Mr. Kalish spent some time in contacting prospective customers for truck trailer flooring in the Northwest and Midwest; however, the response from manufacturers had been disappointing. Despite this reception, smaller local builders of truck trailers were interested, and placed several small orders for MICRO=LAM laminated veneer lumber. The order Mr. Kalish just received was from one of the Midwestern companies he had earlier contacted, thus renewing his belief that the trailer manufacturing industry held great potential for MICRO=LAM laminated veneer lumber as a flooring material.

COMPANY BACKGROUND

The Trus Joist Corporation, headquartered in Boise, Idaho, is a manufacturer of structural wood products with plants located in the Pacific Northwest, Midwest, Southeast and Southwest. It was founded by Art Troutner and Harold Thomas who developed a unique concept in joist design, then implemented a manufacturing process for the design.

[1]This case is reproduced with the permission of its author, Dr. Stuart U. Rich, Professor of Marketing and Director, Emeritus, Forest Industries Management Center, College of Business Administration, University of Oregon, Eugene, Oregon.

The majority of salesmen were assigned to the regional Commercial Division sales offices; four outside salesmen were assigned to the MICRO=LAM Division. The functions of selling and manufacturing were performed at each of the five geographically organized Commercial Divisions; therefore, the salesmen concentrated on geographic selling. The MICRO=LAM Division was more centralized in nature, conducting all nationwide sales and manufacturing activities from Eugene, Oregon.

In 1971, Trus Joist first introduced and patented MICRO=LAM laminated veneer lumber. The product is made of thin $^1/_{10}''$ or $^1/_8''$ thick veneer sheets of Douglas fir glued together by a waterproof phenol formaldehyde adhesive. Under exact and specified conditions, the glued sheets are heated and pressed together. The MICRO=LAM lumber, or billet,[2] is "extruded" from specially-made equipment in 80' lengths and 24" widths. The billets can be cut to any customer-desired length or width within those limiting dimensions. The billets come in several thicknesses ranging from $^3/_4''$ to $2^1/_2''$; however, $1^1/_2''$ and $1^3/_4''$ are the two sizes produced regularly in volume.

MARKETING MICRO=LAM

When MICRO=LAM was first introduced, Trus Joist executives asked an independent research group to perform a study indicating possible industrial applications for the product. The first application for MICRO=LAM was to replace the high quality solid sawn lumber $2'' \times 4''$ trus chords[3] in its open web joist designs and the solid sawn lumber flanges[4] on its wooden I-beam joists (TJI). The findings of the research report suggested MICRO=LAM could be used as scaffold planking, mobile home trus chords and housing components. These products accounted for about 25% of the MICRO=LAM production. Mr. Kalish had also begun to develop new markets for MICRO=LAM, including ladder rails and framing material for office partitions.

When marketing MICRO=LAM to potential customers, Trus Joist greatly emphasized the superior structural qualities of the product over conventional lumber. MICRO=LAM did not possess the undesirable characteristics of warping, checking and twisting; yet it did show greater bending strength and more structural stability. (One ad claimed, "Testing proves MICRO=LAM to be approximately 30% stiffer than #1 dense select structural Douglas fir.") In some applications, MICRO=LAM offered distinct price advantages over its competing wood alternatives, and this factor always proved to be a good selling point. Manufacturers were often concerned about the lead/delivery time involved with ordering MICRO=LAM. Trus Joist promised delivery within one to three weeks of an order, which was often a full two weeks to two months ahead of other wood manufacturers.

The industrial application report had also suggested using MICRO=LAM as a decking material for truck trailers. This use became a reality when Sherman Brothers Trucking, a local trucking firm that frequently transported MICRO=LAM, made a request for MICRO=LAM to redeck some of their worn-out trailers. To increase the durability of the flooring surface, the manufacturing department of Trus Joist replaced the top two veneer sheets of Douglas fir with Apitong. Apitong was a Southeast Asian wood known for its strength, durability, and high specific gravity. This foreign hardwood had been used in the United States for several years because of the diminishing supplies of domestic hardwoods. (See Exhibit B.)

The pioneer advertisement for MICRO=LAM as a trailer deck material had consisted of one

---

[2]MICRO=LAM is manufactured in units called billets, and the basic unit is one billet foot. The actual dimensions of a billet foot are $1' \times 2' \times 1^1/_2''$, and one billet is $80' \times 24'' \times 1^1/_2''$).

[3]Trus chords are the top and bottom components in an open web trus incorporating wood chords and tubular steel webs.

[4]Flanges are the top and bottom components in an all-wood I-beam. Refer to Exhibit A.

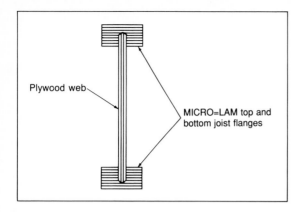

Plywood web

MICRO=LAM top and
bottom joist flanges

EXHIBIT A
End view of an all-wood I-beam.

ad in a national trade journal, and had depicted the MICRO=LAM cut so that the edges were used as the top surface. (See Exhibit C.) The response from this ad had been dismal and had resulted in only one or two orders. The latest generation of advertisement depicting MICRO=LAM as it was currently being used (with Apitong as the top veneer layers) had better results. This ad, sent to every major truck or trailer manufacturing journal as a news release on a new product, resulted in 30 to 50 inquiries which turned into 10 to 15 orders. Approximately 15 decks were sold as a result of the promotion.

Everyone at Trus Joist believed the current price on MICRO=LAM was the absolute rock bottom price possible. In fact, most people believed that MICRO=LAM was underpriced. The current price of MICRO=LAM included a gross margin of 20%. The price of $1\frac{1}{4}''$ thick and $1\frac{1}{2}''$ thick MICRO=LAM was based on the costs of a $1\frac{1}{2}''$ billet. The total variable costs of $1\frac{1}{2}''$ material were multiplied by $\frac{5}{6}$ to estimate the same costs of $1\frac{1}{4}''$ material. There had recently been some discussion over the appropriateness of this ratio. Some of the marketing personnel believed a more appropriate estimate of the variable costs for $1\frac{1}{4}''$ MICRO=LAM would be the ratio of the number of veneers in a $1\frac{1}{4}''$ billet to the number of veneers in a $1\frac{1}{2}''$ billet, or $\frac{14}{16}$. At the present time, the costs of veneer represented 55% of the selling price. Glue cost was

EXHIBIT B
MECHANICAL PROPERTIES OF WOOD USED FOR TRAILER DECKING

Common name of species	Specific gravity	Modulus of elasticity (million P.S.I.)	Compression parallel to grain & fiber strength max. crush strength (P.S.I.)
Apitong	.59	2.35	8540
Douglas fir	.48	1.95	7240
Alaska yellow cedar	.42	1.59	6640
White oak	.68	1.78	7440
Northern red oak	.63	1.82	6760
MICRO=LAM*	.55	2.20	8200

*MICRO=LAM using Douglas fir as the veneer faces of the lumber.
Source: U.S.D.A. Handbook No. 72, *Wood Handbook: Wood as an Engineering Material,* revised edition, 1974, U.S. Forest Products Laboratory.

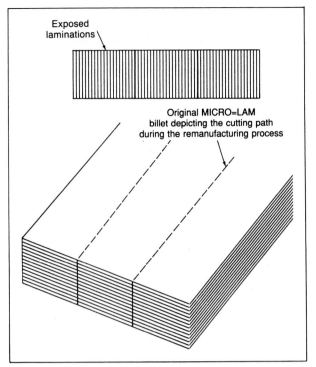

EXHIBIT C
End view of *remanufactured* MICRO=LAM. Original
MICRO=LAM billet depicting the cutting path (– – – –) during
the manufacturing process.

approximately 13¢/sq. foot; fixed overhead represented 14¢/sq. foot; and other variable costs amounted to approximately 12¹/₂¢/sq. foot. The total variable cost was divided by 0.80 to cover all selling and administrative expenses and to secure a profit.[5]

Last year truck trailer manufacturers ordered and used 46 million square feet for installation in new truck trailer construction. This figure was understated because redecking or replacement of worn out floors of trailers had not been incorporated, and there was little organized information to determine what this potential could be.

The problem Mr. Kalish saw with this aggregate data was that it was not broken down into the various segments of trailer builders. For example, not all of the 236 manufacturers who produced trailers used wooden floors. Among those not using wooden floors, for example, were tankers and logging trailers. Mr. Kalish believed the real key to selling MICRO=LAM in this industry would be to determine on what segment of the trailer industry he should concentrate his selling efforts. Mr. Kalish also knew that he somehow had to determine trailer manufacturers' requirements for trailer decking. The Eugene-Portland, Oregon, area offered what he thought to be a good cross-section of the type of trailer manufacturers that might be interested in MICRO=LAM. Some of those firms he had already contacted about buying MICRO=LAM.

---

[5] All cost figures have been disguised.

GENERAL TRAILER COMPANY

Mr. Jim Walline had been the purchasing agent for General Trailer Company of Springfield, Oregon, for the past two and one-half years. He stated, "The engineering department makes the decisions on what materials to buy. I place the orders after the requisition has been placed on my desk."

General Trailer Company was a manufacturer of several different types of trailers: low-boys, chip trailers, log trailers, and flatbeds. General did most of their business with the local timber industry; however, they also sold to local firms in the steel industry.

The flatbeds General Trailer manuactured were 40' to 45' long and approximately 7' wide. Log trailers were approximately 20' to 25' long.

General Trailer manufactured trailers primarily for the West Coast market, although it had sold a few trailers to users in Alaska. On the West Coast, General's major competitors were Peerless, Fruehauf, and Trailmobile, all large-scale manufacturers of truck trailers. Even though General was comparatively small in size, they did not feel threatened, because, "We build a top quality trailer which is not mass produced," as Mr. Walline put it.

General had been using Apitong as a trailer decking material until customers complained of the weight, and expansion/contraction characteristics when exposed to weather. At that time, Mr. Schmidt, the General Manager and head of the engineering department, made the decision to switch from Apitong to laminated fir.

Laminated fir (consisting of solid sawn lumber strips glued together) was currently being used as the material for decking flatbeds, and Pacific Laminated Co. of Vancouver, Washington, supplied all of General's fir decking, so they would only order material when a customer bought a new trailer or needed a trailer redecked. Mr. Walline was disappointed with the two- to three-week delivery time, since it often meant that much more time before the customer's trailer was ready.

Laminated fir in 40' lengths, 11¾" widths and 1¼" thickness was used by General. General paid approximately $2.00 to $3.00 per square foot for this decking.

Even though Pacific Laminated could provide custom-cut and edged pieces with no additional lead time, General preferred ship-lapped fir in the previously noted dimensions, with the top layer treated with a water-proof coating.

The different types of trailers General manufactured required different decking materials. Low-boys required 2¼" thick material, and General used 3" × 12" rough-cut fir lumber. Chip trailers required ⅝" thick MDO (medium density overlay) plywood with a slick surface.

Mr. Walline said they had used MICRO=LAM on one trailer; however, their customer had not been expecting it and was very displeased with the job.[6] Therefore, the current policy was to use only laminated fir for the local market, unless a customer specifically ordered a different decking material. Trailers headed for Alaska were decked with laminated oak, supplied by a vendor other than Pacific Laminated.

Mr. Walline said that if he wanted to make a recommendation to change decking materials, he would need to know price advantages, lead times, moisture content, availability, and industry experience with the material.

---

[6]After purchasing MICRO=LAM, General Trailer modified the material by ripping the billets into 1½" widths and then relaminating these strips back into 12" or 24" wide pieces of lumber. This remanufacturing added substantial costs. Also, the laminations were now directly exposed to the weather. Moisture could more easily seep into cracks or voids, causing swells and buckling. (See Exhibit C.)

SHERMAN BROTHERS TRUCKING

"We already use MICRO=LAM on our trailers," was the response of Mr. Sherman, President of Mayflower Moving and Storage Company, when asked about the trailer decking material they used. He went on to say, "In fact, we had hauled several shipments for Trus Joist when we initiated a call to them asking if they could make a decking material for us."

Mayflower Moving and Storage owned 60 trailers (flatbeds) which they used to haul heavy equipment and machinery. They had been in a dilemma for eight years with the type of materials used to replace the original decks. Nothing seemed to be satisfactory. Solid Apitong was tough but it was too heavy and did not weather very well. Plywood did not provide adequate weight distribution and had too many joints. Often the small wheels of the forklifts would break through the decking, or heavy equipment with steel legs would punch a hole through the decks. Laminated fir was too expensive.

Mayflower Moving and Storage was currently redecking a trailer per week. They usually patched the decks until the whole bed fell apart, then the trailer would sit in the yard waiting for a major overhaul. The trailers by this time needed the cross beams repaired and new bearings, besides new decks.

Mr. Sherman went on to say, "The shop mechanic just loves MICRO=LAM. This is because it used to take the mechanic and one other employee two days to redeck a trailer, and now it just takes the shop mechanic one day to do the same job." Advantages (over plywood and Apitong) of the $2' \times 40'$ MICRO=LAM pieces were: ease of installation, excellent weight distribution due to the reduced number of seams, and reduced total weight of the bed.

Mr. Sherman explained that they usually purchased four or five decks at a time, and warehoused some of the materials until a trailer needed redecking.

Mr. Sherman thought the original decking on flatbeds was some type of hardwood, probably oak, which could last up to five years; however, a similar decking material had not been found for a reasonable price. The plywood and fir decks used in the past eight to ten years had lasted anywhere from one to two years, and some had worn out in as little as six months. After using MICRO=LAM for six months, Mr. Sherman expected the decking to last up to three to five years.

When asked about the type of flooring used in their moving vans, Mr. Sherman emphasized the top care that those floors received. "We sand, buff and wax them just like a household floor; in fact, we take such good care of these floors they will occasionally outlast the trailer." The original floors in moving vans were made out of a laminated oak, and had to be kept extremely smooth, allowing freight to slide freely without the possibility of damaging items of freight with legs. The local company purchased all of its moving vans through Mayflower Moving Vans. The only problem with floors in moving vans was that the jointed floors would occasionally buckle because of swelling.

The fact that MICRO=LAM protruded $\frac{1}{8}''$ above the metal lip[7] which edged the flatbed trailers posed no problem for Sherman Brothers. "All we had to do was plane the edge at 45°. In fact, the best fit will have the decking protrude a hair above the metal edge," Mr. Sherman said. Just prior to this, Mr. Sherman had recounted an experience which occurred with the first shipment of MICRO=LAM. Because the deck was too thick, Mayflower Moving and Storage had about $\frac{1}{8}''$ planed from one side of the decking material. However, the company shaved off the Apitong veneer, exposing the fir. Mr. Sherman said that he laughed about it now, but at the time he wasn't too pleased.

---

[7]Refer to Exhibit D.

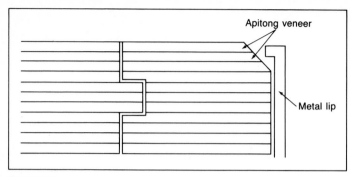

EXHIBIT D
Cross-sectional end view of trailer decking (tongue and groove).

PEERLESS TRUCKING COMPANY

"Sure, I've heard of MICRO=LAM. They (Trus Joist salesmen) have been in here . . . but we don't need that good a material." This was the response of Mel Rogers, head of Peerless' Purchasing Department, Tualatin, Oregon, when asked about the use of MICRO=LAM as a truck decking material. Mr. Rogers, a 30-year veteran of the trailer manufacturing industry, seemed very skeptical of all laminated decking materials.

The primary products manufactured by Peerless (in Tualatin) required bedding materials very different from MICRO=LAM. Chip trailers and rail car dumpers required metal beds to facilitate unloading. Low-boys required a heavy decking material (usually 2″ × 12″ or 3″ × 12″ rough planking) as Caterpillar tractors were frequently driven on them. Logging trailers had no beds.

Approximately 60 decks per year were required by Peerless in the manufacture of flatbeds and in redecking jobs. MICRO=LAM could have been used in these applications, but fir planking was used exclusively, except for some special overseas jobs. Fir planking was available in full trailer lengths, requiring eight man-hours to install on new equipment. Usually, five or six decks were stocked at a time. The estimated life of a new deck was two to three years.

Fir planking was selected for decking applications on the basis of price and durability. Peerless purchased fir planking for $1,000 per MBF. Tradition supported fir planking in durability, as it was a well-known product.

Decking material thickness was critical, according to Mr. Rogers, as any deviation from the industry standard of $1\frac{3}{8}″$ required extensive re-tooling.

Any new decking materials for use in original equipment manufacture had to be approved by the Peerless engineering department. Alternative decking materials could have been used locally if specified by the customer.

Mr. Rogers was certainly going to be a hard person to sell on the use of MICRO=LAM, Mr. Kalish felt. "Why use MICRO=LAM, when I can buy fir planking for less?" Rogers had said.

FRUEHAUF TRUCKING COMPANY

"I'd be very happy if someone would come up with a durable (trailer) deck at a reasonable price," was the response of Wayne Peterson when asked about Fruehauf's experience with decking materials. Mr. Peterson was Service Manager for Fruehauf's factory branch in Milwaukie, Oregon. Fruehauf Corporation, with their principal manufacturing facilities in Detroit, Michigan, was one of the nation's largest manufacturers of truck trailers.

The manufacturing facilities in Milwaukie produced 40-ton low-beds as well as assembled truck bodies manufactured in Detroit. The low-beds were subjected to heavy use, often with forklifts, which required a decking material of extreme strength and durability. Laminated decking materials then available were therefore excluded from this application.

The decking materials used in the truck bodies were specified by the sales department in Detroit, based on customer input. Generally, Apitong or laminated oak was installed at the factory. Any new product to be used in original equipment manufacture had to be approved by Fruehauf's well-developed factory engineering department.

The Milwaukie operation also did about 15 redecking jobs per year. The decking material was specified by the customer on the basis of price and weathering characteristics. The materials used were laminated oak ($11\frac{1}{2}''$ W $\times$ 40′), Apitong (7″ $\times$ $1\frac{3}{8}''$ random lengths), Alaska yellow cedar (2″ $\times$ 6″ T&G), fir planking (2″ $\times$ 6″ T&G), and laminated fir (24″ W $\times$ 40′). Alaska yellow cedar was priced below all other decking materials, followed (in order) by fir planking, laminated fir, laminated oak, and Apitong.

Fruehauf's suppliers of decking materials were as follows: laminated fir—Pacific Laminating, Vancouver, Washington; Alaska yellow cedar—Al Disdero Lumber Company, Portland, Oregon; and Apitong—Builterials, Portland, Oregon. There were no specific suppliers for the other materials.

A minimum inventory of decking materials was kept on hand to allow for immediate repair needs only. Orders were placed for complete decks as needed.

A redecking job typically required 30 man-hours per 7′ $\times$ 40′ trailer, including the removal of the old deck and installation of the new one. Decking materials which were available in full trailer lengths were preferred, as they greatly reduced installation time as well as improving weight distribution, and had fewer joints along which failure could occur.

Use of alternative products, such as composition flooring of wood and aluminum, was not under consideration.

Alaska yellow cedar and fir planking had the best weathering characteristics, while Apitong and laminated oak weathered poorly. Oak and Apitong did, however, have a hard, non-scratching surface that was desirable in enclosed use. When asked about the weathering characteristics of laminated flooring in general, Mr. Peterson responded, "It's all right for the dry states, but not around here."

COMPETITION

There was a large number of materials with which MICRO=LAM competed in the trailer flooring market, ranging from fir plywood to aluminum floors. Trus Joist felt that the greatest obstacles to MICRO=LAM's success would be from the old standard products like laminated fir and oak, which had a great deal of industry respect. For years, oak had been the premier flooring material; recently, however, supplies were short, delivery times long (two months in some cases), and prices were becoming prohibitive. (See Exhibit E.)

Mr. Kalish had found that, in the Northwest, Pacific Laminated Company was one of the major flooring suppliers to local manufacturers. Pacific Laminated produced a Douglas fir laminated product that was highly popular; however, like oak, it was relatively high-priced. Despite the price, Pacific Laminated could cut the product in dimensions up to 2′ wide and 40′ long. Delivery time was excellent for their customers, even with special milling for shiplapped or tongue and groove edges and manufacturing to user thicknesses.

EXHIBIT E
DECKING MATERIAL PRICES

Product	Price	Form
Alaska yellow cedar	$650/MBF	2″ × 6″ T&G 15′ lengths
Apitong	$1.30–$2.00/lineal foot*	1³/₈″ × 7″ random lengths
Fir planking	$1.00/bd. ft.	2″ × 6″ T&G random lengths
Fir laminated	$2.50/sq. ft.	1¹/₄″ × 11³/₄″ × 40′
MICRO=LAM	$1.30/sq. ft.	1¹/₄″ × 24″ × 40′
	$1.50/sq. ft.	1¹/₂″ × 24″ × 40′
Oak laminated	$2.20/sq. ft.	1³/₈″ × 1¹/₂′ × 40′

*Lineal foot = price per unit length of the product.

CONCLUSION

Although Mr. Kalish had had limited success marketing MICRO=LAM to truck trailer manufacturers, he was concerned with the marketing program for his product. Several trailer manufacturers had raised important questions concerning the price and durability of MICRO=LAM compared to alternative decking materials. He knew MICRO=LAM had some strong attributes, yet he was hesitant to expand beyond the local market. Mr. Kalish was also wondering about the action he should eventually take in order to determine the additional information he would need to successfully introduce MICRO=LAM nationally as a trailer decking material. One thought that crossed his mind was the use of a survey questionnaire. He knew that he would also be expected to define the company's marketing strategy for this product. Meanwhile, small orders continued to trickle in.

LANGDALE LUMBER COMPANY[1]

Development of a Marketing Plan for a Recently Acquired Southern Lumber Mill

INTRODUCTION

Stan Sherman, Vice President for Corporate Planning of Clark Industries, Inc., leaned back in his plane seat as a late Friday afternoon flight carried him from Mississippi back to corporate headquarters in a major northern city. He had just spent two rather intense ten hour days at the Langdale Lumber Company, a Southern pine sawmill in a small town in Mississippi which had recently been acquired by Clark Industries, a larger conglomerate. Although Clark Industries had been involved in the pulp and paper field for a number of years, along with a diverse array of other industrial and consumer goods businesses, this was the company's first venture into the lumber business. As Corporate Planning Vice President, it was Stan Sherman's job to oversee the planning activities of the company's various divisions. In the case of newly-acquired firms such as the Langdale Lumber Company, which had never done any formal planning before, this could be quite a task, particularly when neither Mr. Sherman nor anyone else at corporate headquarters had had any working experience in the particular business involved, as in the case of lumber. Just

[1]This case is reproduced with the permission of its author, Dr. Stuart U. Rich, Professor of Marketing and Director, Emeritus, Forest Industries Management Center, College of Business Administration, University of Oregon, Eugene, Oregon.

prior to the Langdale acquisition, Sherman, whose background lay in marketing, had read up on the lumber industry, had collected reports and statistics from government and trade association sources, and had visited the Langdale mill several times. He had also recently talked with several large lumber wholesalers in the metropolitan area where Clark corporate headquarters were located. Right now, however, he was feeling frustrated in his efforts to get local management to use certain basic marketing concepts, such as market segmentation, to draw up a Five-Year Marketing Plan for Langdale.

While Stan Sherman was winging his way northward, Jim Benson, the new Langdale general manager, was sitting in his office at the mill in the late afternoon, after operations had closed down for the weekend, and was pondering over the many meetings he had had with Stan Sherman during the past two days. Mr. Benson's background lay chiefly in manufacturing, but he had also spent some time in log and pulpwood procurement. He had previously worked for a major pulp and paper company, not part of Clark Industries but located in the South. Unlike Stan Sherman, who had lived in the North all his life, Benson had grown up in Mississippi. Although less accustomed than Sherman to the analytical type of work required in business planning, Benson had a good feel for manufacturing operations, and had established a good relationship with the mill workforce. As one of his earlier supervisors at his previous company had remarked, "Jim can make the machines hum, and he has the ability to move among the workers in a mill in the right manner." Benson, who was given profit responsibility for the Langdale mill, currently reported to one of the regional vice presidents of Clark Industries, who was mainly concerned with the company's pulp and paper operations in his region.

Reflecting on the planning job which lay ahead, Jim Benson said to himself, "Basically, the job isn't all that complex. Our marketing plan must answer three questions: Where are we now?; Where do we want to go?; and How can we get there?" What bothered him, however, was the extensive detail work which corporate headquarters seemed to call for, and the time demands this work would place on him and his small staff. He was thinking particularly of his sales staff, which consisted of Mathew Brady, who carried the title of Sales Manager, and Brady's assistant, Miss Alice Perkins, who typed up the sales invoices, handled the sales correspondence, and did some telephone selling herself, chiefly during the infrequent occasions when Brady was away from the office.

Mathew Brady, unlike Sherman and Benson, had been part of Langdale Company's management prior to the acquisition. With Miss Perkins' assistance, Brady had handled the company's sales activities for many years. He had lived in Mississippi all his life, his family was well known in the area, and the Brady family and the Langdale family (the company's previous owners) had been closely connected for several generations. Brady was about fifty, and Sherman and Benson were both about fifteen years younger. Mathew Brady prided himself on the close relationships he had built up with his customers over the years, although it had recently become Stan Sherman's opinion that Brady "often seemed more concerned with protecting his long-time customers than aggressively expanding sales." Brady had attended several of the meetings held during the past two days, although he had spent most of the time listening to the discussions between Sherman and Benson, rather than playing an active role himself. When asked for his views about the need for expanding sales by entering new markets, Brady had replied that he "did not want to go out and talk to new people until he was sure the mill could handle the extra production that might be required and still maintain its reputation for product quality and prompt service." Although the Langdale mill, like many others in the industry, was currently running at below one-shift capacity, Brady felt that when business conditions improved later in the year, the mill would be back up to its optimum level of operation, sustained largely by its regular customers and markets.

COMPANY BACKGROUND

The Langdale Lumber Company, located in Mississippi, had been founded in the early 1930's by Cyrus Langdale to manufacture boards and dimension lumber[2] from Southern yellow pine. During its first 30 years of operation, the company had sold mainly direct to retail lumber yards in Mississippi and neighboring states and, to a lesser extent, to the export market through agents. Many of the sales were by letter and through personal visits, although by the 1960's, the use of telephone WATS lines had largely replaced these earlier selling methods, and the company had expanded by selling to non-stocking (or "office") wholesalers chiefly in the South, but also in the Midwest and a little in the Northeast.

The Langdale Lumber Company, which had been purchased by Clark Industries from the Langdale family about a year earlier, was running one shift and producing at the rate of 100,000 board feet per day. Based on an industry average of 240 workdays per year, this translated into an annual rate of 24,000 Mbf. (Mbf means thousand board feet.) The company's theoretical one-shift capacity was 120 Mbf per day, but the mill had seldom reached that level, even during boom times. The Langdale mill was considered to be in the small to medium-sized range among Southern pine sawmills. Company sales volume was $6,000,000, with a small loss reported, which was not unusual among lumber mills during current depressed industry conditions. Results so far this year had been about the same, but improvement was expected in the second half year, assuming the country started to pull out of the current housing slump.

Besides Jim Benson and Mathew Brady, other supervisory personnel at Langdale included a timber supply forester, a personnel coordinator, a maintenance foreman, sawmill foreman, dry kiln foreman, planing mill foreman, and shipping foreman. Total salaried personnel numbered 12, and total hourly workers numbered 120. Total personnel had been reduced about 10% since the acquisition, as the plant had been considered overstaffed.

The company had two primary manufacturing lines. The band saw line had the capability of producing either boards or dimension lumber, and the chipping saw (or "Chip-n-Saw") line produced primarily dimension lumber.[3] These two production lines gave the mill more flexibility than that available to many competitors which had only one type of production line, usually devoted to dimension lumber. The production costs of these more specialized mills were usually lower than those of a more versatile mill like Langdale.

Over the past year since the mill acquisition, Jim Benson, the general manager, had devoted nearly all of his time to improving the manufacturing efficiency of the mill and installing the many accounting and cost control reports which corporate headquarters now required. Sales and marketing had been left to Mathew Brady, who ran that end of the business much as he had done during his previous 15 years under the old family management. With a certain amount of urging by Stan Sherman, Benson was now starting to get involved in sales and marketing, although he felt that much remained to be done "to bring the manufacturing operation up to speed." For example, a bottleneck sometimes developed in the lumber drying process because of the lack of

---

[2]A board is a piece of lumber less than two inches in nominal thickness and one inch or more in width. Dimension lumber is two inches up to, but not including, five inches thick, and is two inches or more in width. It is also referred to as framing lumber.

[3]A band saw consists of a continuous piece of flexible steel, with teeth on one or both sides, used to cut logs into cants, which are large timbers destined for further processing by other saws (e.g., cut into dimension lumber or boards).

A chipping saw or chipping heading (often called a "Chip-n-Saw" after a particular brand name) mills small logs simultaneously into lumber and chips. The machine chips away the entire outer part of the log and saws the inner part, usually into 2 × 4 dimension lumber.

an adequate cooling ramp leading out of the dry kiln.[4] Similarly, the sorting chain on the planer mill needed to be extended so that the workers could more easily and quickly pull the finished lumber off the moving chain and stack it according to size, length, and grade.[5] The total cost of these two improvements was estimated at about $350,000.

Besides manufacturing, Jim Benson had also spent some of his time developing log supply sources. Although in earlier years Langdale had harvested a substantial portion of its logs from its own timberlands, very little productive timberlands were included in the acquisition, and the company was now more dependent on outside raw material sources than many of its larger competitors. Log costs constituted approximately half of total lumber production costs. To get better product recovery from the logs, the company sold the chips which developed from its chipping saw production line to a nearby pulp mill as raw materials for pulp manufacture.

INDUSTRY BACKGROUND

End Markets

The major markets for Southern pine, and U.S. softwood lumber in general, were residential construction, non-residential construction, repair and remodeling (or home improvement), exports, and industrial uses.

The largest single market for softwood lumber was residential construction, particularly conventional, site-built, detached single-family homes. It was estimated that a site-built single-family house used on the average 10,500 board feet of lumber; a multi-family home used 4,200 board feet; and a mobile home, 2,000 board feet. About two-thirds of the lumber used in homes was framing lumber, and 90 percent of all single-family homes were built with wood frames. The relative importance of the residential construction market varied from year to year, reflecting the strength of the housing market. In recent years, the repair and remodeling market had taken on increased importance. Finally, in terms of regional demand, the South's share of new housing starts continued to increase and reached 58% during the first five months of the current year. Within the South, certain metropolitan areas were currently experiencing strong housing demand even during the severe national housing slump. These areas included Houston and Dallas/Ft. Worth, Texas; Tampa and Orlando, Florida; Oklahoma City and Tulsa, Oklahoma; and Shreveport, Louisiana.

By mid-year, it was clear that earlier projections of housing demand and lumber demand had been too optimistic, although longer-term projections still looked fairly good. Most of the forecasts which Mr. Sherman had recently seen now predicted 1 million to 1.1 million starts rising to 1.3 million next year. Some housing analysts were also challenging the conventional belief that a huge pent-up demand for single-family homes would be unleashed as soon as mortgage rates dropped to 13 or 14% from the current 17%. Revised Census Bureau data showed that the housing stock of the nation was 92 million dwelling units, more than 2 million units greater than previously calculated. Many of these additional units had resulted from the conversion of schools, warehouses, and other non-residential buildings, as well as the repartitioning of older houses, a trend which was expected to continue.

On the demand side, the 25- through 39-year-olds, a key buying group for housing, was increasing. However, federally subsidized housing, which accounted for an average of 240,000

---

[4]A dry kiln is an oven-like chamber in which lumber is seasoned by applying heat and withdrawing moist air.

[5]A planer mill is where rough lumber is surfaced or dressed by a planing machine for the purpose of attaining smoothness of surface and uniformity of size on at least one side or edge.

housing starts per year throughout the 1970's, had been drastically curtailed and seemed unlikely to be a major factor in the future. Finally, the purchase of homes (particularly single-family homes) as an investment had become less attractive because of the availability of new, high-yielding money market instruments, as well as the fact that home prices, which had been increasing at nearly twice the rate of inflation during the second half of the 1970's, had been falling in real terms since then.

Channels of Distribution

Channels of distribution for U.S. domestic markets and for overseas markets are shown in Exhibits 1a and 2a, and brief descriptions of the various channel members are given in the Notes to these two exhibits (Exhibits 1b and 2b). Of the various types of wholesalers in the domestic market, the merchant wholesalers, both stocking and non-stocking, had traditionally been the dominant channel for the distribution of lumber and other construction materials.

Among the merchant wholesalers, those that carried inventory had become the dominant channel in recent years, and many of the non-stocking (or "office") wholesalers had gone out of business. The major reason for this trend was that the high cost of working capital had caused the retail building materials dealers to push the inventory carrying function back to the wholesale level. Retailers preferred to buy in smaller amounts closer to their time of sale, making their purchases from nearby stocking distributors, either independent merchant wholesalers or large manufacturer-owned distribution branches. Lumber mills selling to the large distribution yards of these two types of channel members shipped chiefly straight carloads of lumber rather than mixed cars, which were sold to the non-stocking or office wholesalers.[6] Another trend that was taking place (as portrayed in Exhibit 1a) was a "lateral" channel of distribution from both stocking and non-stocking wholesalers to manufacturers' sales branches and sales offices, and to a lesser extent, to cooperative buying offices. Finally, there were increasing numbers of wholesalers who had become "remanufacturing wholesalers" that bought a product from the mill, altered it, and then resold it to retailers or industrial users. These products ranged from millwork items (door and window parts and decorative trim) to pressure-treated lumber.

A major trend at the retail level was the increasing importance of home improvement centers (or home centers) and the mass merchandisers that served the rapidly growing repair and remodeling market. They catered particularly to the homeowner or do-it-yourself (DIY) trade, but also sold to contractors and builders. They purchased their goods mainly from stocking merchant wholesalers, as well as from manufacturers' sales branches with distribution yards. Some of the major lumber items they handled were pressure-treated lumber for fencing and other outdoor use, such as 8-foot 4 × 4's, and shorter lengths and wider widths of pine boards for shelving and cabinets, such as 1 × 12's appearance (higher quality) grades.

In the overseas markets, the most common channel used followed this pattern: manufacturers to export merchant wholesalers to overseas sales agents to overseas merchant wholesalers to retailers to final overseas markets. (See Exhibit 2a.) As in the case of domestic markets, although not nearly as strong a trend, a number of major manufacturers sold through their own captive wholesale distribution branches which sold to retailers and to end-use markets. Of the many channel members involved, the key role was generally played by the overseas sales agent who

---

[6]A straight car is a loading of lumber and other wood products consisting entirely of one type of item and usually limited to one species, grade, and width. Therre may be a variation in length, such as in a random loading of dimension lumber.

A mixed car consists of a variety of items, sizes, species, etc. Mixed car items generally command a higher price than a comparable single item in a "straight" car.

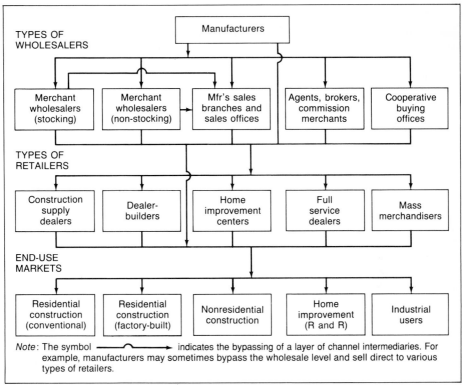

TYPES OF WHOLESALERS

| Manufacturers |

| Merchant wholesalers (stocking) | Merchant wholesalers (non-stocking) | Mfr's sales branches and sales offices | Agents, brokers, commission merchants | Cooperative buying offices |

TYPES OF RETAILERS

| Construction supply dealers | Dealer-builders | Home improvement centers | Full service dealers | Mass merchandisers |

END-USE MARKETS

| Residential construction (conventional) | Residential construction (factory-built) | Nonresidential construction | Home improvement (R and R) | Industrial users |

Note: The symbol ━━━⌒━━► indicates the bypassing of a layer of channel intermediaries. For example, manufacturers may sometimes bypass the wholesale level and sell direct to various types of retailers.

EXHIBIT 1a
Channels of distribution for lumber—domestic markets.

made daily price offers to importers on commodities based on Telex-confirmed prices with the exporters (i.e., the manufacturers or their U.S.-based channel members). The agent arranged for shipments from the exporter, and followed up to make sure deliveries were made and payment received.

Pricing

The process of price determination of most lumber items took place in a nationwide (actually U.S.-Canadian) auction market. The lumber wholesaler played a key role in matching what the mills were asking with what the dealers or other purchasers were willing to pay, and in that way determining the price at which a sale could take place. The mill's offering price was influenced by its backlog of unfilled orders, by how much stock it had on hand, and by the supply of raw material or logs which were available. The price the dealer was willing to pay depended on his inventory and on the strength of market demand. This huge lumber auction involved thousands of telephone transactions. The daily flood of buyer-seller negotiations meant a highly volatile price structure, almost immediately responsive to supply and demand pressures. However, instead of a single market price determining market equilibrium, as portrayed in economic theory of price competition, there was more likely to be some variation in the prices of the same items sold to the same general categories of purchasers. These price differentials stemmed from imperfect knowledge of the market—for example, small local wholesalers making infrequent purchases versus large distributors plugged into a constant information stream of telephone inquiries and orders. The differing prices were also caused by variations in the order mix shipped by different mills, such as straight cars versus mixed cars, overgraded versus undergraded lumber, and regular orders versus those requiring special handling or packaging.

EXHIBIT 1b
NOTES ON CHANNELS OF DISTRIBUTION FOR LUMBER—DOMESTIC MARKETS

Types of wholesalers

Merchant wholesalers are independent firms which purchase products from manufacturers for the purpose of resale to dealers or retailers or industrial users. They take title to the goods they purchase and may or may not carry stock.

Agents, brokers, and commission merchants are similar to merchant wholesalers, except that they do not take title to the products but only perform the sales function, and typically do not carry stock.

Manufacturers' sales branches and sales offices are establishments maintained by the manufacturers for marketing their products at the wholesale level. Captive wholesale distribution centers are also included in this category. If the sales branches or distribution centers sell primarily to household consumers or to local builders, however, they are not included in this category but in the appropriate retailer category. Finally, a mill salesforce, selling mainly the mill's output to establishments at the wholesale level, would not be considered as a sales branch or sales office, but rather as part of the mill operation.

Cooperative buying offices are maintained by a group of dealers or retailers who pool orders so that they can buy in quantity from mills and drop-ship partial carloads to the various yard locations of the dealer-members.

Types of retailers

Construction supply dealers carry large inventories of bulky materials such as lumber, plywood, hardboard, and gypsum which they sell at near-wholesale prices to big-volume builders and contractors. These dealers may be involved in the manufacture of millwork items, such as cabinets and window frames, as well as roof trusses and wall panels.

Dealer-builders sell building materials and also engage in either custom or speculative building of homes, farm buildings, major room additions, and general remodeling, using either their own or subcontracted crews.

Home improvement centers, also called Home Centers, are dealers or retailers who sell, chiefly to homeowners, a wide range of building materials and other products used for adding to, repairing, renovating, and maintaining the home.

Full service dealers carry a complete line of building products, including appliances and kitchen cabinets, and serve both homeowners and builders. Service is emphasized, including design, cost estimating, and financing for small home builders.

Mass merchandisers are retail institutions from outside the building materials field, including department stores, mail order/department stores, and discount stores. They may have home improvement or building supply departments in their regular stores, or may have chains of separate building supply centers.

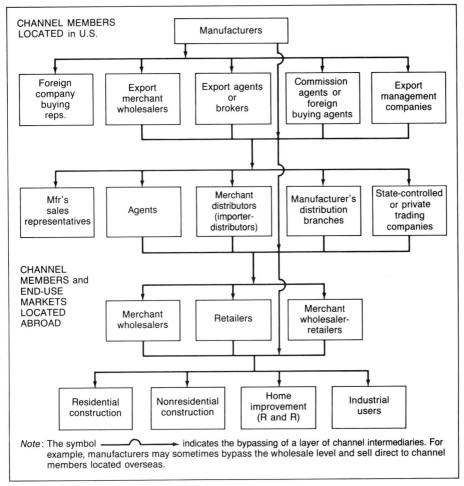

EXHIBIT 2a
Channels of distribution for lumber—overseas markets.

Position of the South

The 12 Southern lumber-producing states were in a favorable position to capitalize on the growing worldwide demand for lumber and wood products. In contrast with the West, where over 60% of commercial forest lands was in government ownership, in the South 10% was government-owned and 90 percent privately-owned, either by forest products companies or by other forest land owners. Cuttings on private holdings in the Northwest were being made at a rate faster than timber growth, but at the same time there was a flat trend in federal timber sales. In the South, growth rate was greater than harvest rate because of the rapid growth and early maturity of Southern pine, as well as more intensive forest management. It was estimated that the South grew timber at a rate one-fourth greater than the national average. Nearness to the major market areas of the Midwest and Northeast was another advantage of the Southern mills, as compared with those on the West coast, particularly with the rapid rise in freight rates. The faster increase in population

EXHIBIT 2b

NOTES ON CHANNELS OF DISTRIBUTION FOR LUMBER—OVERSEAS MARKETS

Channel members located in U.S.

Export merchant wholesalers are the same as merchant wholesalers in the domestic market, but specialize in selling to channels and markets in the export trade.

Export agents or brokers are the same as agents and brokers in the domestic market, but specialize in selling to export channels and markets.

Commission agents or foreign buying agents are "finders" for foreign firms wanting to purchase U.S. products, and a commission is paid to them by their foreign clients.

Foreign company buying reps are foreign company direct representatives, such as those sent here by Japanese trading companies.

Export management companies (also called EMC's) act as export departments for several manufacturers of noncompetitive products. They solicit and transact business in the name of the manufacturers they represent for a commission, salary, or retainer plus commission. Some may also take title and carry the financing for export sales.

Channel members located abroad

Agents represent American exporters in the principal foreign lumber-buying ports. They develop sales with foreign importers, such as merchant wholesalers or industrial users. They do not take title to the goods, but are paid by commission.

Manufacturers' sales representatives take responsibility for a manufacturer's goods exported to a particular country or market area, and are paid by commission. They are more under the control of the manufacturer-exporter than are most categories of agents.

Merchant distributors, unlike agents, take title to the goods they purchase from U.S. manufacturers or U.S.-located channel members. They resell the goods either direct or through wholesalers or retailers to end-user markets abroad. (Merchant distributors are also called importer-distributors.)

Manufacturers' distribution branches have their own shipping terminals and storage facilities in foreign ports from which they sell their own mills' production as well as that of other U.S. mills. Their sales are to foreign wholesalers, retailers, or end-users.

State-controlled trading companies are found in those countries which have state trading monopolies, where business is conducted by a few government-sanctioned and controlled trading entities. *Private trading companies* are privately-owned merchant importers. In some countries, such as Japan, they may exercise considerable control over both the importation of goods and their distribution within the country.

Merchant wholesalers take title to the goods purchased and may act as stocking wholesalers or drop shippers, and sell to retailers or direct to end markets.

Retailers typically buy from distributors or wholesalers and sell to building contractors, industrial users, and home owners.

Merchant wholesaler-retailers are firms involved in both wholesaling and retailing, and may own retail chains which sell to building contractors and to the do-it-yourself market.

in the Sunbelt states, with the South's increasing share of national housing starts noted earlier, also meant a stronger regional market for the Southern lumber mills.

Most mills in the South were designed for the high-speed production of dimension or framing lumber, which constituted 85% of total Southern pine lumber, with 13% consisting of boards, and the remaining 2% mostly timbers (i.e., lumber at least 5 inches in least dimension). Dimension lumber was used primarily for structural members and components such as floor and roof trusses, floor and ceiling joists, and roof rafters and studs in homebuilding. Boards were used in home construction for paneling, siding, and mouldings or trim to finish off around window and door openings and walls. They were also used for shelving, cabinets, furniture parts, and ladders, and the lower grades were found in a wide variety of industrial uses such as crates and pallets. In earlier years, flooring had been a major market for boards, but with the popularity of wall-to-wall carpeting starting in the 1950's, plywood and particleboard had largely replaced lumber floor boards. Finally, the growing repair and remodeling market had caused an increase in the demand for boards. A major portion of the high quality, "appearance" grades of boards were manufactured in the West, with Ponderosa Pine the leading species. Boards were also produced from other species in the West and Northeast, as well as from Southern pine.

Besides boards, another type of lumber in current demand by the home improvement market was treated lumber, which was used particularly for fences and decks. Uses in other markets included wood foundation systems, bridges, and piers. Increasing amounts of treated lumber were also being exported to the Caribbean region. The wood treating industry favored Southern pine as it readily accepted preservative chemicals. In fact, nearly 50% of all Southern pine lumber produced was treated. The most frequently treated items were #2 grade 2 × 4's and 2 × 6's in 16-foot and shorter lengths, as well as 4 × 4's in 8-, 10-, and 12-foot lengths. Another reason for the increased popularity of treated lumber was the decreased availability of rot-resistant redwood and cedar, the traditional species for fences and decks and similar outdoor uses.

The South got about 70% of its total softwood lumber needs from Southern mills, a percentage which had remained fairly steady for the past 25 years. The second major source was Canada, which had recently surpassed the U.S. Western states as a supplier. In fact, in the case of Georgia and Florida, over half of lumber consumption needs came from Canada, mainly in the form of dimension lumber. (Canadian mills had some advantages over U.S. mills, particularly those in the West, because of more favorable government timber sale policies, lower ocean shipping rates, lower rail transportation costs across Canada and into U.S. markets, and a favorable exchange rate for U.S. purchasers.)

In the case of Mississippi, which typically accounted for about 10% of total Southern softwood lumber production, 84% of the state's softwood lumber consumption needs (slightly over 500 million board feet) came from Southern mills, 11% from Canada, and 5% from the U.S. West. Although a major portion of their production was shipped out of the state, the Mississippi mills still supplied a large amount of the needs of the Mississippi market. The imports from Canada into that state were chiefly of Douglas fir and other Western species.

In the overseas export market, Southern pine lumber was expected to play an increasing role. World lumber consumption, excluding North America, was expected to increase by 30% over the next 2 decades. Currently, the world's leading exporters were Canada, Scandinavia, and Russia. However, U.S. softwood growth rates, particularly in the South, far exceeded those of Canada and Russia, and Scandinavia was already at the limit of sustainable yield production. Major export markets for Southern pine were Western Europe and the Caribbean, that is, the Caribbean island nations and the countries around the rim of the Caribbean basin. Southern pine lumber shipped to

Europe was primarily clear (knot-free) material, used for paneling, furniture parts, and joinery.[7] A major portion of the pine shipped to the Caribbean was used for structural purposes, particularly in roof structures.

PRODUCT LINE

Langdale's current product line was broken down as follows: boards, 35%; dimension, 55%; flooring and siding, 6%; and SAPS and Prime, 4%.[8] The last two categories were the highest priced and most profitable items. As for the first two categories, boards were usually higher priced than dimension lumber of the closest similar size and grade. Board prices also fluctuated less widely than dimension prices. However, the sawing and planing process for boards took longer than for dimension. In addition, since boards were sawed to nominal one-inch thickness versus 2-inch thickness for dimension, there was greater raw material wastage in the form of sawdust when producing boards. Thus, manufacturing costs were high for boards, and the relative profitability of the two types of lumber varied according to their price relationship. Other factors determining the production emphasis on boards versus dimension were the types of logs available from local logging contractors and wood-lot owners, competition from other mills, customer commitments, and overall market demand.

With a strong commitment to both boards and dimension lumber, the Langdale Company's product line was more complete than that of most of its competitors. Similarly, most of the company's shipments were in mixed carloads, rather than straight cars, enabling the company to charge higher prices to those customers who preferred mixed cars, such as office wholesalers and dealers. Product quality, in both boards and dimension, sometimes enabled Langdale to get slightly higher prices than many of its competitors in certain market uses, including boards used in finished work, such as cabinets, shelving, furniture, and window facings. In dimension lumber, wood truss manufacturers were a market which called for higher quality lumber, with strength a major requirement.[9] Southern pine was good for trusses because of its strength, and the grades required were #1, and #2 Dense, the first of which was made by Langdale.

DISTRIBUTION, SALES, AND PRICING

Langdale's current distribution channels were as follows: wholesalers, 65%; dealers or retailers, 15%; direct to industrial users, 10%; and export wholesalers or agents, 10%. Some 20 years earlier, the percentages through wholesalers versus retailers had been just reversed. A major reason for the change was the increased importance of telephone selling and the use of WATS lines.[10] Other reasons for the increasing importance of wholesalers for Langdale and for the industry generally were the shift of the inventory carrying function from retailer to wholesaler (described earlier), and the fact that the wholesaler could, in one telephone call, supply the retailer

---

[7]Joinery is a term used in Europe to denote the higher grades of lumber suitable for such uses as cabinetry, millwork, or interior trim.

[8]Flooring and siding are higher quality boards usually requiring more manufacturing steps than boards used for other purposes. For example, flooring is a tongue-and-groove piece of lumber, with the basic size 1" × 4". SAPS and Prime are high quality export grades of Southern yellow pine boards, shipped mostly to Europe.

[9]A truss is an assemblage of wooden members which forms a rigid framework to support a roof or other part of a building.

[10]A WATS (Wide Area Telephone Service) line enables a caller to phone anywhere in a designated area for a flat monthly fee. Many lumber wholesale firms use a WATS line in lieu of regular long-distance lines; some have toll-free incoming WATS lines that customers may use.

with all of his regular lumber needs in terms of species, sizes, types, and grades, thus saving much time that the retailer might have to spend dealing with many different mill salesmen. Wholesalers had also become more aggressive and were shaving the traditional 5% markup which they had taken on mill prices when they sold to retailers.

The majority of the company's sales to wholesalers were to office wholesalers (non-stocking merchant wholesalers). Those sales that went to stocking wholesalers were usually to the smaller, local ones, rather than to the big distribution yards which serviced retailers in larger areas. It was these large distribution yards which had shown the fastest growth in recent years. Direct sales to industrial accounts included manufacturers of furniture, bleachers, toys, and laminated beams, as well as a little to a wood treating firm. Other industrial users, such as truss manufacturers, were reached through wholesalers. The majority of the company's customers were in the home state of Mississippi, with some located in the adjoining states of Alabama, Tennessee, and Louisiana, and relatively few elsewhere.

Although during its early years Langdale had been very active in the export trade, this was currently not a market that it was actively pursuing. It sold to this market through U.S. export wholesalers, and a few foreign-based agents. The most important of these had been an agent located in a major European lumber-importing country which had recently taken on the export business of a large U.S. forest products firm and had apparently lost interest in selling Langdale products. If the company decided to expand its export sales, it would probably have to seek new export wholesalers and/or new overseas agents. It was generally accepted in the industry that success in selling to the export market required sorting out the higher grade items in a mill's production, tight control over appearance and quality, rigid adherence to clearly drafted product specifications which often involved sizes and grades different from those in the domestic market, and first-hand knowledge of the trade customs, usage patterns, and government regulations of the various countries involved. It also helped if the U.S. exporter was willing to enter into long-term forward price commitments, since that was what foreign importers preferred. Since this represented a departure from the auction-market type of fluctuating price pattern to which U.S. firms had long been accustomed in the domestic market, few companies were willing to change. Most mills, including Langdale, tried to keep a 2- to 4-week order file, that is, not to give price quotes for any period more than 4 weeks ahead.

The company's pricing policy, following a long-established tradition of previous management, was to price somewhat higher than most competitors and not to engage in aggressive price cutting during a falling market, but rather to try to hold price and curtail production if inventory levels started rising too much. Mathew Brady, the Langdale sales manager, felt that such a policy was, in the long run, best for the company and for his established customers. However, Stan Sherman, Planning Vice President from corporate headquarters, had questioned the desirability of this policy during the recent meetings at the Langdale mill. "We've got to make greater effort to maintain our sales volume," he said, "and if the market slides, there are two ways to do this: 1) lower price, or 2) expand our market through broadening our customer base, both geographically and through emphasizing new channels and new end-users." During a depressed market such as the company was currently experiencing, Sherman recommended lowering prices on a selective basis, in order to get new customers in new areas. Brady disagreed with this proposal. "If our inventory of particular items is getting too high, I would prefer to offer a one-time special sale to all our customers. To cut prices to new customers in other areas while at the same time charging higher prices to our long-established customers would raise a real moral issue with me. Furthermore," he continued, "these new customers would probably not stay with us for very long, but would go back to their established suppliers as soon as the market strengthened."

DEVELOPMENT OF A MARKETING PLAN

The Five-Year Marketing Plan, which Langdale Company management was expected to draw up with the general assistance of Stan Sherman, was to be part of the overall Five-Year Long-Range Plan for Langdale. The main goal for the Langdale Company, which had been set by corporate headquarters, was to increase production capacity to 45,000 Mbf over the next 5 years. To accomplish this, capital expenditures of between $500,000 and $750,000 had been tentatively planned. These expenditures included improvements in the dry kiln and planer mill noted earlier, as well as improvements in the log yard, and the chipping saw and band saw lines. The relative amount of expenditures on these two production lines depended on the future emphasis to be given to dimension lumber versus boards, as well as on the quality and sizes of the logs which could be procured.

How to sell the increased output of the mill over the coming five years was an issue which Stan Sherman felt had not really been addressed as yet. The purpose of his visit had been to get Langdale management moving ahead on drawing up a Five-Year Marketing Plan, which he believed should be a key component of the overall Five-Year Long-Range Plan to be submitted to corporate headquarters for approval by the first of October. As Sherman had explained to Benson and Brady, the basic approach in marketing planning for Langdale, or for any company for that matter, should be to start out by identifying market segments in the industry which might be meaningful to the company, and to see to what extent the company was selling to these segments. In this way, a plan or program could be laid out detailing which segments should be emphasized, which ones deemphasized, and which ones, previously overlooked, should be cultivated. Both Benson and Brady agreed that this seemed like a good idea in principle; however, they did not appear to be convinced that a formalized approach such as this was necessary for a mill the size of Langdale. Furthermore, given the limited management personnel at Langdale, and current restrictions on additional hiring, there simply wasn't the time to do the analytical work and paperwork required for such a complete marketing plan.

There had been considerable discussion during the recent meetings of the sorts of data needed to draw up a complete marketing plan based on a market segmentation analysis. Some of the overall industry data had already been compiled by Stan Sherman. (These data consisted chiefly of the industry background exhibits and comments thereon already portrayed in the case.) Some of the industry trends were of particular significance to Langdale, and would have to be analyzed in more detail. More company sales information would also be needed in order to see where Langdale stood in relation to industry trends. Besides gathering the data, there was also the question of how to analyze it.

Prior to its acquisition by Clark Industries, the only data processing being used by the Langdale Company was a bank service to process its payroll. Now, using the data processing facilities of the parent corporation, the Langdale Company was planning to computerize other parts of its operations, including order entry, inventory, production, and sales analysis. An input terminal and printer had been ordered to provide the company with the devices necessary to control its data entry and report printing needs. As part of the current marketing planning efforts, Jim Benson, with the help of Sherman and Brady, was starting to determine how the company's sales invoices should be redesigned in order to secure the data necessary to do the sorts of sales analyses needed to improve the firm's customer service and marketing efforts. Currently, the invoices contained the following information: date of sale, name and address of customer, types of items purchased, number of units purchased, price charged per unit, and total dollar amount due. Stan Sherman pointed out that a lot could be learned about what market segments were being served simply by analyzing past company invoices. Jim Benson, however, preferred to wait a couple of months, when the new data processing system was supposed to be in place, before spending much time on such an analysis.

Another possible source of data which Stan Sherman felt would be useful in improving the company's marketing efforts was a telephone log. During his recent visits to several large lumber wholesalers, Sherman had noticed that some of the firms had required that their traders (those buying and selling lumber) keep such logs, or daily telephone call reports. These reports contained such information as the following: names of customer accounts (or supply sources) called, calls completed, price quotes made, orders bought (or sold), new accounts called, and new accounts sold. Mathew Brady's reaction to such a phone log, however, was that he could keep all this information in his head, that he was too busy talking on the phone to write up such a log, and that even if such call reports were written out, they would be misleading to anyone trying to do sales analyses from them. The reason for this latter objection was that every customer was different. For example, some wanted to be called only at certain hours on certain days of the week, whereas others welcomed calls from him, or called themselves, almost any time; some always liked to dicker on prices and items, whereas others quickly decided whether to accept or refuse an offer. Jim Benson had not yet decided how he felt about the telephone log idea.

CONCLUSION

As he neared the end of his flight, Stan Sherman decided to draft a memo over the weekend to be mailed to Jim Benson on Monday. This memo would first summarize what he perceived as the areas of common agreement reached during the recent meetings at the mill. The desirability of spending capital to increase the efficiency of the mill was one of these areas, and one which seemed to rank top priority with Benson in particular. The need to improve the Langdale Company's current profit situation and to increase future sales were two other issues on which Sherman, Benson, and Brady agreed, although how best to go about doing this was an issue not yet resolved.

After describing these areas of general agreement, Sherman felt he should next reemphasize that some kind of market segmentation approach was the logical way to go, not only to improve current profit performance, but also to reach the long-term sales goals that would be needed to sell out the mill's capacity during the coming years. He wondered if he should suggest several different ways to segment the lumber market and leave it up to Benson and Brady to choose the way, or ways, most applicable to Langdale. An alternative approach would be to lay out exactly what steps should be taken to analyze the market according to several different dimensions. This analysis would take the form of specific segmentation projects, a description of the benefits to be gained from each project, and timetables for their completion.

Finally, Sherman had not yet decided which other people, either within the Langdale mill or in other parts of Clark Industries, should receive copies of his memo to Jim Benson. He felt that both Benson and Brady were valuable men to the Langdale Company and to Clark Industries. However, something had to be done to "get Langdale on the march" toward developing a meaningful Five-Year Marketing Plan which was due at corporate headquarters on October first, just three months away.

OREGON PINE LUMBER COMPANY[1]

Marketing, Distribution, and Manufacturing Problems of a Small Lumber Mill

INTRODUCTION

"I remember when this highway was lined with green and dry mills," remarked Jim Sutter, Oregon Pine Lumber's President and Sales Manager. "When this mill was built, about 1946, there wasn't much of anything else but mills around here. When we took over in 1973 there were several mills operating nearby. Now we're one of the few mills still running. For my partners and me this is the realization of a dream. We wanted to work for ourselves, and when the opportunity finally presented itself, we couldn't say no. We all left jobs with other forest products companies and took over this operation. We didn't want anything big and fancy—just a small partnership, not a big corporation. Now the cost of timber, transportation, and credit are just about putting an end to all of it."

In the background the hum of the mill's saws, debarker, and finishing equipment was punctuated occasionally by the roar of a chainsaw or lumber carrier. Several of the mill's 40 nonunion millworkers and one or two of its 10 woodland crew darted back and forth, preparing to wind down another day on the job. The last semi-trailer of the day pulled out for California with a load of green Douglas fir timbers lashed securely on board. A yellow Southern Pacific flat car sat idle on a siding across the yard from Jim. Jim's two partners, Ross Gilbert and Charlie Olsen, joined him by the office. As they surveyed the small, neatly stacked piles of green Douglas fir dimension stock, they reminisced about their eight years in business together. They mused about how they swung the sale of the mill for $165,000 down to $1.1 million in debt. They were pleased that they had retired their original debt within six years even though economic conditions had forced them to shut down for 15 months in 1974 and 1975.

Ross Gilbert, the company's Secretary-Treasurer, was obviously concerned about the company's current financial condition. "We're overextended on federal timber sales which we bid when lumber prices were high and strong. Even with the federal set-aside programs for small mills, we are paying a premium for our timber. We've got great relations with our bank and our bonding company, but we're likely to run into trouble if we can't sell our lumber at the prices we estimated when we committed ourselves to those timber sales."

Charlie Olsen, Log Department Manager, listened as Jim Sutter spoke of changes in the market for green Douglas fir timbers and dimension lumber. Jim said, "Even with our ability to cut special orders and special lengths, we're probably going to have to curtail our operations if the markets don't get stronger. The phone has been ringing a lot lately, but everybody is just fishing. The export business was taking up slack for our lost business in the Midwest and South, but lately things seem to be slowing down. There doesn't seem to be a lot of agreement about the strength of the export markets in the future."

As the hum of the mill machinery died down, Matt Kilkenney, the Mill Superintendent, joined the others to discuss the scheduling of orders for the next day. Collectively the group wondered what the future held for the mill.

[1]This case is reproduced with the permission of its author, Dr. Stuart U. Rich, Professor of Marketing and Director, Emeritus, Forest Industries Management Center, College of Business Administration, University of Oregon, Eugene, Oregon.

RAW MATERIALS

Oregon Pine purchased all of its timber from U.S. Forest Service and Bureau of Land Management (BLM) timber sales. Currently, 90 million board feet of timber were under contract to be cut over the next three years. Since only Douglas fir logs were manufactured, other species in the timber sale such as Hemlock, White fir and Cedar were separated at the logging site and sold direct to mills in the valley. Occasionally, mixed species were delivered to the mill. These logs were rebucked (sawed to length) and upgraded to add value, then sold on the open log market.

Because Oregon Pine employed less than 500 people, the company was allowed to bid on SBA set-aside sales.[2] Once a sale was put up for bid, Oregon Pine foresters cruised the timber to provide management with specific information on the quality and volume of timber being auctioned. Sometimes the U.S. Forest Service prospectus contained erroneous information that the unsuspecting purchaser based his bid on, thus resulting in financial distress to the firm.

Oregon Pine hired contract loggers to harvest and haul the timber from the woods to the mill. Charlie Olsen, the log department manager, was responsible for contract administration and coordination of logging activities to meet the daily production requirements of the mill. Because of the small log storage pond, Mr. Olsen and Mr. Kilkenney had to closely coordinate their logging and production schedules.

Douglas fir stumpage prices in the last decade had escalated rapidly. In search of less expensive high-quality logs, Oregon Pine had expanded its procurement radius to 270 miles. At full mill capacity, delivered log costs (including transportation and stumpage costs) ran as high as $30,000 to $40,000 per day.

MANUFACTURING

The Oregon Pine sawmill was located on 13 acres of land leased from the Burlington Northern Railroad. In addition to a rail siding adjacent to the mill, Interstate 5, a major connecting link between Canada and Mexico, was four miles away by state highway. The mill ran one eight-hour shift per day, cutting approximately 100,000 board feet of green surfaced and rough-sawn Douglas fir lumber in this time period.

Logs were delivered by contract haulers to the mill and stored in a small log pond where they were sorted to provide Mr. Kilkenney with high-quality, properly-sized logs. From the log pond, logs were moved to a debarker and converted into timbers, cants, and dimension lumber by a vertical bandsaw head rig.[3] Timbers, crossarms, and cants were trimmed and edged before they reached the sorting chain, where they were graded, stamped, and prepared for shipment. The falldown was resawn and trimmed into dimension lumber, then edged and planed to smooth the surfaces before it reached a green chain at the far end of the mill. Here it was sorted by length and width, graded, and placed in bins. As the bins became full, a lumber carrier moved them to a small lumber storage area where their contents were strapped and loaded for shipment. High-quality green lumber was often stored in a covered shelter behind the main office to prevent checking or warping in hot weather.

Export clears and timbers were inspected for defects to insure that each product met or exceeded grading specification. Oregon Pine's reputation in the domestic and international markets was based on the high-quality Douglas fir used, and the finished product they manufactured. Their order file was filled with repeat customers who didn't mind paying a premium for

[2]Small Business Administration (SBA) set-aside sales are reserved by the U.S. Forest Service exclusively for small firms who can't bid competitively with large integrated and better capitalized firms.
[3]For these and other technical terms, see Exhibit 1, "Glossary of Terms."

EXHIBIT 1
GLOSSARY OF TERMS

---

band saw   A saw consisting of a continuous piece of flexible steel, with teeth on one or both sides, used to cut logs into cants and also to rip lumber.

cant   A large timber cut from a log and destined for further processing by other saws.

clears   Lumber free, or practically free, of all blemishes, characteristics, or defects.

dimension   Lumber that is from two inches up to, but not including, five inches thick, and that is two or more inches in width; classified also as framing, joists, planks, rafters, etc.

edger   A piece of sawmill machinery used to saw cants after they come off the head rig, squaring the edges and ripping the cants into lumber.

export clear   A high grade of lumber produced for shipment to overseas markets.

falldown   Those lumber items of lesser grade or quality produced as an adjunct to the processing of higher-quality stock.

green chain   A moving chain or belt on which lumber is transported from saws in a mill. The lumber is pulled from the chain by workers and stacked according to size, length, species, and other criteria.

green lumber   Moisture content of lumber has not been reduced due to either air or kiln drying.

head rig   The primary saw in a sawmill operation, on which logs are first cut into cants before being sent on to other saws for further processing.

hog (hogged) fuel   Fuel made by grinding waste wood in a hog (a machine for grinding wood into chips). Used to fire boilers or furnaces, often at the mill at which the fuel was processed.

merchantable   An export grade that describes a piece of lumber suited for general construction use.

planer   A machine with cutting knives mounted on cylindrical heads which revolve at high speeds to surface the faces and edges of rough lumber.

timber   1. Standing trees; stumpage. 2. A size classification of lumber that includes pieces that are at least five inches in least dimension.

---

excellent service and the assurance of a structurally superior product. Jim and Matt were proud of the fact they could cut any size product up to 13″ × 13″ and 32′ long, and fill most orders the same day they were received.

Operating costs, net of residuals and stumpage, averaged $64/MBF. Residuals sold in the form of planer shavings, sawdust, and hog fuel lowered this cost by $14/MBF.

PRODUCTS

Oregon Pine cut old-growth Douglas fir into clears, merchantable, and dimension lumber. Export clears were used primarily for cabinets, moldings, venetian blind slats, window trimmings, and other decorative wood product uses. Domestic clears had a variety of uses, including utility pole crossarms. Merchantable referred to export lumber that was good in appearance, strength, serviceability, and had no serious defects. Merchantable lumber was used in residential and commercial construction. Dimension lumber (2″ to 4″ thick) was primarily used for light framing and construction in the domestic market.

Jim Sutter indicated that they would essentially cut anything within the mill's capabilities that brought a premium price. This was reflected in Oregon Pine's product line. In search of a price premium, the company would cut 4″ clears for the export market. The company's ability to cut to a 32′ length also offered an opportunity for a premium price. Most of the company's competitors, within one hundred miles, cut instead 3″ clears and only up to 24′ lengths. About 15% of Oregon Pine's output was in dimension lumber. (A breakdown of the company's year-to-date product line for both domestic and export markets is presented in Exhibit 2.)

EXHIBIT 2
LUMBER SALES, JANUARY THROUGH NOVEMBER*

	Board feet	Percent of total	Average sales price per Mbf (thousand board feet)
Clears			
1 × 2 & wider	682,309	3.5%	$440.76
2 × 3 & wider	954,717	5.0%	523.13
2½ × 4 & wider†	158,805	0.8%	810.35
3″ & thicker†	1,576,217	7.9%	702.43
Dee &/or reject†	1,111,791	5.5%	442.21
Misc. (kiln dried)	38,453	0.1%	570.00
	4,522,292	22.8%	563.78
Crossarms, small squares, beams, & timbers			
Crossarms (incl. rej.)	3,070,610	15.5%	617.12
4 × 4 to 4 × 12 (#2 & Btr.)	4,231,552	21.4%	280.63
4 × 14 & wider (#2 & Btr.)	11,022	· · ·	584.17
6 × 6 & wider (#2 & Btr.)†	2,564,729	12.9%	376.23
8 × 8 & larger (#2 & Btr.)	151,862	0.7%	507.44
#3 (Util) S4S & Rgh.	1,131,611	5.7%	127.28
#4 & Econ.	203,325	1.0%	70.39
	11,364,711	57.2%	377.41
Dimension			
2 × 4 & wider (#2 & Btr.)	2,168,771	10.9%	214.50
2 × 4 & wider (#3)	642,125	3.3%	126.84
2 × 4 & wider (Econ.)	276,239	1.3%	72.93
	3,087,135	15.5%	183.60
Miscellaneous			
1 × 2 & wider	642,894	3.2%	167.52
2 × 2 & wider	264,950	1.3%	195.34
	907,844	4.5%	175.64
Totals	19,881,982	100.0%	$/MBF 380.50

*Notes: All products cut to ALS (American Lumber Standard) standards. The most commonly used species of framing lumber (except Southern pine) are graded as follows: Standard and Better; #2 and Better; Utility; #3; and Economy. S4S means surfaced (planed) on four sides.
  †Export sales.

MARKETS

Oregon Pine sold to both the domestic and the overseas markets. Both Jim and Ross felt that, geographically, their domestic market had decreased in size as a result of high rail transportation rates and railroad shipping incentives aimed at accommodating large producers. Thus, the company's current domestic market consisted of the following areas: Oregon; Washington; Western Idaho; California; Reno, Nevada; and Seward, Nebraska. Oregon Pine's export market included Western Europe (U.K., Switzerland, Greece, Italy) and Australia. Italy was an important market with primary interest in Douglas fir clears, while merchantable Douglas fir appealed more to the Australian market.

PRICES

Jim did not consult any price list when attempting to make a sale. He felt that a salesman needed to have the flexibility to change prices in order to keep up with a fluctuating market. This flexibility was what Jim felt was one of Oregon Pine's distinctive competencies over the larger mills (where organizational structures often did not facilitate rapid price adjustments to changes in the market). Oregon Pine's pricing policy was to price at what the market would bear. Jim's years of experience in lumber sales enabled him to quote a price that covered variable costs and that he felt was competitive with the rest of the industry. He admitted, however, that he received less than 2% of the orders on which he quoted. While the buyer might express interest in the quoted price, he was very likely to shop around for the lowest possible price offered.

Oregon Pine's year-to-date sales, through November, were $7.5 million.

CHANNELS OF DISTRIBUTION

To serve the domestic and export markets, Oregon Pine relied upon a network of wholesalers and stocking distributors. The company did none of its business by consignment.

In the previous eleven months, Oregon Pine's sales in the export market had topped $1,878,000. Oregon Pine's sales manager estimated that between 25 and 30% of the company's annual production reached the export market through 22 channel intermediaries. Among these were the international distribution networks of several of the country's largest integrated forest products companies. Two export wholesalers and one integrated forest products company accounted for 45% of Oregon Pine's export business. (Exhibit 3 presents a generalized view of traditional export distribution channels.)

Oregon Pine sold 70 to 75% of its annual production in the domestic market. Wholesalers and stocking distributors[4] in the domestic market handled approximately 50% of the company's annual production for resale purposes. The remaining 20% was nearly evenly distributed direct to industrial end users and remanufacturers.

Jim Sutter avoided direct retail sales channels and seldom sold direct to contractors. He felt that direct retail channels in particular involved unnecessary complications. He preferred to

---

[4]Stocking distributors sell through their own salesforce or sales agents. Their product lines are generally quite broad. Their customers are retail building supply centers and retail lumber yards.

EXHIBIT 3
TRADITIONAL EXPORT DISTRIBUTION CHANNELS

Exporter or shipper ⟶	*Agent* ⟶	*Importer* ⟶	*Merchant* ⟶	*End user*
Liaison between the domestic producers and the agent or importers in foreign locations; exporters may solicit business from producers or producers may initiate the contact.	Frequently the most knowledgeable source in the foreign location; responsible for lining up importers as contacts for exporters, makes shipping arrangements, quotes prices, and checks on status of orders.	Like a domestic wholesaler or stocking distributor; grants credit, carries inventories; may provide for further processing of lumber; may simply act as a drop-shipper; sells mostly to merchants.	May be a combination retail and wholesale operation; sells to contractors, industrial users, do-it-yourself retail outlets, homeowners; may do millwork and cabinetry.	Industrial, construction, and homeowner markets.

conduct business on wholesale terms (2/10 net 30) under an F.O.B. mill pricing policy. He believed that such conditions guaranteed prompt payment of receivables, and the channel intermediaries bore the costs of insuring and holding inventories. Offloading these inventory carrying costs onto the distributors, he felt, was critical during periods of high interest rates and low product prices.

TRANSPORTATION

A Southern Pacific rail spur and a state highway bounded the mill site on two sides, thus giving the company quick access to either rail or highway carriers. Under the F.O.B. mill pricing system, the distributor selected the appropriate transportation mode and the mill loaded the order.

Oregon Pine's modal mix, 20% rail and 80% truck, resulted from a large number of less-than-carload orders, its green lumber product line, the mill's quick-order-filling sales approach, and the lumber's destination. Oregon Pine's use of truck transportation was greater than that of most other coast mills in the same size class. However, it followed the trend toward increased reliance on truck transportation for all Oregon mills.

Rail transportation costs were in a state of flux, due to the recent federal legislation aimed at lessening the Interstate Commerce Commission's regulatory powers. Because the "deregulation" was still new, many carriers were experimenting with new rate structures. Shipments from Oregon Pine originated under the Southern Pacific's new carload point-to-point rates. Under the per carload rates with weight ceilings, dry lumber producers enjoyed a cost advantage over green lumber producers such as Oregon Pine. The dry lumber producers could pack flatcars and boxcars with more board feet of lumber and remain within the weight limitations, thus achieving a lower transportation cost per thousand board feet than that for green lumber. Also, under the new system, the nearer an origin point was to a destination, the lower the per carload rate. Therefore, the greatest cost advantages under this new rate system accrued to shipments of dry lumber from producers located closer to the destination point than their competitors producing either green or dry lumber.

CONCLUSION

Toward the end of the conversation among the company managers, Jim turned to Ross and said, "None of our children seem interested in taking over the business, and we're all thinking about retiring in a few more years. I don't think we should be looking much further ahead than the next three or four years."

Ross agreed, but quickly added "We would like to keep our sales volume up so that the business remains a going concern and continues its history of good service. I know the hassles of doing business, especially when all the regulations don't seem worth the returns, but we've got a really good operation here."

Matt Kilkenney spoke up: "The mill's in good shape. With a little regular maintenance we could run for another ten years, just so long as we don't try to make it a high-volume operation."

Jim said, "We don't have to do that, Matt. The big boys rely too much on us as a small specialty mill. We round out their product lines in ways that they can't. Without us they might lose some big accounts, especially some in foreign countries. I just wish I had a better idea about how we could sell more to customers who pay a premium for our products. I'd also like to figure out how we could stay flexible, but be better able to balance our domestic and foreign sales. We've been pretty good about switching to take advantage of premiums in either market, so far, and I hope we can continue. Sometimes, though, I feel like we may be missing something."

"Our total October sales were down $115,000 from the prior month, and prices continue to decline. Industry-wide production still exceeds demand and the pressure on prices is all downward. The market isn't orderly. Volume producers and dimension mills are pressing us in our best markets. We've been selling a larger volume of high grades, but I'm not sure this will continue. I'm concerned because the demand forecasts for the next six months don't indicate a turnaround. Why don't we review our basic business strategies and consider making some revisions to help us get through the next couple of years?"

PREMIUM HARDWOODS: PLANNING FOR THE FUTURE[1]

Strategic Planning for a West Coast Lumber Company

Fred Cohen had a new office, a new job, and a new challenge. As the result of a corporation-wide restructuring, he was asked to head the Premium Hardwoods Division of Pacific Building Products Corporation. Premium Hardwoods is a relatively new but large producer of hardwood lumber and dimension parts located in the Pacific Northwest.

Before heading Premium Hardwoods, Mr. Cohen held several positions with the Corporation—the majority of which were in the Corporation's west coast softwood lumber operations. In his previous positions, Fred initiated cost reduction programs that were mildly successful. However, he spent much of his time solving problems with logging contractors and the union that represented the mill workers.

While Premium Hardwoods has been successful, Fred is interested in making a name for himself within the Corporation by improving the performance of the Division.

PACIFIC BUILDING PRODUCTS CORPORATION

Pacific Building Products was established in 1894 and has become a large, diversified corporation with annual sales in excess of 3 billion dollars. The Corporation is involved in the production of a wide variety of wood-based products including softwood and hardwood lumber, hardwood dimension parts, various types of structural and nonstructural panels (plywood, oriented strandboard, particleboard), chips, pulp and paper. In addition, the Corporation is involved in the sale of chemicals, treated wood products, and office supplies (to complement the paper products line).

Solid and reconstituted wood products account for approximately 40% of the corporation's sales. Hardwood lumber and dimension produced by Premium Hardwoods represent less than one-eighth of solid and reconstituted products sales. Paper products account for another 40% of sales and the remaining 20% is attributable to the various other products.

Since Pacific Building Products initially produced only softwood lumber, the company operated in a boom or bust environment, heavily dependent on building activity. This resulted in a corporate philosophy that emphasized operating with a minimum investment in fixed assets and timberlands, and a small work force. However, during the last decade, company executives have become increasingly concerned with the supply of timber and have deviated from this philosophy by purchasing additional timberlands in Oregon, Washington and several southeastern states.

---

[1]This case was developed by Robert J. Bush, Assistant Professor, Department of Wood Science and Forest Products, Virginia Polytechnic Institute and State University, with help from Stan Wise and John Punches. The works of M. E. Plank, C. Buhler and D. G. Briggs provided information for some of the exhibits in this case. The helpful comments and suggestions provided by Steven A. Sinclair are gratefully acknowledged. Copyright © 1991 Robert J. Bush.

The Corporation has a strong reforestation program that is geared toward the reestablishment of conifers on logged lands. In spite of this program, company foresters find that many logging sites in the west are dominated by hardwood species such as red alder. In fact, company timberlands in the northwest contain approximately 1.1 billion board feet of hardwood sawtimber.

Fifteen to twenty years ago, executives with the Corporation viewed red alder and other western hardwood species as weeds. However, growing demand for alder convinced executives that it is a viable commercial species. As a result, Premium Hardwoods was established as a profit center to make use of this resource.

PREMIUM HARDWOODS

The previous president of Premium Hardwoods, William Thompson, ran the division from its formation five years ago until he accepted early retirement during a recent corporate restructuring. At that time, Mr. Cohen took over. Under Mr. Thompson's direction the company sold western hardwood lumber in standard National Hardwood Lumber Association (NHLA) grades.

Paul Morris is in charge of production operations. Paul earned a B.S. degree in Wood Technology from a large eastern university and grew up working at his father's hardwood mill. His first job after graduation was in one of the Corporation's large southern pine sawmill operations. When Premium Hardwoods was formed, Paul (in part because of his experience in a hardwood mill) was asked to transfer to the west coast and run the production operations of the new division.

Kathy West is in charge of sales and marketing for Premium Hardwoods. Kathy holds a B.S. degree in forestry and an MBA. Her first job after graduation was as a procurement forester at one of Pacific Building Products' major mills. However, she later took a sales position and eventually was transferred to Premium Hardwoods.

Ms. West's staff includes a sales manager, and 6 inside salespeople. Premium Hardwoods has never used outside salespeople or technical support personnel since they believed that NLHA grades of lumber were well understood by their customers. Kathy's office is located in Eugene, Oregon, as are the offices of the sales staff and management personnel.

Premium Hardwoods operates three sawmills in the Pacific Northwest with a total annual production capacity of approximately 148 million board feet. The mills have planing and kiln drying capacity and all three utilize single-band headsaws and gang resaws. Logs are debarked and sawn on two sides until all the #1 shop and better material is removed. The resulting cant is then sent to the resaw.

Lumber is produced in lengths of 8 and 10 feet and thicknesses from one to three inches (4/4 to 12/4). The company also owns a remanufacturing facility which purchases western hardwood lumber from within the company (70% of total purchases) and outside sources (30%). The lumber is made into dimension parts which are sold to furniture and cabinet manufacturers.

The company produces hardwood lumber from several west coast species including oak, maple, cottonwood and red alder. However, over 95% of production is red alder. The better lumber is sold in standard NHLA grades to furniture, cabinet, and moulding manufacturers in the western U.S. or used internally in the remanufacturing facility. Lower quality lumber is sold primarily to pallet manufacturers under contracts which range in length from 6 months to 2 years. Sawdust is used to help fuel boilers which provide steam for the dry kilns.

Premium Hardwoods is the largest producer of western U.S. hardwood lumber, accounting for approximately 50% of total production. However, the company controls only a small share of the U.S. hardwood lumber market since western U.S. hardwood lumber production is small in comparison to eastern production. For example, total U.S. production of hardwood lumber last year was approximately 11.5 billion board feet. Mills in the eastern and central United States accounted for 11.2 billion board feet (97%) of this total.

The majority of lumber production (75%) is marketed directly to industrial lumber users such as furniture, cabinet, and pallet producers. Brokers market approximately 10% of production and 10% is used internally. The remaining portion of production is marketed to distribution yards or directly to retailers.

RED ALDER

Red alder (*Alnus rubra* Bong.) is the most common hardwood species on the west coast. Sawtimber volume in Oregon and Washington is 24.4 billion board feet (International $\frac{1}{4}$ inch rule). Unlike many softwood species, which are concentrated on public lands, 80% of red alder sawtimber volume is located on private lands.

Annual growth of red alder in the Pacific Northwest is estimated to be 2 to 4 times greater than the volume harvested. However, this excess growth is due, in part, to the high cost of harvesting the relatively small logs. Environmental regulations regarding logging near streams (where alder stands are often found) increase the cost of harvesting and contribute to this underutilization.

Red alder has a rotation age of 20 to 50 years and a lifespan of approximately 80 years. Relatively little of the present growing stock on commercial forest lands is younger than 20 years and the proportion of growing stock consisting of trees 17 inches or larger in diameter is estimated at 25%. Exhibit 1 provides the grade yield mix by red alder log diameter. The average log processed by Premium Hardwoods has a diameter of 14 inches.

Low grade red alder logs are most often converted to pulp chips and higher grade logs are sawn into lumber. In the domestic market, select and shop grades of lumber are used by the furniture (including upholstered) and cabinet industries. Exhibit 2 provides a breakdown of Pacific Northwest hardwood (the majority of which is red alder) use by value.

Demand for alder has been growing for some time. The Northwest Hardwoods Association states that there has not been a decline in demand for alder products since 1955. Today, much of the demand has been for alder chips in export markets (primarily to paper manufacturers in Pacific Rim countries). In fact, a few companies have closed their sawmill operations and concentrated on producing chips for the export market. Domestic demand for alder lumber is primarily on the west coast, relatively little is shipped to the eastern United States. However, eastern hardwood lumber is routinely shipped to west coast lumber users.

Red alder is a uniform, fine textured wood that has properties similar to yellow-poplar and soft maple (Exhibit 3). Unlike east coast hardwood lumber, red alder lumber has traditionally been sold kiln-dried and surfaced (abrasive or knife planed). NHLA lumber grades for red alder (Exhibit 4) differ from many grades for eastern hardwood species in that the latter is typically graded on the poorest face in the rough condition while the former is typically graded on the best face after surfacing.

The strong chip market and the resulting demand for alder logs helped to increase the average market price of sawlogs (12–14 inch diameter, 24–40 feet long, FOB mill) from approximately $190/MBF to nearly $260/MBF last year. Prices for the best grade of red alder lumber are comparable to prices for similar grades of hard maple and birch. When compared to the *Weekly Hardwood Review* price index (a composite price index for a variety of lumber products) they appear to be more stable (Exhibit 5). Premium Hardwoods has found that customers often "price shop" among the various western hardwood lumber producers. Both customers and producers rely on industry price reporting publications when negotiating a sale.

EXHIBIT 1
GRADE AND VALUE RECOVERY PER THOUSAND BOARD FEET LUMBER TALLY (MLT) BY
VARIOUS DIAMETERS OF WOODS-RUN, MILL-LENGTH RED ALDER LOGS

Diameter		Grade, %					$/MLT
	Select	No. 1 shop	No. 2 shop	No. 3 shop	Frame	Pallet	
7	1	3	4	5	2	85	170
8	2	9	7	6	5	72	195
9	4	14	9	6	6	61	220
10	6	17	11	6	7	53	246
11	9	20	11	5	8	46	271
12	13	22	11	5	3	41	296
13	17	23	11	5	8	36	322
14	21	25	11	4	8	32	347
15	25	26	10	3	7	29	372
16	29	26	10	3	7	26	397
17	33	27	9	2	6	23	423
18	38	27	8	1	5	21	448
19	42	27	6	1	5	19	474
20	47	27	5	0	4	17	499

PLANNING FOR THE FUTURE

Business is quite good at present and Premium Hardwoods has had no trouble selling its standard
NHLA grades of alder lumber. However, Fred thinks that performance can be improved and is
concerned with the company's ability to profit during an economic downturn. One of the
company's three mills, a 50,000 MBF/Yr. operation in Ironta, Washington, has been the least
profitable over the last few years. Fred thinks that changes should begin with this mill. He asked
Paul and Kathy for suggestions on how the Ironta operation could be changed to help reach his
goals for Premium Hardwoods.

EXHIBIT 2
PACIFIC NORTHWEST HARDWOOD USE BY VALUE

End use	% of total value
Fine furniture	19
Upholstered furniture	20
Institutional furniture	3
Cabinets	18
Chips	17
Pallets	20
Plywood	2
Other	1
Total value = 98.0 million dollars	

EXHIBIT 3
PHYSICAL AND RELATIVE WORKING PROPERTIES OF RED ALDER AND SELECTED
OTHER SPECIES AT 12% MOISTURE CONTENT

	Species				
Property	Red alder	Yellow-poplar	Soft maple (silver)	Pine (longleaf)	Douglas fir (coastal)
Specific gravity	.41	.42	.47	.59	.48
Modulus of elasticity, megapascals	9500	10,900	7900	13,700	13,400
Modulus of rupture, kilopascals	68,000	70,000	61,000	100,000	85,000
Volumetric shrinkage % (green to OD)	12.6	12.7	12.0	12.2	12.4
Ease of machining	Very good	Very good	Very good	Good	Good
Resistance to splitting	Excellent	Excellent	Fair	Fair	Fair
Screw and nail holding	Fair	Fair	Good	Excellent	Good

EXHIBIT 4
NATIONAL HARDWOOD LUMBER ASSOCIATION GRADES AND GRADE REQUIREMENTS
FOR RED ALDER FACTORY LUMBER

Grade	Required cuttings, %	Width, inches	Length, feet
Selects and better	83⅓ (clear)	4+ (max. 5% of 3 in.)	4 and longer
No. 1 Shop	66⅔ (clear)	4+ (max. 5% of 3 in.)	4 and longer
No. 2 Shop	50 (clear)	4+ (max. 5% of 3 in.)	4 and longer
No. 3 Shop	33⅓ (sound)	3+	4 and longer
Frame grade	83⅓ (sound)	4+	7 and longer

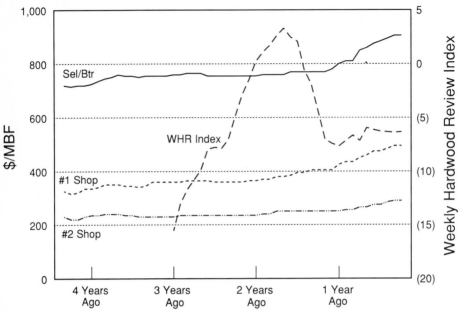

EXHIBIT 5
Prices for various grades of 4/4 kiln-dried red alder and the *Weekly Hardwood Review* hardwood price index.

Kathy's Proposal

Kathy suggested that instead of selling alder lumber using the standard NHLA grading system, the company develop its own grade system. These company specific or proprietary grades could be developed to better meet the needs of their customers and would be produced at the Ironta mill. Kathy envisions the following grades:

Premium grade. This grade would be the best lumber available. Narrow boards and boards with most types of defects would not make this grade. Premium grade would be similar but not identical to the NHLA select and better grade.

Moulding grade. As the name implies, this grade would be designed for moulding and millwork producers. Boards would be long and essentially clear on one face. Moulding grade would be selected from what is now #1 shop grade and some narrow select and better lumber.

Cabinet grade. This grade would be developed with the intent of maximizing the clear cuttings yield for cabinet manufacturers. It would consist primarily of #1 shop grade material and have a high percentage of cuttings which are clear on two faces. However, cabinet grade lumber could be shorter than moulding grade and need not be as clear since it will be cut into shorter lengths by the cabinet company.

Upholstery grade. Upholstery grade would be selected for structural integrity rather than freedom from visual defects. Boards would generally be longer than cabinet material but not as long as moulding grade. Upholstery grade would be selected from what is now #1 and #2 shop material.

Strip grade. Strip grade would be designed for exposed frame parts in cabinets. The pieces would be narrow (under 6 inches), relatively short, and essentially clear.

Kathy thinks there may be a need to develop additional grades for markets such as the pallet industry. Various combinations of length, width, planing, and grade would be sorted from production to best meet the customer's specific needs. Over twenty-five separate sorting operations would probably be required.

Kathy pointed out that the key to this scheme would be to offer customers better value. While the initial price of the new grades might be higher than for the grades customers presently use, the new grades would result in better yields and lower cost per usable part since they would be tailored to the specific needs of the user. Additional sorting by length and width would reduce the variability within and between loads of a particular grade. Consequently, the new grades would be more consistent than NHLA grades, would facilitate greater operator efficiency in the rough mill, and would require fewer setup changes. This would promote efficiency in the user's manufacturing process and increase the value of the product to the customer.

Value could also be increased by abrasive rather than knife planing some grades since abrasive planing eliminates some roller splitting, torn grain, and miss. The company presently owns only knife planing equipment and an abrasive planer would cost approximately $60,000.

An advantage of proprietary grading, Kathy suggested, would be to avoid price competition. Customers could no longer bargain based on the prices offered by Premium Hardwoods' competitors since no other company would offer the same grades. However, extra employees would be required at the mill to conduct the additional sorting and handling. Additional employees would almost certainly be required for outside sales and to provide customers with technical information about using the new grades. Kathy thinks that advertising and promotional materials will also be needed. Using the company's accounting practices, Kathy developed some financial projections for her proposal (Exhibit 6).

Paul's Proposal

Paul had a different approach. He noted that red alder is accepted by the Western Wood Products Association (WWPA), a major trade association, and the American Lumber Standards Committee (ALSC) in the STUD grade of construction lumber. He recommended that Premium Hardwoods begin producing alder studs.

Since alder is prone to warp, he thinks that the Saw-Dry-Rip (S-D-R) process should be used. Saw-Dry-Rip has been shown to be quite effective in eliminating warp in studs manufactured from aspen, yellow-poplar and alder. When using S-D-R the log is first live sawn into flitches. The flitches are then dried, ripped to width, and planed. Paul thinks that existing equipment could be

EXHIBIT 6
ESTIMATED FINANCIAL DATA

	Paul's proposal: studs	Kathy's proposal: proprietary grades	Now: NHLA grades
Avg. selling price, $/MBF	230	420	365
Avg. variable cost, $/MBF	212	382	340
Fixed costs, $	612,000	680,000	575,000
Projected prod. and sales, MBF	80,000	40,000	50,000

modified to produce studs using the S-D-R process at a cost that is competitive with Douglas-fir or southern pine studs.

Producing alder studs would require that more material be kiln-dried. However, the existing kilns would be adequate (with minor modifications) for the increased production since a faster, high-temperature schedule can be used with the S-D-R process (18 hours as compared to approximately 100 hours using a conventional schedule).

Paul predicted that producing alder dimension lumber using the S-D-R method would yield 91% STUD grade lumber, 8% ECONOMY grade, and 1% would be rejected. Lower quality (and lower priced) logs would be purchased for the stud mill. The studs would be used in the construction of homes and other light frame buildings.

Paul suggested four advantages to this plan:

1 Stud production would reduce the company's susceptibility to downturns in the furniture and cabinet markets by diversifying into the construction market.

2 Studs could be made from lower grade logs which would otherwise yield primarily low grade lumber. The price per thousand board feet (MBF) for kiln-dried and precision end trimmed (PET) studs (approximately 230 $/MBF for Douglas-fir) can exceed that of #3 Shop red alder. Even at a slightly lower price, Paul thinks that increased production efficiency when producing studs would make up the difference.

3 Alder timber is located, primarily, on private lands. Consequently, it will not be involved in the environmental policy conflicts which presently affect public lands and may limit the supply of softwood timber. If limited supplies increase the price of softwood timber, alder studs could have a cost advantage.

Paul also developed financial projections for his proposal. These figures, along with figures for the current Ironta mill operation, are provided in Exhibit 6.

SUMMARY

Fred thought that both ideas had merit and wanted additional information. He asked for an analysis of each proposal including: advantages and disadvantages relative to his goals for Premium Hardwoods, the feasibility of each proposal, and the reality of assumptions. He was particularly concerned with how the resources and expertise of the company would fit with the markets addressed by the two proposals as well as the factors which drive the markets.

Fred also had some ideas concerning possible changes. He thought there might be some advantage to producing dimension parts at the Ironta mill. This could involve closing the sawmill and putting a dimension operation in the existing building or adding a building and producing both lumber and dimension parts at the same location. He wondered about possible efficiencies that could be gained if the existing dimension facility were moved to the Ironta location. Since the demand for alder chips was strong, Fred also considered the idea of converting the Ironta mill to a chip production facility.

Since Fred would be meeting in one week with Corporate executives to discuss strategic planning for Premium Hardwoods, he needed the analysis and a recommended course of action by then.

APPENDIX B: Major Company Biographical Sketches

Boise Cascade
1 Jefferson Square
Boise, ID 83728

COMPANY PROFILE

Boise Cascade is a major integrated forest products firm with assets in all four major geographic regions of the United States. Annual sales by the early 1990s exceeded $4.3 billion and were split between three major product groups: paper products 55 percent, office products 22 percent, building products 22 percent (1 percent in other).

BUSINESS STRATEGY

Robert Hansberger, the President of Boise Payette Lumber Company, at age 36, arranged a merger with another lumber producer of similar size, the Cascade Lumber Company. The merger of these two Idaho lumber companies in 1957 formed the new Boise Cascade Corporation. Boise Cascade Corporation, under Hansberger's leadership, grew at a rapid rate through a long series of acquisitions during the late 1950s and 1960s.

Boise Cascade's initial acquisition strategy was that of integrated utilization. That is, Boise grouped complementary timber conversion operations together in order to achieve the maximum return from each log. In a typical sawmill as much as 50 percent of the round log is converted to waste materials as the log is converted to square lumber. In the past, many firms had simply burnt this waste. However, by utilizing this waste to produce other forest products, such as paper or particleboard, Boise Cascade could achieve the maximum return from its logs.

Through additional mergers in the 1960s Boise Cascade also became a significant producer of paper and packaging products. It also integrated forward in the wood products business and ventured into home building and land development. Hansberger noted at this point in time that his company was in the "shelter business." Boise Cascade further integrated its operations to the retail level within the building products industry. It was one of the very few major integrated forest products firms to have a significant building products retail group.

Hansberger, in the late 1960s, was caught up in diversification fever and proceeded to turn Boise Cascade into a conglomerate. One of the more interesting acquisitions was the Cuban Electric Company which Boise has had claims against the Cuban government for since 1960. For a while it seemed this strategy was working and Boise Cascade became one of the glamour companies of the late 1960s. However, by the 1970s this strategy had almost proved disastrous and Hansberger was forced to resign.

In 1972 the Board of Directors chose John B. Fery as the new Chairman and Chief Executive Officer of Boise Cascade. Fery is credited by many with rescuing Boise Cascade from near disaster by selling off many of the earlier diversified assets acquired by Hansberger. The income generated by selling these nonforest products assets was reinvested back into the core forest products businesses of Boise.

Fery instituted a sophisticated planning system for Boise Cascade with 5- and 10-year blueprints along with comprehensive business reviews. These became the basis for strategic discussions at Boise Cascade headquarters. Boise spent nearly $3 billion to move the corporation into what Fery believed were young markets with high growth potential. Boise moved rapidly into white paper production where demand was being strengthened by the office computer and copy machine. Fery also lessened Boise's dependence on wood and building materials which are very vulnerable to the ups and downs of the housing market.

In the 1970s, Boise Cascade was largely a lumber company with wood products accounting for half of company sales. The company is still a large lumber and structural panel producer; however, by the early 1990s paper and office product operations for Boise Cascade accounted for over three-quarters of sales. Boise Cascade's clear strategy for the future is to concentrate on focused growth in the paper, packaging, and office products businesses. It is a recognized leader in high-value white papers and in office products distribution. Boise Cascade operates a string of office products distribution centers which serve all 50 states. These distribution centers have given it a strong position in the office products market and provide a unique marketing outlet for Boise's fast-growing white paper production.

Boise Cascade has announced that it does not plan to grow much in building products. However, better efficiency in existing mills is clearly a strong theme. In 1983, Boise Cascade owned 87 retail building material centers, which served primarily the small remodeling contractors and do-it-yourself consumers. To support these retail building material centers, it operated 15 wholesale distribution units. These facilities were located in high-growth areas of the western and southwestern United States. However, by 1988 only nine wholesale distribution units remained. The retail centers were closed or sold. Lowe's purchased 25 stores in 1985, expanding their market area into Texas and Oklahoma, and the last 20 retail stores were sold in 1987.

Boise Cascade had a three-point strategy during the 1980s:

1 Become a more efficient producer and distributor of paper, paper products, office products, and building products
2 Develop and maintain a distinctive competence by adding value to its products and services, thereby gaining a competitive edge
3 Focus growth in high-potential segments of its business

In recent years, a total commitment to quality has been adopted as part of the company strategy to augment this three-point strategy.

MAJOR PRODUCT LINES AND MARKET POSITIONS

Boise Cascade is a major producer of uncoated free sheet, newsprint, uncoated ground wood, and wood building products. It also operates one of the largest office products distribution systems. Boise Cascade owns 2.9 million acres of timberland and controls another 3.3 million acres through leases, contracts, and licenses.

Champion International
One Champion Plaza
Stamford, CT 06921

COMPANY PROFILE

The new Champion was formed in 1984 with the acquisition of St. Regis Corporation. This newly combined company commands strong market positions across a variety of paper grades including coated and uncoated white papers, newsprint, and bleached paper board. Additionally it owns Nationwide Papers, a wholesale paper distribution firm, and has a large lumber and plywood business along with 6.4 million acres of timberland.

BUSINESS STRATEGY

When Andrew Sigler became Champion's Chairman in 1974 the corporation was suffering from overdiversification, which had resulted in a hodgepodge company created by his predecessors. The company had been caught up in the jargon and faddism of the 1960s and attempted to turn Champion into a company that would "serve the total home environment" (*Forbes,* March 5, 1979).

Sigler's commitment was to trees and he set about busily undiversifying the company. He sold 18 separate companies, which provided him an additional $400 million of capital. This capital along with retained earnings was used to purchase Hoerner Waldorf, a strong packaging firm located in St. Paul, Minnesota. Additional capital was used to upgrade the production facilities of the firm.

In 1984, Champion made its biggest acquisition ever, St. Regis Corporation. This resulted in a significant debt load and by early 1985 the company had established a goal of paying down $1 billion of acquisition-related debt by the end of 1986. This resulted in more selling, including six folding carton plants, two recycled paperboard mills, and envelope manufacturing, laminated and coated products, and flexible packaging businesses. In addition, a pulp and linerboard mill, the Wheeler Construction Products business, distribution yards and industrial supply operations, and various other pieces of the corporation were sold. Stone Container purchased the brown paper packaging business, and the U.S. Plywood operations, which had been originally purchased in 1972, were spun off in a leveraged buyout by the management.

Over $1.1 billion was raised through these divestiture operations to pay down the debt of acquiring St. Regis. The impact of this divestiture program was evident in total corporate sales, which dropped from approximately $5.8 billion in 1985 to $4.4 billion in 1986. However, net income rose from $163 million in 1985 to over $200 million in 1986. Champion's goal, as stated in its 1987 annual report, is to substantially increase earnings by becoming a low-cost manufacturer of quality products with service second to none. It was certainly a leaner and more profitable corporation in the late 1980s than it was in the mid-1980s. Its current product mix is focused on high-growth white paper such as printing, writing, and publication papers.

PRODUCT LINES

During 1986, Champion reorganized into four product line business units: printing and writing papers; publication papers; newsprint; pulp and kraft and forest products. The paper and pulp side of this mix contributed three-quarters of company sales and its forest products contributed one-quarter of company sales in 1986. Champion's largest business unit, the printing and writing papers unit, produces such products as coated and uncoated grades of white paper for forms, bond, envelopes, brochures, catalogues, and annual reports. Champion is a major producer of uncoated free sheet and in the late 1980s was the largest producer of coated groundwood paper.

The publication papers group produces paper used primarily in magazines, catalogues, and periodicals. These products are manufactured primarily at three Champion mills in Maine, New York, and Minnesota. Newsprint, pulp, and kraft operations contribute strongly to sales in most years.

The forest products unit contributed over $1 billion in annual sales during the late 1980s. This unit is responsible for managing lumber mills, plywood plants, and 6.5 million acres of domestic timberlands.

Champion operates a Weldwood subsidiary in Canada (mostly in British Columbia). Weldwood operates plywood, lumber, and OSB mills, and has a string of sales offices and distribution centers.

Champion also has a pulp and paper operation in Brazil.

FUTURE GROWTH

Champion's clear strategic focus is to remain a major player in the printing and writing papers, publication papers, and newsprint markets.

Chesapeake Corporation
Box 2350
Richmond, VA 23218

BACKGROUND

Chesapeake Corporation is an integrated paper and forest products company headquartered in Virginia, organized in 1918. Chesapeake is a major manufacturer of kraft paperboard and paper, bleached hardwood pulp, and commercial and industrial tissue products. It converts paperboard and other products into corrugated containers, displays, and other packaging products. Chesapeake also converts timber into lumber and has a significant stake in the treated-lumber marketplace. During the 1980s, Chesapeake grew rapidly and more than tripled its sales.

PRODUCT LINES

Chesapeake has four core businesses: kraft products, tissue products, containers, and treated-wood products.

Kraft Products

Chesapeake's overall paper product line is the dominant segment of corporate sales. This segment accounts for over 50 percent of total corporate sales.

The West Point, Virginia, facility, the original paper mill in the company, has concentrated on producing kraft paper products specializing on a family of trademarked products, TRI-LITE, C-WHITE, and TRI-KRAFT. TRI-KRAFT is a three-ply kraft paper which Chesapeake believes will give it a major technological edge in the market.

Tissue Products

During 1985, Chesapeake acquired from Philip Morris two additional paper mills, Wisconsin Tissue and Plainwell Paper. The Wisconsin Tissue facility uses recycled paper as its furnish and has developed a marketing strategy based on a full line of disposable products for the commercial and industrial away-from-home tissue markets. This is a rapidly growing niche market and has been a successful one for Wisconsin Tissue and also for other firms such as Fort Howard Paper.

Plainwell Paper, a producer of printing and specialty papers, was sold in late 1987.

Containers

Chesapeake Corporation owns container and box plants located from Ohio to North Carolina to New York. This is a very competitive market with many competitors. The container group's marketing strategy seeks to make Chesapeake a trendsetter rather than a follower.

Wood Products and Treated-Wood Products

Chesapeake moved into this segment heavily during the 1980s, being primarily a paper firm prior to that point. Its production facilities at one point included both hardwood and softwood sawmills, a softwood plywood facility, a manufactured housing division (Home Craft), and wood-treating facilities. The manufactured housing division, the plywood plant, and a sawmill were all sold in 1987. The strategic decision was to concentrate on treated-wood products. With modern wood-treating plants in North Carolina, Virginia, Maryland, and Pennsylvania, the company is a major player in northeastern markets. Chesapeake produces a decay-resistant treated product and a brand-named decay- and moisture-resistant product, WOOD + PLUS.

SUMMARY

Chesapeake Corporation, a medium-sized wood products firm, is attempting to transform itself from a company based primarily on commodity products to a company whose products serve clearly identified markets where they have developed specific strengths. It is working hard at developing innovative products to complete this specialty product marketing strategy. In addition, it is devoting energy to servicing its customers in ways which meet or exceed the customer's expectations. Chesapeake indicates it is more interested in long-term strategic results and is less interested in quarter-to-quarter increases and earnings per share. The degree of commitment it has to these goals can only be borne out through time and experience in the marketplace. However, during 1989 it was named one of America's ten best-managed companies by *Business Week* magazine.

Georgia-Pacific
133 Peachtree Street NE
Atlanta, GA 30303

In 1927, a Virginian named Owen Cheatham borrowed $12,000 and purchased a wholesale lumberyard in Augusta, Georgia. It wasn't long before Mr. Cheatham was able to entice his college friend, Robert Pamplin, into joining him in this new business. Together these two men built the company now known as Georgia-Pacific.

The early years of Georgia-Pacific were marked by the Great Depression and World War II; however, by 1946 the company owned five lumber mills in Alabama, Arkansas, Mississippi, and South Carolina, along with its original lumberyard in Augusta. In addition, it had established sales offices in New York and Portland, Oregon.

Beginning in the late 1940s Georgia-Pacific entered into an era of tremendous growth. In the late 1940s the company moved into the plywood industry by purchasing and building several west coast plywood mills. This was considered quite risky at that time because plywood was a very new product with an unsure market potential. Beginning in the 1950s Georgia-Pacific moved aggressively into the pulp and paper field and began to acquire some of the last remaining large tracts of old-growth west coast timberlands. The corporation, at that time, borrowed very heavily to acquire these lands and then through accelerating its harvesting schedules produced enough cash to pay its short-term creditors and bring its long-term debt down to a manageable level.

The corporate office, during these early years, moved from Augusta, Georgia, to Portland, Oregon. However, beginning in the 1960s Georgia-Pacific made a strong move back into its original base of the southern United States. Large tracts of land were acquired in the Crosset, Arkansas, area and in other deep-south states. True to form, in 1963 Georgia-Pacific built the first southern pine plywood plant. And by the end of that first year it had an additional four plants running, which gave it about a 1-year lead over its major competitors. Georgia-Pacific has continued its emphasis on plywood and other specialty panels to this day and remains the world's largest plywood manufacturer.

Throughout its history Georgia-Pacific has tried to emphasize market positions in fast-growth fields. Early on it chose to emphasize plywood production over that of lumber and also to emphasize the rapidly expanding pulp and paper segment of its business. Georgia-Pacific, like Weyerhaeuser, decided to emphasize its distribution of building products to the building materials retail industry, believing this gave it a strong advantage in the marketplace. Today, Georgia-Pacific's distribution division is the largest wholesale distribution system for building products in North America.

In 1987, Georgia-Pacific consolidated its leadership in building products distribution by acquiring the privately held U.S. Plywood Corp., which had been spun out of Champion International in a leveraged buyout by its managers. U.S. Plywood had the third largest distribution system behind Georgia-Pacific and Weyerhaeuser. Georgia-Pacific uses this extensive distribution network as its eyes and ears in the marketplace. It has a strategy of integrating backward from distribution, with recent examples in molding and roofing products.

Perhaps the largest setback to the growth of Georgia-Pacific Corporation resulted from a 1972 Federal Trade Commission ruling in which the company was deemed to have illegally restricted competition within the softwood plywood industry. The company was forced to sell 20 percent of its assets centered in Louisiana. As a result, the Louisiana-Pacific Corporation was born and was immediately the sixth largest domestic producer of lumber. In addition, the Georgia-Pacific Corporation was prohibited from purchasing any additional softwood plywood producers in the United States for 10 years.

In the late 1950s, Georgia-Pacific diversified into the chemical industry. This began as a result of the company's desire to manufacture the resins necessary for its plywood plants. However, the more recent corporate goals, as articulated by Chief Executive Officer T. Marshall Hahn, have been to liquidate all but the most profitable of the chemical operations and invest the proceeds of the sales into modernizing the core forest products business of the corporation.

In the early 1980s, Georgia-Pacific began construction of a new corporate headquarters in Atlanta, Georgia. It was felt desirable to move the corporate headquarters from Portland to Atlanta to be nearer to the major part of Georgia-Pacific's assets. By the mid-1980s, the corporation owned 4.6 million acres of timberlands in North America, of which 65 percent was in the southern United States, 16 percent in the Pacific northwest, and 19 percent in the eastern United States and eastern Canada.

In 1990, Georgia-Pacific acquired Great Northern Nekoosa, a major U.S. producer of corrugated packaging and a major paper distributor and envelope producer. At the beginning of the 1990s, this clearly places Georgia-Pacific as the largest U.S. forest products corporation.

Grossman's Inc.
200 Union Street
Braintree, MA 02184-2997

BACKGROUND

Grossman's Inc. is one of the new-old firms in the home center marketplace. Grossman's Inc. became a separate company again on November 19, 1986, as it was transferred out of the Evans Products bankruptcy proceedings. However, the history of Grossman's dates back to its founding in 1896 by the grandfather of the current Chief Executive Officer, Mike Grossman. Grossman's was run for many, many years as a family concern and was a pioneer in the cash-and-carry era of lumberyards.

In the mid-1960s, Evans Products Corporation began to acquire building materials retailers. Their first acquisition was in 1965, when they acquired the Moore's chain headquartered in Richmond, Virginia. Later in 1969, Evans acquired Grossman's retail chain and formed a retail group headed by Mike Grossman in 1970. In 1979, Evans acquired Lindsley's chain for its retail group. Although the retail side of Evans had done well for a number of years, the corporation as a whole was doing very poorly in the early 1980s and on March 11, 1985, Evans Products declared bankruptcy and filed for protection under the Chapter 11 clause of the Federal Bankruptcy Code. After a considerable period of uncertainty and time for other retail home centers to gain competitive advantage, Grossman's did emerge from the Evans bankruptcy procedure in November 1986 as a separate company.

SALES AND FACILITIES

Grossman's is one of the larger retail home center chains, operating 156 retail stores at year-end 1989. Sales were roughly 80 percent to the do-it-yourself (DIY) customer segment and 20 percent to the professional segment of the market.

Grossman's operates basically two kinds of retail stores. The first is aimed primarily at the do-it-yourself marketplace, with the typical store in smaller markets having approximately 49,000 square feet of selling space of which 24,000 square feet is under roof with the rest in an outdoor lumberyard. In larger markets the company facilities may be as large as 100,000 square feet of selling space of which 60,000 square feet is under roof and the remainder in an adjacent open-air lumberyard. The second type of retail store Grossman's refers to as a "dual-yard." These dual-yards cater to consumers, builders, and contractors. They are an outgrowth of the traditional lumberyard. A typical dual-yard has approximately 45,000 square feet under roof of which about 12,000 square feet is retail showroom space and the balance is warehouse storage. There is an additional outside storage area of 40,000 to 60,000 square feet. Dual-yards also have sales staffs which call on builders and contractors, and provide a delivery service for those customers.

STRATEGIES

Although the bankruptcy proceedings allowed its competitors to gain an advantage in some markets, Grossman's has been fighting back. A store improvement program is well under way. This is a multiyear program to remodel, enlarge, and in many cases relocate stores for the company. In its recent annual reports, Grossman's highlights the following five strategies:

1 Concentrate on existing markets where the company is already successful
2 Continue the capital improvement program to upgrade existing stores

3 Broaden the selection of products carried to establish category dominance and extend new merchandising programs throughout the chain

4 Strive to better serve customers

5 Improve financial performance

In 1989, Grossman's sold its Moore's and Northwest Divisions to raise cash to pay down long-term debts.

Hechinger Company
3500 Pennsy Drive
Landover, MD 20785

Price is only one piece of the marketing puzzle. Customers want a fair price, but they want much more. They want choice. They want quality. They want compelling presentation. They want information and ideas. And they want service. Most of all, I believe, they want value.

John Hechinger, Jr., 1986

COMPANY PROFILE

Hechinger Company is known widely in the industry for at least two reasons. One is its unique slogan, "The World's Most Unusual Lumberyard," and the second is because it is recognized as a leader in DIY home products retailing. Hechinger was founded in 1911 and by the early 1990s was operating over 100 home centers with annual sales exceeding $1.2 billion.

Hechinger's growth through the 1980s was very strong. The decade of the 1980s closed with a sprint of growth as Home Quarters and Triangle Building Centers were acquired, along with valuable leases on prime retail locations.

MARKETING STRATEGIES

As with most home center retailers, Hechinger developed a strong geographic base. Its primary base has been in the Washington, D.C., Maryland, Virginia area, with other centers of strength in Pennsylvania and the piedmont of North Carolina. Expansion in the late 1980s and early 1990s moved it beyond this original base to encompass the northeast and southeast.

Hechinger has twin strategies for growth in the 1990s: capitalize on the proven appeal of the Hechinger stores and become a significant participant in the warehouse home center segment through its Home Quarters division. Hechinger has a special sensitivity to the needs of women and to the fact that women are the motivators of most do-it-yourself projects as well as becoming active do-it-yourselfers themselves in growing numbers. Hechinger is one of the few home center chains which has a significant proportion of women at its top level of management and in key positions. Also, it has made a strong search to include women in its store manager training programs and professional staff.

While carrying a wide variety of products, Hechinger uses the age-old retailing principle of trying to get customers to upgrade their purchases to the higher-quality and higher-priced items. Hechinger's is known for providing top-quality merchandise to its customers. Hechinger's offers its own credit service and invests heavily in advertising and promotion on a year-round basis. It is sophisticated in its use of computerized marketing research and census information, economic indicators, in-depth consumer surveys, and exhaustive contacts with local planning officials to develop new store locations.

The Home Depot
2727 Paces Ferry Road
Atlanta, GA 30339

BACKGROUND

The Home Depot is a pioneer building products retailer for the warehouse home center industry. Bernard Marcus and Arthur Blank were the original founders. They were both fired in 1978 after a management shakeup at the Handy Dan Home Center chain. After their dismissal from Handy Dan, they formed The Home Depot, which began to operate with an entirely new philosophy for building materials retailing.

Their new philosophy was to be the dominant factor in every market their company serves. The key to their success, according to them, has been upon entering a new market to make a substantial commitment: that is, opening new stores, providing excellent customer service, creating highly visible promotions, and thereby "growing" the entire market. Their primary target is the do-it-yourself market, with a secondary emphasis on the home remodeling market.

THE WAREHOUSE CONCEPT

The warehouse concept pioneered by The Home Depot is unique. Warehouse home centers are 65,000 to over 100,000 square feet of selling space, with an additional 10,000 to 15,000 square feet of outdoor selling area. In these large stores, all the materials and tools needed to build a house from scratch are stocked, along with anything necessary to landscape its grounds. These stores generally function as their own warehouses and typically stock over 25,000 to 30,000 different items. They advertise heavily using a low-price theme while at the same time providing a tremendous selection of building materials and brand name merchandise.

This size of operation is in direct contrast to a typical home center, which has an average store size of approximately 25,000 square feet and may stock 10,000 to 15,000 different items.

CURRENT STATUS

The Home Depot has grown tremendously from its beginning in 1978 to become one of the largest home center chains today. It has shown a remarkable ability to manage this rapid growth and has increased its earnings as well. However, The Home Depot does face some significant competition in various markets from other warehouse home centers such as Builders Square and Homeclub.

MARKETING STRATEGY

The Home Depot has an interesting marketing strategy of concentrating its operations in specific geographic markets. It typically chooses a large metropolitan area and opens several stores within that area. This gives it tremendous efficiencies of scale in both transportation and advertising. It also allows the company to become the dominant factor in that market. The company strongly believes that it can "grow the market" in all its geographic areas. It believes it does not have to take business away from traditional hardware stores and other traditional home improvement outlets, but rather that it can create new do-it-yourselfers out of those who have never done their own home improvements before. The company attempts to accomplish this through strong advertising and holding customer clinics to develop such skills as electrical wiring, carpentry, and plumbing, to name a few.

International Paper
Two Manhattanville Road
Purchase, NY 10577

COMPANY PROFILE

Incorporated in 1898, International Paper is one of the world's largest integrated papermaking firms. Its product lines are extensive, including uncoated white paper, coated paper, market pulp, containerboard, bleached packaging, industrial paper, envelopes, timber, wood products, and other specialty products. International Paper is also vertically integrated toward its markets, with building products distribution centers and paper distribution centers.

HISTORICAL

International Paper was quick to apply new technology to achieve a dominant position in the paper industry during the early part of this century. However, by the 1960s its rate of return on net worth and its growth in sales were much lower than competing paper companies. Then rumors on Wall Street that a group of investors was attempting to buy sufficient shares of International Paper stock to control the company shocked the directors out of their complacency.

Under the leadership of Edward Hinman and Frederick Kappell (a former chairman of AT&T) International Paper began a drive in the late 1960s to expand and diversify. International Paper moved into the consumer tissues and disposable diapers market, purchased a producer and distributor of specialty health care products, and also spent over $1 billion to increase International Paper's capacity by 25 percent. All of this served to leave International Paper strongly in debt after having almost no debt as late as 1965.

To better manage this situation, Paul Gorman took over the helm of International Paper in early 1971. Gorman was apparently chosen because of his strong financial abilities. He immediately began to close and/or sell the unprofitable recent acquisitions and to modernize International Paper's financial systems. With these new controls in place, International Paper began to manage more on the basis of profit and not just as an attempt to be the largest paper company.

After Gorman's retirement in 1974, the new chairman, J. Stanford Smith, signaled a new strategy, indicating that International Paper was no longer just in the paper or even the forest products business but rather in the land resource management business. This thinking led International Paper to acquire new companies to provide it with the expertise it needed to develop oil deposits which had been discovered on the company's large land holdings. In addition, a new emphasis was placed on solid-wood products in the mid to late 1970s, when the markets for lumber and other wood products were strong.

In 1979, at the time that Dr. Edwin A. Gee assumed the chairmanship of International Paper, another strategy change was signaled. Gee moved to consolidate International Paper's operations back to its central core of basic forest products. The oil and gas subsidiaries were sold, along with other assets not directly related to the core business. The cash raised through these sales was plowed back into a capital spending program to upgrade International Paper's aging paper mills. Since 1979, International Paper has spent approximately $6 billion in modernizing its facilities. International Paper's latest chairman, John Georges, has continued to upgrade and modernize the core businesses.

STRATEGIES

International Paper has been sticking to its knitting since the early 1980s, with a stated goal of becoming one of the low-cost producers of white paper, packaging, containerboard, and bleached board (*Business Week,* July 28, 1980). To reach this goal, International Paper converted some of its paper mills and modernized them to produce uncoated white papers. At the same time, International Paper announced its intention to move out of the newsprint business and has either converted newsprint mills to produce other paper products or sold them.

In 1986, International Paper purchased Hammermill Paper. Its paper lines fit well within International Paper's strategy to boost sales of value-added products with premium prices in less cyclical markets. The Hammermill Bond name in business papers should be very valuable for International Paper.

International Paper has upgraded its wood products segment also. It is one of the top five largest producers of southern pine lumber. International Paper has focused on specific markets such as export lumber, white pine paneling, CCA-treated (chromated copper arsenate) products, and MSR (machine-stress-rated) lumber. In 1988, it acquired the Masonite subsidiary from U.S.G. to give it the premier producer of hardboard products. To make itself less attractive as a takeover target, International Paper manages its 6.2 million acres of timberlands through International Paper Timberlands, Ltd., a timberland partnership.

International Paper became much more market-oriented in the last half of the 1980s, stating that its highest priority was to improve the quality of every product and service it offers.

James River
P. O. Box 2218
Richmond, VA 23217

COMPANY BACKGROUND

James River Corporation is one of the world's largest paper companies. Headquartered on the banks of the James River in Richmond, Virginia, it is an integrated manufacturer and converter of paper and paper-related products, including pulp, plastic products, and coated film. Through its subsidiaries, the company processes basic raw materials—wood, wood pulp, synthetic fibers, and plastic resins—into finished products such as towel and tissue papers, disposable food and beverage service items, and folding cartons, as well as a wide array of communication papers and specialty industrial packaging papers.

CORPORATE STRATEGY

Since its founding in 1969, James River has pursued a strategy of acquiring specialty paper manufacturing operations. Acquiring operations and investing in their modernization has proved to be a successful strategy which has enabled James River to significantly expand its business and the diversity of its product line. However, perhaps more importantly, it has been able to plan these acquisitions to continue to focus on the company's strengths. Its specific strategies as articulated in recent Annual Reports are characterized by:

- Market focus
- High-value-added products
- Stability of demand
- Substantial market share
- Broad product lines

By pursuing markets with these characteristics, James River has become a highly successful corporation, moving from a standing start in 1969 to a corporation that by the early 1990s had roughly $6 billion in sales. Unlike other major competitors, it has built a major paper firm on a very small timberland base, although timberland acquisitions in some areas changed this a bit in the late 1980s.

James River, in the late 1980s, had only 10 to 15 percent of its assets positioned abroad. Its Chief Executive Officer, Brenton Halsey, has stated the corporation's desire to be the premier worldwide pulp, paper, packaging, and related products company. This will clearly require expansion overseas. Halsey has made it clear that he hopes to take on Scott Paper in the struggle for the European tissue market (*Forbes,* November 28, 1988).

A new approach to its work force is also apparent at James River. It is experimenting with operating plants with a 100 percent salaried work force. In some cases, the United Paperworkers International Union was invited to help set salaries and participate in the experiments. This was in clear contrast to the anti-union stance of other papermakers.

PRODUCT LINES

Product group	% of 1990 Sales
Sanitary paper products	27.8
Disposable food and beverage service products	16.4
Specialty industrial and packaging papers	13.8
Food and consumer packaging	16.8
Communication papers	21.4
Market pulp and other	3.8
Total	100.0

James River's product line was clearly enhanced by its acquisition of the paper activities of Crown Zellerbach. This strategically important acquisition reinforced James River and its strategic markets without dilution into other paper markets which were not deemed strategic by James River. By concentrating its efforts in the markets listed in the table above and by achieving a value-added product line in markets which exhibit relatively stable demand such as the tissue market and disposable food and beverage container markets, James River has been able to weather economic changes and grow at the same time. It produces many of the brand names seen at your local grocer: Dixie (cups and plates), Vanity Fair (napkins), Northern (bathroom tissue), Brawny (towels).

Kimberly-Clark
World Headquarters
Box 619100
D. F. W. Airport Station
Dallas, TX 75261-9100

Kimberly-Clark moved its World Headquarters to Dallas in 1985. However, it maintains a large operation headquarters at its former home office in Neenah, Wisconsin, and a new operations headquarters in Roswell, Georgia, near Atlanta.

COMPANY PROFILE

Kimberly, Clark and Company was established in 1872 as a partnership in Neenah, Wisconsin, with its first product being newsprint produced from cotton and linen rags. Over the next several decades it added a producer of manila wrapping paper and also built additional pulp and paper-making complexes. In the 1920s, following the use of crepe cellulose tissue for wound dressings in World War I, it introduced new products—products which are now household words. In 1920, Kotex feminine pads were introduced and in 1924 Kleenex tissues. In 1928, the name was formally changed to Kimberly-Clark Corporation. During the 1940s and 1950s, Kimberly-Clark had tremendous expansion into both consumer tissues and newsprint operations. During the 1950s and 1960s, it moved heavily overseas and by 1985 was operating 66 overseas paper machines in 20 countries.

Today, Kimberly-Clark Corporation manufactures and markets a wide range of products (most of which are made from natural and synthetic fibers) for personal care, health care, and other uses in the home, business, and industry. These products include disposable diapers, facial tissue, household towels, bathroom tissue, feminine pads, tampons, incontinence products, industrial wipes, surgical gowns, packs, and wraps. The products are sold under a variety of brand names such as Kleenex, Huggies, Kotex, New Freedom, Delsey, High-Dry, Depend, Kimwipes, and Kimguard. The corporation also produces and markets papers requiring specialized technology and development or application, as well as traditional papers and related products for newspaper publishing and other communication needs. It also operates Midwest Express Airlines.

COMPANY STRATEGY

According to the *Wall Street Journal*, Kimberly-Clark has been transformed from a staid paper company to a plucky consumer-product competitor selling many durable brand names, largely due to the work of Darwin E. Smith, the company chairman since 1971 (*Wall Street Journal*, July 23, 1987). Smith restructured the corporation by closing four large coated paper mills which were not always profitable, and selling off other mills and timberland as well as two sawmills in northern California. Smith envisioned the new direction of the firm as moving into the fiercely competitive consumer-product industry to soften the wild swings of the basic commodity paper business. Kimberly-Clark already controlled as much as one-third of the market for facial tissues through its Kleenex brand, and Kotex had more than half of the sanitary napkin market. However, Smith thought that future growth would come mostly from new businesses, and the arena in which he decided to compete was disposable diapers. An earlier foray into the disposable diaper market with a product named Kimbies was a failure, as Procter & Gamble's Pampers brand overtook it. In 1978, Smith brought out a new disposable diaper, Huggies, very similar to Kimbies but with a new hourglass shape and a new more absorbent filling. With a fresh name and extensive market support, the brand took off and accounts for roughly 25 percent of total corporate sales. Current studies show that each child under the age of 2 years goes through 2300 diapers a year, which makes a market of over $4 billion in the United States alone.

Kimberly-Clark did not rest on its success with Huggies, however; it continued to work with its Kleenex brand as well, which now commands about half of the market. In addition, Kimberly-Clark's adult incontinence products also have about 50 percent of that market. Some analysts say from 1987 to 1991 this particular market has grown from approximately $150 million to over $500 million in total sales.

Kimberly-Clark has, in recent years, invested large sums of money in research and development to keep its consumer product line the best in the industry. In addition, significant sums of money have been spent on cost containment and production improvement at its various pulp- and paper-producing facilities, and Smith has also begun a process to "flatten" the organization of Kimberly-Clark by removing unnecessary layers of supervision and pushing the decision-making process much closer to the operating levels.

PRODUCT LINES

Kimberly-Clark divides its basic businesses into three different categories. Class I includes the household and institutional tissues, feminine, infant, and incontinence care products, and industrial and commercial wipes. This class represents roughly 80 percent of total corporate sales. Class II products include business papers, tobacco papers, newsprint, printing papers, and other specialty papers. Kimberly-Clark remains a major producer of newsprint and thin papers. Class III products include aircraft services, air and truck transportation, and other miscellaneous products. The dominant portion of this class is Midwest Express Airlines, which is owned by Kimberly-Clark.

Louisiana-Pacific
111 SW 5th Avenue
Portland, OR 97204

A 1971 Federal Trade Commission complaint against Georgia-Pacific ultimately resulted in Louisiana-Pacific Corporation. On January 5, 1973, Georgia-Pacific spun off $327 million of assets to create Louisiana-Pacific. Harry A. Merlo, a former Georgia-Pacific executive, was named president of the new corporation. Merlo wasted no time getting started and in 1973 the company made no less than 17 acquisitions, investing over $21 million in capital improvements and adding 1800 employees.

On January 5, 1973, Louisiana-Pacific stock began trading on the New York Stock Exchange. Later in that same year the stock was split two for one. Net sales in that first year of operation were approximately $417 million. By 1974, company ownership of timberlands had climbed to 625,000 acres and the company had the capacity to supply 2.5 million seedlings on an annual basis to reforest cutover company lands.

By 1979, the annual sales of Louisiana-Pacific had reached $1.3 billion but declined to $1.1 billion in 1983 as a result of the weakness of the building products market. Louisiana-Pacific has remained primarily a lumber and structural panel producer. In 1973, 92 percent of corporate sales was from building products and 8 percent was from pulp and paper operations. By 1989, the distribution of sales between building products and pulp and paper operations had changed modestly, with building products contributing 87 percent of net sales and pulp and paper products contributing 13 percent. Of the major integrated forest products firms, Louisiana-Pacific has the largest exposure to the building products market.

Louisiana-Pacific has two principal market areas in which it has a strong leadership position. The first involves the structural composite panel market. Louisiana-Pacific had a stated company goal of having 1 billion square feet of capacity in this market segment by 1985. This goal was met and exceeded, with capacity at the beginning of the 1990s at nearly 3 billion square feet.

Louisiana-Pacific entered the structural composite panel market with an oriented waferboard sold under the brand name of Waferwood. It was positioned in the marketplace as a lower-cost substitute for plywood. Later it introduced an oriented strandboard product with a new brand name, Inner-Seal, and a new resin system. Louisiana-Pacific is the clear volume leader in this product line.

Louisiana-Pacific is also a major plywood producer. And, in the panel product line, it also produces medium-density fiberboard, particleboard, and hardboard.

Another area in which Louisiana-Pacific had stated a goal of market leadership was in the southern pine lumber industry. The company stated in 1981 that its goal was to have 1 billion board feet of southern pine lumber capacity by 1985. The company pursued that goal by acquiring southern pine sawmills and by constructing new mills in the southern states. In 1986, it had acquired Kirby Forest Industries with its 650,000 acres of timberland, sawmills, and plywood plants. This pushed the company over the goal of 1 billion board feet of southern pine lumber production.

Other products round out the product line. These include western softwood lumber, redwood, windows and doors, fiber gypsum panels, and market pulp.

Louisiana-Pacific has largely concentrated on the development of its manufacturing base and has forgone the development of an internal distribution system for its building products. However, in 1979 it did acquire, from Lone Star Industries, a chain of 15 building material centers in southern California. These were later sold and Louisiana-Pacific operated only eight building products distribution centers by 1989.

Lowe's Companies, Inc.
P. O. Box 1111
North Wilkesboro, NC 28656

BACKGROUND

Lowe's 1987 annual report described its company's beginnings as a renegade retailer of heavy hardware and building materials back in the days of the post–World War II building boom when Carl Buchan bypassed wholesalers to establish the tradition of "Lowe's Low Prices." The company notes that some Lowe's customers can still remember buying refrigerators and building supplies directly from a freight car on the railway siding serving the North Wilkesboro store. From these modest beginnings, Lowe's grew rapidly, becoming a publicly traded company on October 6, 1961. Currently it is one of America's largest specialty retailers of building materials and related products for the do-it-yourself home improvement and home construction markets.

STORE TYPES

The early Lowe's stores were designed to sell building materials to professional contractors in the business of residential construction. This early customer base most influenced the appearance and attributes of Lowe's stores well into the 1970s. The typical Lowe's store evolved as a free-standing building located near railroad tracks. It had a small retail floor displaying very limited inventory, and a lumberyard out back. Many services were offered for building contractors, but the stores were ill-suited for the do-it-yourself retail customers. With the growth in the do-it-yourself home improvement and home repair industry in the 1970s, Lowe's began a program of transition to move itself from a contractor-based lumber and building materials business to a do-it-yourself home improvement center. The last phase of this transition began in 1980 with Lowe's RSVP program (Retail Sales Volume and Profit). Through this program, Lowe's sales floors in the retail stores were enlarged and given an attractive new look designed to better appeal to the retail customer.

Even with the RSVP program, the older Lowe's stores still only averaged 5,000 to 7,000 square feet of indoor selling space. Clearly an attractive home center cannot be accommodated in such a small space. Lowe's own annual reports stated the minimum effective size for home center inventory presentation is roughly 20,000 square feet of indoor showroom area. As of the end of fiscal 1987, only one-third of Lowe's stores met that particular size criterion. Lowe's has been expanding the average size of its retail stores and has experimented with 60,000-square-feet prototype stores. It has a Phoenix program for older retail stores, which is designed to expand on existing stores by moving a single wall with minimal disruption of existing business. These various programs have been successful, with the average square footage of a Lowe's retail store expanding from 10,626 square feet in 1983 to an average size of over 20,000 square feet by the beginning of the 1990s.

BUSINESS STRATEGIES

Lowe's is somewhat unique among its competitors in that it focuses on three distinct sets of customers, pursuing all three from the same store. These three customer segments are the building contractor business, the consumer durables business, and the do-it-yourself home center business. Originally, the building contractor business was the dominant segment for Lowe's total sales; however, by the early 1990s the do-it-yourself business had grown larger than the contractor segment. Consumer durables also had strong growth and Lowe's expects the do-it-yourself and consumer durables business to represent 65 percent to 75 percent of total sales in the near future.

In the home center business, most sales growth comes from additional selling space. This increased selling space can be gained either by building new stores or enlarging and retrofitting existing sales locations. From 1985 to 1987, Lowe's constructed 58 new stores and retrofitted or relocated 36 additional stores, for a total of 94 new and retrofitted stores. During this same time frame, Lowe's corporate sales increased almost $400 million. The 94 new and retrofitted stores accounted for 96 percent of this 2-year sales gain, with the 201 other stores accounting for the remaining 4 percent. These numbers dramatically show the necessity to add retail space for sales growth.

Lowe's also has expanded through small acquisitions. One ill-fated acquisition occurred in 1985 with the purchase of 25 stores from Boise Cascade in the Texas and Oklahoma oil patch. Due to the poor economic conditions in this region, 15 of the stores acquired had been closed by fiscal year-end 1987.

As an aside, for those interested in the home center business, Lowe's annual reports are practically small encyclopedias on the home center industry. They are simply chock full of useful information. Lowe's was a pioneer and early leader in the development of the Do-It-Yourself Research Institute, which serves to provide market information and market research on the rapidly growing do-it-yourself consumer market segment. Much of this information can be found in the company's annual reports.

MacMillan Bloedel Ltd.
1075 West Georgia Street
Vancouver, British Columbia
Canada V6E 3R9

COMPANY PROFILE

MacMillan Bloedel is one of North America's major forest products companies, with integrated operations in Canada and the United States in addition to operations in the United Kingdom and continental Europe. Approximately two-thirds of MacMillan Bloedel's timberlands and productive assets are in the Canadian province of British Columbia. The products of MacMillan Bloedel and its affiliated companies which are marketed throughout the world include lumber, panel boards, kraft pulp, newsprint, groundwood printing papers, fine papers, containerboard, and corrugated containers.

BACKGROUND

The history of MacMillan Bloedel dates back to the year 1919 when H. R. MacMillan and M. L. Meyer formed the H. R. MacMillan Export Company Ltd. This new firm secured a very large order for lumber. This single large order resulted in the formation of the Associated Timber Exporters of B.C., which was a consortium of coastal mills which banded together as the best and perhaps only way of filling the order for MacMillan Export Company Ltd. This resulted in a partnership of sorts, with MacMillan Export Company Ltd. providing sales and shipping expertise and the Associated Timber Exporters providing the sawn lumber. However, as years passed the lumber producers began to feel that MacMillan was not passing on a fair proportion of profits to their mills. As a result, the sawmills formed their own lumber sales organization, Seaboard Lumber Sales Ltd. This action forced MacMillan to integrate backward and purchase production facilities, resulting in the development of an integrated forest products company. In 1951, H. R. MacMillan Export Company merged with Bloedel, Stewart and Welch (a large lumber and timber firm) to form MacMillan Bloedel Ltd.

PRODUCT LINES

MacMillan Bloedel is a major integrated producer of paper and forest products. Its major building product lines include lumber, plywood, and Parallam. Perhaps the most exciting development to occur in the building product line of MacMillan Bloedel in a number of years has been the development of the Parallam product. Parallam is a composite, parallel strand lumber which can substitute for large sawn timbers, laminated veneer lumber, glue laminated beams, and steel beams in some instances. Additionally, the company has introduced PSL 300, which is an engineered parallel strand lumber, for use as an industrial product in applications such as core material for window and door frames.

MacMillan Bloedel's pulp and paper product lines include newsprint, both standard and groundwood specialties, and market pulp. Its containerboard and packaging group produces linerboard, corrugated medium, and corrugated containers.

STRATEGY

MacMillan Bloedel has had a long history of looking toward overseas markets for its production. This has resulted somewhat from necessity but has placed MacMillan as a premiere firm in the world markets for forest products. It has strong and unique ties with both the Japanese market and the markets in Great Britain and continental Europe. MacMillan Bloedel has shown an increasing willingness to produce specialty products designed for these export markets while at the same time committing significant company resources to unique product innovations such as Parallam.

Potlatch
1 Maritime Plaza
Box 3591
San Francisco, CA 94119

COMPANY PROFILE

Potlatch Corporation began as a lumber company early in this century. Since the 1950s, pulp and paper products have gradually increased their share of the company's sales and operating income. Pulp and paper now represent over 75 percent of company sales. Most of Potlatch's manufacturing facilities and land base are centered in three states: Idaho, Minnesota, and Arkansas. Potlatch is a medium-sized integrated forest products firm with annual sales in the early 1990s exceeding $1 billion.

BACKGROUND

Potlatch Corporation was founded in Potlatch, Idaho, in 1903. It was started by the same Frederick Weyerhaeuser who, along with his partners, founded Weyerhaeuser Company. In fact, some 200 Weyerhaeuser family members still hold shares in Potlatch and four of Frederick Weyerhaeuser's direct descendants sat on the Potlatch Corporate Board in the mid-1980s.

In 1950, the company built a pulp/paper mill at Lewiston, Idaho, to diversify its operations and reduce the cylicality of its wood products business. Since this time, it has concentrated on pulp-based products. In the 1950s, Potlatch acquired the Southern and Bradley companies in Arkansas, and in the 1960s added tissue production to the Lewiston mill, merged with Northwest Paper Company in Minnesota, and acquired the Ozan Lumber Company in Arkansas. In the late 1970s and early 1980s Potlatch built two of the early OSB mills in the United States in Minnesota.

When Potlatch's Chairman of the Board and Chief Executive Officer, Richard B. Madden, was hired in 1971, he faced a problem which had unfortunately become common in many U.S. corporations of that time frame. Potlatch was in 20 different businesses as diverse as corrugated containers and modular buildings. It was having a difficult time managing such a diverse group of operations and Madden wisely sold off most of the old diversification scheme and began to focus on four basic product lines: bleached pulp and paperboard for packaging, coated printing papers, private label tissue, and lumber and structural panels.

Beginning in 1976, as cash flow was strong Potlatch spent more than $1 billion over the next decade to develop the company's 1.5 million acres of timberland and to build and improve mills to better utilize the timber. Potlatch survived the recession of the early 1980s by instituting a company-wide cost-cutting program, trimming employees, and successfully negotiating with labor unions in Idaho for wage concessions. Some market analysts have credited Potlatch with being one of the first forest products companies to really hunker down in the early 1980s.

Even with Madden's strong performance, Potlatch was the target of a failed takeover bid in the mid-1980s.

CURRENT OPERATIONS

Potlatch Corporation has regional timberland holdings of 1.5 million acres centered in Arkansas, Idaho, and Minnesota. It operates sawmills, plywood plants, OSB plants, a particleboard plant, bleached kraft pulp mills, bleached paperboard mills, printing paper mills, tissue and toweling mills, and packaging plants.

Potlatch has had extremely good success with its branded OSB product Oxboard, as well as with other specialty products. However, the main corporate strategy for the last several years has been to consolidate and trim internal functions including reductions in the salaried work staff. Its objectives are to continue to tighten the organization and improve profitability while retaining the ability to operate its facilities efficiently and provide good customer service.

PRODUCT MIX

The product mix is focused in four areas: coated free sheet, bleached paperboard and packaging, private label tissue, and wood products.

Potlatch was the fourth largest U.S. producer of coated free sheet in 1988, and it continues to upgrade and modernize its facilities.

Potlatch had about a 10 percent share of the bleached paperboard market in 1988, ranking it fourth among U.S. producers. It operates strategically located plants to convert paperboard into milk and juice cartons. Also, it exports significant volumes of paperboard to Japan and Australia.

Potlatch is the only major household tissue supplier committed solely to generic and private label tissue. Potlatch holds a 20 percent share of this market in the United States and a 50 percent share in the western states. This unique strategy allows Potlatch to avoid direct competition at the high end of the market with James River, Scott, Procter & Gamble, Fort Howard, and Georgia-Pacific.

Scotty's Inc.
P. O. Box 939
5300 North Recker Hwy.
Winter Haven, FL 33882

COMPANY PROFILE

Scotty's is Florida's leading retailer of home improvement and building supplies.

Merchandise typically sold at Scotty's includes lumber, plywood, paneling, roofing, insulation, cement, paint, hardware, tools, electrical fixtures, plumbing supplies, cabinets, wall and floor coverings, lawn and garden supplies, and automotive products. The company has historically derived approximately two-thirds of its revenues from the do-it-yourself consumer segment of the market and the remainder has come from builders, commercial enterprises, and export and wholesale customers. Lumber, plywood, and other basic building materials have typically accounted for 40 to 50 percent of annual sales.

RETAIL STORE TYPES

Beginning in the mid-1980s, Scotty's began an extensive new store program. The first of these new store types introduced in 1984 was the Scotty's Hardware Store. This was designed to reach rural and neighborhood markets which are ill-suited for the larger full-line stores and to allow Scotty's to further penetrate larger market areas where full-line stores are widely separated. A typical Scotty's Hardware Store is 7500 square feet in size and sells about 9000 items. The merchandise assortment includes basically everything carried at the larger full-line stores except for lumber and basic building materials.

In 1985, Scotty's introduced a new format of full-line stores which have since proven to be quite popular and competitive in the marketplace. These larger Scotty's outlets typically have 49,000 square feet of selling space of which 25,000 square feet is an air-conditioned shopping area. The format is somewhat flexible so that stores can be built in larger or smaller sizes as appropriate for a specific location. The full-line stores stock more than 16,000 merchandise items on modified warehouse racking. Nearly all of the Scotty's full-line stores are located on sites of approximately 5 acres, easily accessible from major highways and close to major shopping areas.

STRATEGY

Although Scotty's is Florida's number one home center company this has not allowed it to rest on past accomplishments. It has a strategic plan to bolster its position as Florida's preeminent retailer of home improvement and building supplies and to enter additional markets. It is continuing to open more full-line stores and Scotty's Hardware Stores in Florida, its principal market area, and in southern Georgia. In addition, it has constructed full-line stores in Mobile, Alabama, to initiate its move into that state. By clustering its stores in a relatively narrow but high-growth geographic base, Scotty's enjoys significant efficiencies in its warehousing and distribution operations, and perhaps more importantly in its advertising. By having numerous stores in the same geographic area, significant economies of scale can be realized in advertising due to the fact that one advertisement may serve to draw customers to a number of stores rather than just a single store in a given area. This allows the advertising expense to be spread over numerous stores, thus creating significant efficiencies.

Scott Paper Company
Scott Plaza
Philadelphia, PA 19113

COMPANY PROFILE

Scott Paper is the world's leading manufacturer and marketer of sanitary tissue paper products. Through its Scott worldwide group, a broad range of sanitary paper products, as well as nonwoven products and personal cleansing soap and dispensing systems, is marketed.

The S. D. Warren subsidiary produces coated and uncoated printing, publishing, and specialty papers, principally for U.S. markets. Scott's Natural Resources Division is responsible for the company's 2.8 million acres of woodlands in the United States and Canada, and directs its land management pulp and forest products marketing and minerals activities.

Scott Paper is a participant in a number of joint ventures including Canso Chemicals Limited and Mountain Tree Farm Company. In addition it operates a number of joint ventures in Latin America, the Pacific Rim, and Europe.

COMPANY STRATEGY

Philip Lippincott, Scott's Chairman and Chief Executive Officer, was quoted in a July 5, 1982, article in *Forbes* as saying that when he joined Scott in 1959 it was the Cadillac of the industry. Scott had almost a monopoly on toilet tissue and pretax margins were 16 percent. However, when Lippincott took over as Chief Executive Officer in 1982 Scott's pretax margins were at 7 percent and it ranked almost at the bottom of forest products companies in terms of profitability.

The trouble began in 1965, when Bounty paper towels were introduced by Procter & Gamble. Two years later, Procter & Gamble brought out its premium Charmin bathroom tissue and went for Scott's throat. Scott failed to act and in the mid to late 1970s it was outflanked again by generic producers at the low end of the market led by Georgia-Pacific. A worsening company performance and a blunted takeover bid by Brascan Ltd. brought in Lippincott as the new Chief Executive Officer in 1982.

Lippincott initiated a new strategic plan which called for the company concentrating its resources on the best parts of its business and the parts that it knew the best, and to eliminate those parts of Scott which did not fit its long-term strategy. In addition, the company worked hard to change its work environment and culture and to significantly improve its financial performance. This strategy has been largely successful. Scott has positioned itself in two of the most attractive segments of the paper industry, tissue products and coated printing and publishing papers. Within the tissue segment, Scott is the world leader and plans to move from strictly tissue into a growth area of selected personal care and cleaning product markets which is a natural extension of their leading position in the sanitary tissue paper products area. Scott has stuck to the less competitive economy end of the tissue market and has looked at less competitive overseas markets.

Within the coated paper segment, its S. D. Warren subsidiary has been following an aggressive strategy to build upon its traditional strength in heavyweight coated grades used for annual reports, book publishing, and similar applications while at the same time becoming a major factor in the much faster growing lightweight grades, which are used in magazine publishing, etc. S. D. Warren is a strong name in the printing and publishing paper industry, having been in business for over 132 years. During that time it has developed a low-cost production base, strong distribution, and a substantially integrated asset base.

In each of these two basic market segments, Scott's strategy has been to produce value-added and distinctive products, resulting in strong market franchises as well as better profit margins. While Scott has been growing strongly, its growth has been guided very carefully. It is expanding only in areas where it believes it can enjoy or develop a competitive advantage. It is seeking areas which match with existing businesses and where it can demonstrate management excellence. In other words, it is building on its key strengths, its integrated asset base, and the Scott and S. D. Warren reputations for quality and value, including the market franchises it has established in sanitary tissue paper products and coated paper. The company's current worldwide sales and distribution strengths are also important.

FUTURE OPPORTUNITIES

In addition to adding to its base businesses, Scott is expecting significant growth to occur from the broader area of personal care and cleaning products worldwide. It is examining products and services for personal body care. These products and services include adult incontinence items, liquid soap dispensing systems, premoistened wipes, and products and services for the cleaning or sanitizing of environmental surfaces including bathroom bowl cleaners, bathroom seat covers, and other cleaning disposables. It believes growth in this area will accelerate in the future.

Scott is unique in that it is positioned worldwide in the sanitary tissue paper products market. This is important because the worldwide growth in unit volume, at nearly 5 percent a year, is about twice the growth rate of the United States. This faster international growth rate is expected to continue because annual consumer tissue products consumption in most international markets, on a pounds per capita basis, is only a fraction of the U.S. per capita consumption. However, competition is also heating up, with James River moving into the European tissue market.

Stone Container
150 North Michigan Ave.
Chicago, IL 60601

COMPANY PROFILE

Stone Container began as a family-run business in 1926. It was originally a jobber of paper bags, wrapping paper, and twine. Today, it is the nation's largest producer of corrugated boxes, paper bags, linerboard, and kraft paper. Stone also has strong positions in newsprint and market pulp and is a significant wood products manufacturer.

BASIC BUSINESS STRATEGY

Roger Stone, grandson of the company founder, joined the firm in 1957 and became its President in 1975. The turning point for the management of Stone Container may have come in 1979 when Boise Cascade offered to buy Stone for approximately $125 million. It was a very tempting offer and caused much concern at the largely family-owned firm.

By the early 1980s, Stone realized that there was a growing demand for linerboard (the material that corrugated boxes are made from) and that this demand would soon bump up against a fixed supply. The demand for linerboard has continued to grow and as there are few substitutes for corrugated containers there are few substitutes for linerboard. Stone also knew that supply was fairly fixed in the short term because new mills take up to 3 years to build and very few new mills had been announced in the early 1980s.

About this same time the market for linerboard had declined and many firms were trying to get out of this business. Stone Container capitalized on its competitors' disenchantment and was able to buy existing capacity from other firms. This was not only cheaper than building, but also allowed Stone to increase its market share without adding to the total supply which would surely push prices down. In 1983, Stone Container acquired Continental Groups Forest Products Division. In 1986, it acquired Champion's brown paper business and in 1987 it purchased all of Southwest Forest Industries. These deals and others quintupled Stone's annual capacity at a cost of only $123,000 per ton, approximately one-fifth of what it would cost to build new plants. In addition, the Southwest purchase also marked Stone's entrance into the newsprint and wood products business.

The biggest leap came in 1988 as a strategy was formulated to make Stone Container a major force in European and North American papermaking in the three core areas of packaging, newsprint, and market pulp. As this strategy was developed, it became clear that acquiring Consolidated-Bathurst, a major Canadian producer of newsprint and owner of the largest newsprint producer in Great Britain and the largest packaging company in Germany, would be an excellent match to the strategic plans. Consolidated-Bathurst was acquired in March 1989, increasing Stone's 1989 sales by 42 percent over 1988 and also strongly increasing the company's debt.

Temple-Inland Inc.
Drawer N
Diboll, TX 75941

Temple-Inland is one of the newest major independent forest products companies in the United States. It began operations on January 1, 1984, following its spinoff from Time, Inc.'s forest products operations which were part of Time's communication activities. Temple-Inland was formed from two subsidiaries of Time, Inc., which had long operating histories: Temple-Eastex and Inland Container.

TEMPLE-EASTEX

T. L. L. Temple acquired 7000 acres of timberland in Texas and established the Southern Pine Lumber Company in 1893. By 1894, the Southern Pine Lumber Company had its first sawmill operating in Diboll, Texas, cutting approximately 50,000 board feet of southern pine per day. During the early years of this company, it was primarily a producer of basic lumber products for the construction and furniture industries. In the late 1950s and early 1960s the company moved into the production of particleboard, gypsum wallboard, and other building materials, along with entering the mortgage, banking, insurance, and construction businesses.

In 1964, the name of the Southern Pine Lumber Company was changed to Temple Industries, Inc., and the company's initial 7000 acres of timberland had grown to more than 450,000 acres.

In the early 1950s, Time, Inc., entered into a joint venture with the Houston Oil Company and established the East Texas Pulp and Paper Company. In 1956, Time purchased Houston Oil's ownership in the East Texas Pulp and Paper Company and in 1973 it acquired Temple Industries and merged it with Eastex Pulp and Paper Company to form Temple-Eastex, Inc.

INLAND CONTAINER

Herman C. Krannert is responsible for the heritage of Inland Container Corporation. Krannert started the Inland Box Company in Indianapolis in 1925 and soon acquired a second box plant in Middleton, Ohio. Then in 1938 the company was reincorporated as Inland Container Corporation. Up until 1946, Inland was primarily a multiplant box converter relying solely on outside sources for its paper supply. That same year, through a joint venture with Mead Corporation, the Georgia kraft Company was formed and a new linerboard mill was constructed in Macon, Georgia. Georgia kraft, 50 percent owned by Inland Container, operated three linerboard mills with five paper machines, along with plywood and lumber mills, and owned approximately 1 million acres of timberland in Alabama and Georgia. In 1978, Inland Container Corporation was acquired by Time, Inc., to be operated as a subsidiary.

TEMPLE-INLAND

The formation of Temple-Inland Corporation from the Time subsidiaries of Temple-Eastex and Inland Container produced an integrated forest products firm with sales of approximately $1.2 billion in 1983. Temple-Inland has four main areas of operation: paper, containers, building materials, and financial services. The paper activities of Temple-Inland are concentrated in two paper grades; containerboard and bleached paperboard, both of which serve as raw materials for their container-producing operations.

Temple-Inland has a diversified building products product line, including southern pine lumber and plywood, fiberboard, particleboard, wall paneling, and gypsum wallboard. In addition to manufacturing operations, the company also owns building materials retail distribution centers in Texas and Louisiana. Its financial services group is comprised primarily of mortgage banking, real estate development, and savings and loan activities.

Effective January 4, 1988, Mead and Temple-Inland divided the assets of their Georgia kraft joint venture. Temple-Inland acquired a 2000-ton-per-day linerboard mill, a sawmill, and 400,000 acres of timberland from the dissolution.

Temple-Inland has a substantial timberland base consisting of 1.8 million acres of timberland in east Texas, Louisiana, Alabama, and Georgia.

Temple-Inland's Chief Executive Officer and Chairman has stated the firm's strategy very clearly:

> Most of the products sold by Temple-Inland are variations of commodity-based items and are readily available from other suppliers. Although we continue to diversify our product mix, we accept the fact that the success of the Company is dependent on excelling in commodity-based sales and maintaining a low-cost structure.
>
> Today, many companies like to talk about low cost. At Temple-Inland, our concern for low cost goes beyond just manufacturing expense. We want to win all ties, and that means being competitive in more than just price. We remain committed to providing superior service and delivering quality products manufactured at the lowest possible cost.

TJ International
9777 West Chinden Blvd.
Boise, ID 83714

COMPANY PROFILE

Trus Joist Corporation was founded in Boise, Idaho, in 1960 by two entrepreneurs, Harold Thomas, a businessman, and Art Troutner, an architect and contractor. Troutner invented a unique structural truss, blending the light-weight nailable characteristics of wood with the strength of steel. This new product had a superior weight-to-strength ratio when compared to other competing products on the marketplace. The new product was named TJL Trus and was originally designed to support the roofs and floors of light commercial buildings such as stores, banks, and small office buildings.

From the initial TJL Trus, the company developed a complete family of structural products consisting of five series of open-web trusses, six series of I-beam wood trusses, and the high-strength manufactured lumber substitute MICRO=LAM. MICRO=LAM is a laminated veneer lumber product.

In 1987, *Forbes* placed Trus Joist on its list of the top 200 small corporations in America. In 1988, Trus Joist changed its name to TJ International. It now operates under two divisions: Trus Joist Corporation and Design Master Corporation, a producer of wood windows.

MARKETING STRATEGIES

The structural building products that TJ International has produced over the years are designed to compete typically with solid-sawn lumber in larger sizes such as 2×10s and 2×12s. As these larger sizes of lumber become more expensive and of lower quality, TJ International believes it is in a position to take advantage of these changes in the timber resource.

Because its initial product line, the TJL Trus, was aimed at the light industrial and commercial market, a direct sales force was established to market this product line. In an effort to capture more of the residential construction market, TJ International has a large network of stocking lumber dealers spread across the United States and Canada. This program has been very successful. It seems builders are willing to pay more per lineal foot of TJI Joist because they find that labor and material savings provided by the product reduce the total cost of construction.

Many of the residential housing sales have gone to the multifamily market. Multifamily housing construction has been the key to the success of the TJ International residential program during the early years of its growth. Because the effort required to sell products for a large condominium or apartment complex is often the same as required to sell for a single-family house, the TJ International sales effort purposefully targeted multifamily construction as the more efficient opportunity.

TJ International summarizes its strategies in specialty structural products as remaining the low-cost producer and the largest producer as well as maintaining unsurpassed customer service and premium quality, and constantly evolving specialty building products.

DIVERSIFICATION

TJ International has attempted to diversify to expand its opportunities in the marketplace. Two of these diversification attempts have failed: a joint-venture subsidiary, Advanced Power Structures, and a wholly owned subsidiary, T. J. Controls.

Norco Windows, Inc., and Dashwood Industries Ltd. were acquired in 1987. Both are high-quality wood window manufacturers and appear to be a good match with TJ International's existing product line and company culture. The window operations are now organized in a separate division, Design Master Corporation, and represented over 30 percent of total company sales in 1989. TJ International's most recent acquisition was a major interest in MacMillan Bloedel's Paralam operations in the United States.

TJ International's acquisition criterion has been delineated as: Candidates must have an established profitable, sound management and the potential to grow with TJ International capital. But, most important, they must have an obsession for customer service.

Union Camp
1600 Valley Road
Wayne, NJ 07470

CORPORATE HISTORY AND STRATEGIES

Union Camp Corporation has been a medium- to large-sized pulp and paper operation for a number of years. In the 1930s and 1940s, it was known as the Union Bag and Paper Corporation, highlighting its early emphasis on bags.

In recent years, the Union Camp Corporation developed a strategy of having very modern, large, cost-efficient manufacturing facilities. Union Camp tends to build new plants rather than acquire existing ones and has concentrated its productive capacity in a few very modern efficient operations. This is especially important in paper products, where the economies of scale are so significant. As a result of this strategy, in many of the paper market segments where the company operates, it is an acknowledged, efficient, low-cost producer. In more recent years, it has placed new emphasis on product quality, customer service, and higher-value products.

PRODUCT LINES

Packaging and Converted Products

Union Camp's container division produces corrugated and preprinted corrugated containers and plastic and paper bags. In addition, it produces folding cartons, consumer packaging, and school supplies.

Paper and Paperboard

Union Camp's paper and paperboard operations are centered at four major locations. White fine papers are produced at Eastover, South Carolina, and Franklin, Virginia. The Franklin operation also manufactures coated and uncoated bristols. A large, modern kraft linerboard operation is in Montgomery, Alabama, and a new facility in Savannah, Georgia, produces various grades of kraft paper and paperboard. Union Camp's broad line of products in this category includes forms, envelopes, tablet and copy papers, uncoated offset, coated and uncoated bristols, kraft paper, linerboard, and saturating kraft papers.

Chemicals

Another large segment of Union Camp's operations revolves around the chemical division. The raw materials for this division are generally from pulping by-products. This division produces such products as fatty acids, esters, rosin and rosin derivatives, high-performance resins, terpene products, and flavor and fragrance chemicals.

Building Products

Union Camp has a small but significant building products segment producing southern pine plywood, particleboard, and dimension lumber. Its long-term strategy for building products is to continue to upgrade from commodities to higher-value-added products and to develop more volume in markets that are less sensitive to housing. In order to do this, it has been searching out market niches where innovation or the application of a special skill or process gives it a competitive advantage. It has been successful in doing this in its plywood business, having now developed large industrial application markets for most of its plywood production.

Woodlands

Union Camp owns or controls approximately 1.7 million acres of woodlands in Alabama, Florida, Georgia, North Carolina, South Carolina, and Virginia.

EMPHASIS ON MARKETING

Throughout recent annual reports, Union Camp emphasizes that its organization is more aware of the need to be "customer-driven." This is reflected in its operations and commitment to develop specialty products for its customers. The drive for specialty, value-added products coupled with an efficient low-cost production base places Union Camp in a strategically advantageous position.

Weyerhaeuser
Weyerhaeuser Building
Tacoma, WA 98477

BACKGROUND

By 1900, Frederick Weyerhaeuser had already been active in the lumber and timber industry for 43 years. He led a group of midwestern lumbermen to Tacoma, Washington, in that year to organize a new lumber company. This new company was incorporated as Weyerhaeuser Timber

Company. When you mention a forest products or lumber company, probably the name that first comes to people's lips is Weyerhaeuser. In fact, since its founding six of the nine men who have led the firm have also borne the Weyerhaeuser name.

Weyerhaeuser first purchased 900,000 acres from the Northern Pacific Railroad for $5.4 million. Weyerhaeuser was joined by partners James and Matthew Norton and William Laird, who put up $1.2 million of the $5.4 million. These early partners put in funds for many other ventures that Weyerhaeuser organized including Boise Payette (now Boise Cascade) and Potlatch. In 1902, Weyerhaeuser purchased its first sawmill in Everett, Washington. The company expanded slowly, opening the Weyerhaeuser Sales Company in St. Paul, Minnesota, in 1919, and acquiring its first sulfite pulp mill in 1931 in Longview, Washington. Weyerhaeuser was an early leader in the industry in promoting the integration of utilization by purchasing pulp mills to use the residues available from the company's sawmilling operations. Weyerhaeuser was also an early leader in the development of tree farms and the reforestation of cutover forest land. In addition, Weyerhaeuser was one of the first major forest products firms to institute a research and development program.

Weyerhaeuser remained in the Pacific northwest until the mid-1950s, at which time it moved into the southeast and south central United States with a strong acquisition program. Timberlands were first acquired in 1956 in Mississippi and Alabama, with subsequent timber acquired in North Carolina and Virginia in 1957. One of the largest acquisitions ever of southern timberlands came in 1969 when Weyerhaeuser acquired Dierk's Forests Inc., adding 1.8 million acres of timberlands in Arkansas and Oklahoma, along with various forest products mills.

Weyerhaeuser has tended to be a leader in the forest products industry, instituting such slogans in early advertising as "We're the tree growing company" and "High yield forestry." Beginning in mid-1961, George Weyerhaeuser, then the Executive Vice President, announced a basic shift from commodity selling of lumber to end-user marketing of wood products. This signaled Weyerhaeuser's move into the consumer market with specialty products and a decreasing emphasis on the commodity grades of lumber. The company's huge reserves of prime high-quality timber supported this move and allowed it to produce many high-quality specialty items such as prefinished paneling, glue-laminated trusses, preprimed siding, and molded wood products. This same product policy was extended to the pulp and paper division. Here emphasis was shifted to a wide variety of printing papers and finished paper products. The theme has been carried forward even to this day and Weyerhaeuser is now the largest U.S. manufacturer of private-labeled disposable diapers.

While pursuing this policy of end-user marketing, Weyerhaeuser became, according to the *Wall Street Journal,* the tenth largest home builder in the United States in 1983 and one of the country's largest mortgage bankers. Weyerhaeuser developed a system of wholesale customer service centers with a stated goal of becoming a leading force in the wholesale marketing of building products in the 1980s. Weyerhaeuser's advertising theme at the time, "100% customer satisfaction," was keyed upon making each of these customer service centers a separate profit center with the manager having considerable autonomy to purchase and provide appropriate product mixes for the local building markets.

Beginning in the late 1950s, Weyerhaeuser began to diversify overseas, and by the end of the 1980s a major portion of net corporate sales was from overseas sales.

Weyerhaeuser alone among major U.S. forest products producers has sufficient timberlands to support 100 percent of its wood and fiber needs. While this can be viewed at different times as either an advantage or a disadvantage, it does free the corporation from heavy dependence on expensive government timber contracts in the Pacific northwest and the problems associated with purchasing small amounts of privately owned timber in the southeast. The company owns approximately 5.7 million acres of commercial timberland.

PRODUCT LINE

Weyerhaeuser has a diverse product line. On the building products side it is the leading U.S. producer of softwood lumber and a major producer of plywood, oriented strandboard, and hardwood lumber. Its paper and fiber group is one of the biggest marketers of northern softwood bleached market pulp and is a major producer of bleached paperboard, bleached kraft, and newsprint. Private-label disposable diapers are also a big business.

FUTURE STRATEGIES

The year 1989 was the beginning of a major restructuring for Weyerhaeuser. Resources were concentrated back onto the main core product lines to accelerate their growth, improve productivity, lower cost, strengthen product quality, and strengthen market position.

Willamette Industries, Inc.
3800 First Interstate Tower
Portland, OR 97201

BACKGROUND

Willamette Industries is an integrated forest products company manufacturing bleached pulp, fine paper, paperboard, corrugated containers, business forms, bags, copy paper, lumber, plywood, particleboard, and medium-density fiberboard. Willamette owns or controls 1 million acres of timberland in Oregon, Louisiana, Tennessee, Arkansas, Texas, and the Carolinas.

Willamette has grown from three sawmills in Oregon in 1951 to a nationwide corporation with over 70 plants by the early 1990s. During this time frame, the firm has been headed by the father-son team of William Swindells, Sr., and William Swindells, Jr. Over the last few years, Willamette has concentrated on reducing production costs, improving vertical integration, improving efficiency of operations, and expanding its marketing efforts, largely in that order.

PRODUCT LINES

Paper Products

Willamette has a reasonably broad line in certain paper areas, typically emphasizing downstream integration of its pulping facilities. An example of this is in the area of bleached pulp. It is a major producer of bleached hardwood market pulp. However, it also uses this pulp to produce a line of fine papers which then are typically used by its business forms division. Its business forms group operates plants located throughout the United States serving customers through a network of sales and distribution centers. In 1985, it entered the copy paper market, introducing a new line of copy paper brand named Willcopy. This allowed the company to further integrate its operations using existing raw material and papermaking capacity to provide another product for sale through its paper distribution centers.

Another major paper product area is unbleached pulp, which is used in the manufacture of linerboard, corrugated medium, and bag paper. These products are then taken and manufactured into corrugated containers and grocery bags. Carrying this one step further, Williamette's corrugated container plants produce an upscale product with six-color printing and so forth to capture the growing market for preprinted boxes used as sales and advertising media within the retail store.

The company is heavily involved in fighting the challenge of plastic sacks and has introduced a paper bag called the Easyloader. The Easyloader is 3 inches shorter than the traditional paper bag and is designed to speed up check-stand operations in grocery stores.

Building Materials

Willamette is a major player in the particleboard marketplace. It is adding value to its particleboard production by using a variety of colors and patterned overlays on the particleboard. Willamette also produces medium-density fiberboard.

Willamette is a major U.S. plywood producer and a strong softwood lumber producer.

Williamette's building products lines are marketed largely through independent wholesalers and distributors throughout the nation.

BUSINESS STRATEGY

Vertical integration is the centerpiece of Willamette's business strategy. It converts most of its basic products into high-value products: sawmill chips to pulp, pulp to linerboard, linerboard to corrugated products, and corrugated products to six-color printed boxes. This allows Willamette's earnings to be more stable than some of its competitors which produce mostly basic commodities.

WTD Industries, Inc.
2 Lincoln Center, Suite 601
10220 Southwest Greenburg Road
Portland, OR 97223

COMPANY PROFILE

WTD Industries was founded in 1983 and had sales in its first year of almost $14 million. The company went public in October 1986, selling 1,850,000 shares of common stock, and by 1987 had sales of $176 million. WTD Industries, Inc., is organized as a holding company whose subsidiaries produce softwood and hardwood lumber, softwood veneer, and by-products at mills in Oregon, Washington, and Montana. The lumber products are marketed under the TreeSource brand name primarily in the United States and Canada. Recently WTD Industries has entered into negotiations with potential partners to build a 300,000-ton-per-year kraft pulp mill in that region to utilize the sawmill by-products. WTD's plan is to export 90 to 95 percent of the pulp to Pacific Rim countries. By the late 1980s, WTD Industries ranked among the top five U.S.-based lumber producers. In early 1991, due to weak markets and other factors, WTD declared bankruptcy under Chapter 11.

BUSINESS STRATEGIES

WTD Industries has a strategy of growing through acquisition. It locates and reorganizes mills that appear to have a high potential for profitable operation. Potential mill acquisitions are evaluated under the following criteria: timber supply, transportation, labor availability, product opportunities, manageable environmental concerns, opportunities for improving operational efficiencies, and potential high return on investment.

WTD is a competitive processor of lumber and related products that takes a nontraditional approach to manufacturing. For example, WTD operates using a highly decentralized management structure, holding each mill strongly accountable for the profitability of its operations. Local managers are given maximum flexibility to operate each mill property within the local constraints it faces. Weekly bonus compensation encourages hourly workers to increase their per-shift production output by rewarding them for regular attendance, safety, and productivity.

WTD also has a rather unique timber acquisition strategy. During fiscal 1987, 78 percent of its logs was purchased on the open market, with only 22 percent coming from public timber. By continuing to purchase the majority of its logs on the open market, it is able to buy logs at a price that is more in line with current markets. WTD owns mills, not timberlands. However, it does currently have 25,000 acres of timberland owned mostly in Oregon. Its philosophy is to own some timberland where necessary to support a mill but not to speculate on timber.

WTD recognized the economies of scale which could be achieved by centralized marketing, sales, transportation, and credit activities. As a result, TreeSource, Inc., was formed as the lumber sales arm of WTD. All lumber manufactured by WTD is sold under the registered service mark of TreeSource. WTD has invested in and upgraded its mills and equipment as needed. It evaluates investments in plant equipment based on a projected return on investment of 2 years or less. The company's stated objective is to maintain and improve mills in a manner that matches its high-production manufacturing strategy. It strives to avoid the excessive capital cost of expensive high-technology equipment that purports to eliminate production employees, preferring to rely on the flexibility and productivity of its workers.

FUTURE

WTD's strategy worked well in the 1980s. However, by the late 1980s trouble with tight and expensive timber supplies, weak lumber prices, and high debt from rapid expansion began to take its toll. The year 1991 opened with WTD in bankruptcy proceedings. The company's future is unclear.

GLOSSARY

agents Independent channel intermediaries who do not take title to goods, but rather arrange for their sale and are paid a commission based on the selling price.

air-dried Dried by exposure to the atmosphere without external heat applied—may be in the open or under cover.

appearance grade A grade of lumber intended mostly for exposed use in housing or other light construction where good appearance is required.

baby square Roughly a 4 × 4-inch piece of lumber, often prepared for the Japanese market; fine-grained woods like hemlock are preferred.

board foot Basic unit for pricing and measuring lumber—1 board foot is equivalent to a board 1 inch thick, 12 inches long, and 12 inches wide (board feet calculations are based on nominal sizes).

brand name A word or group of letters used to identify a specific manufacturer's product/ service.

breakeven point The number of units sold where the firm's revenues equal its incurred costs (both fixed and variable), i.e., no profit is made.

bristols A general term referring to a group of stiff, heavy papers; common uses are for index cards and postcards.

broker A type of agent who arranges the sale of a product for a commission.

bulk carrier A ship that carries loose or irregularly shaped commodities such as logs, grain, lumber, and plywood.

cash discount A discount offered off of the selling price sometimes to encourage early payment of invoices—most commonly 2 percent if paid within 10 days.

channel intermediaries The middlemen in the channels of distribution, i.e., agents, brokers, wholesalers, etc. That portion of a channel of distribution between the producer of a product or service and the consumer.

channel of distribution   The series of firms or individuals that moves goods and services from producer to customer.

chemical pulping   A process that uses chemicals to break down wood into separate fibers.

chipboard   A term used very loosely to describe a wide array of particleboards, fiberboards, or oriented strandboards.

chromated copper arsenate (CCA)   A water-based wood preservative.

coated publication papers   Shiny, slick paper used for magazines, other periodicals, catalogues, etc. Coating is usually clays and talcs bonded to the paper surface.

commodity product   A product made to a standard set of specifications by many firms to serve a broad market.

com-ply   A panel or sometimes lumber product manufactured with a particleboard core and veneer faces.

cooperative   In the forest products industry, usually a group of retailers that run their own group buying service and fund joint promotion efforts. But it could be any group of independent firms that band together for group selling or buying.

corrugated medium   A kraft paper used as the corrugated middle in corrugated box manufacture.

cost leadership   A strategy by which a firm seeks competitive advantage by being a lower-cost producer than its competitors.

cumulative quantity discounts   Reductions in price based on the total quantity purchased over a set time period.

derived demand   Demand for industrial products which is derived from the products which utilize the industrial product. For example, the demand for pallet lumber is derived from the demand for pallets.

differentiation   A strategy by which a firm seeks competitive advantage by creating something different or unique about its product or service.

dimension lumber   Usually softwood construction lumber from 2 inches up to, but not more than, 5 inches thick and 2 inches or more in width.

dimension parts   Lumber manufactured to rough, semifinished, or finished parts for the manufacture of furniture, cabinets, and other items.

distribution yard   A wholesaler which inventories and distributes lumber, plywood, and other products to industrial and retail accounts.

diversification   Moving into new products or even totally new businesses.

do-it-yourself (DIY)   That part of the R & R (repair and remodeling) market which is comprised of individuals who do their own repairs and projects.

economies of scale   As more units of an item are produced, the per unit cost of manufacture goes down.

elastic demand   If prices for a good or service are dropped, the amount demanded will increase enough to increase total revenue.

elastic supply   The quantity supplied will increase if prices are raised.

exporting   Selling to foreign markets.

focus   A strategy by which a firm seeks competitive advantage by concentrating on the needs of a specific customer segment and serving it better than its competitors.

free on board (FOB)   That point at which title and responsibility for transportation charges pass to the buyer.

freight forwarders   Almost in the category of transportation wholesaler, they perform many functions in export trade, and can combine small shipments of many firms into more economical shipping arrangements.

furniture gallery A showroom space within a store or an entire store controlled by a furniture producer to display and sell its products.

Furniture Market A market where furniture producers exhibit their products to solicit orders from furniture retailers and wholesalers.

GNP (gross national product) The market value of the goods and services produced in a year, usually measured within political boundaries such as a nation, province, or state.

guerrilla marketing A special form of a focus strategy especially well suited for smaller firms.

hardboard A thin fiberboard panel manufactured from wood fibers consolidated with heat and pressure, similar to medium-density fiberboard but made to heavier densities.

hardwood lumber Lumber produced from angiosperms such as oaks, maples, cherry, walnut, yellow poplar, ash, gum, birch, beech, etc.

home center A building materials retailer catering to individual DIY customers as well as professional contractors.

inelastic demand If prices are lowered for a good or service, the quantity demanded would increase but total revenues would fall.

inelastic supply Price increases have limited impact on increasing the supply.

kiln-dried Dried in a kiln to a set moisture content.

kraft paper A strong paper produced by the kraft process typically used in bags and packaging. A common example is the brown grocery bag.

laminated veneer lumber (LVL) A composite lumber produced from veneer bound together with resin under heat and pressure. Veneer grain is oriented to the long axis of the LVL.

letter of credit A letter issued by a bank for the buyer of certain goods which allows the seller to draw funds from the bank for payment upon certain conditions being met.

linerboard Usually a kraft paper used as the outside wall layers in corrugated box manufacture.

loss leader pricing Setting very low prices (even below cost) usually on only a few items to get customers into a retail store or to call a wholesaler.

machine-stress-rated (MSR) Products (usually softwood lumber) are evaluated piece by piece by machines which measure certain strength criteria. The measured ratings are marked on each piece.

manufacturers' representatives Agent middlemen who sell noncompeting products from several producers for a commission on what they sell.

marginal cost The change in total cost that results from producing one additional unit.

marginal revenue The change in total revenue resulting from the sale of one additional unit.

market A group of current and potential customers with similar needs for a product or service.

market pulp Pulp not consumed at the site of production and offered for sale on the market.

market segment A somewhat homogeneous group of customers that have similar needs for a given product or service.

market segmentation The process of identifying market segments.

marketing concept The notion that an organization should aim all its energies at satisfying its customers at a profit.

marketing mix The controllable components of marketing that a firm uses to satisfy a given target market. At a minimum, this includes price, product, promotion, and distribution.

marketing plan A written document which lays out an organization's marketing strategy.

markup The percentage of the selling price which is added to the cost to get the selling price.

mass marketing A marketing approach which uses the same marketing mix for all customers.

mechanical pulping A mechanical process of breaking down wood into separate fibers, used for less expensive papers.

merchants Channel intermediaries who take title to goods, then offer them for resale.

millwork Door and window parts, decorative trim, molding, etc.

modular housing A type of housing which has its major components assembled in a factory. The components are shipped to the building site, then combined to form the finished structure.

monopolistic competition A market structure where products are not homogeneous and sellers compete with each other.

national accounts sales force Salespeople who sell direct from manufacturer to large accounts/ customers.

newsprint An uncoated paper made from mostly mechanically produced pulp used for newspapers and other printing.

nominal size The commonly named sizes of lumber usually expressed to the nearest inch, like a 2×4. Not to be confused with the actual size, which is usually less. For example, the actual dimensions of a 2×4 are $1^{1}/_{2} \times 3^{1}/_{2}$ inches.

office wholesaler A wholesaler that trades a product without taking physical possession of it.

oligopoly A market structure with only a few producers and many buyers where the producers generally recognize their interdependence on each other.

OSB (oriented strandboard) Veneer-like thin strands of wood glued together to form panels like plywood.

pallet A low platform which keeps a load of product off the floor, allowing a forklift to move it easily.

paper merchant A paper wholesaler.

penetration pricing A low-price strategy often used with new products to build market share quickly and discourage competitors.

plywood A flat panel comprised of veneer sheets bonded together with resin under heat and pressure. The veneer sheets have their grain angles at $90°$ to each other in an alternating pattern.

precision-end-trimmed (PET) Lumber trimmed to a uniform, precise length with square ends.

price The amount of money needed to acquire a given good or service.

price fixing Sellers illegally getting together to set a price.

price leader A seller who legally sets a price that other sellers in the industry follow.

product A physical item, a service, an idea, an organization, etc., which satisfies a customer need.

product differentiation The process of developing and promoting differences in a given product or service to distinguish it from competing products or services.

product life cycle The stages a new product goes through from introduction to growth to maturity to decline.

product-market A market with similar needs served by products which are close substitutes.

product positioning Shows where current or proposed products are rated on various attributes in a given market by the customers.

publicity An unpaid method to communicate ideas about products or services.

pure competition A market structure where products are very similar, made by many producers, and bought by many buyers. Additionally, market knowledge is widely available, and barriers to both producers and buyers are low.

quantity discounts Discounts offered to encourage large-volume orders.

R list A set of export lumber grade rules for some western lumber species published by the Pacific Lumber Inspection Bureau.

random-length lumber Lumber offered for sale on a mixed-length basis.

ready-to-assemble (RTA) furniture Furniture shipped unassembled which customers assemble at their residences or places of business.

reload center A storage and transshipment facility, usually accepting large rail shipments, then reloading to trucks for further shipment.

repair and remodel (R & R) market   Refers to expenditures by both DIY customers and professional contractors for the repair and remodeling of residential and commercial buildings.

skimming price   A high-price strategy usually used during the introduction and growth stages of a new product to recoup the initial investment quickly.

softwood lumber   Lumber produced from gymnosperms such as pines, firs, hemlocks, spruces, cedars, etc.

specialty product   A product tailor-made to serve the needs of a small customer segment or niche.

Standard Industrial Classification (SIC)   A uniform system of classifying companies based on their economic activity.

stocking wholesaler or wholesaler   Channel intermediary who buys goods, maintains inventories, and resells to industrial and retail accounts.

structural panel   A term referring typically to panels like plywood and oriented strandboard which have known physical properties and are designed for use in construction situations.

stud   A vertical framing member cut to precise length at the mill—usually 2 × 4 inches or 2 × 6 inches in cross section.

surfaced four sides (S4S)   Lumber which has been surfaced on all four sides.

target market   A homogeneous group of customers which an organization wishes to appeal to.

target return pricing   Setting prices to achieve a given rate of return on sales, assets, net worth, etc.

total-product concept   The notion that a product is more than a physical item—it includes price, service, quality, credit, dealer reputation, and many other factors.

trade association   A group of companies with common interests and/or producing a common product which band together for more effective dealings with government agencies, more effective promotional efforts, and other reasons.

uncoated free sheet   A paper mostly made from chemical pulp (i.e., mostly *free* from mechanical pulp, no more than 10 percent mechanical pulp) used for office bond, computer paper, and a wide variety of other printing and writing papers.

uncoated groundwood   A paper similar to newsprint but of somewhat higher quality, used for directories, phone books, newspaper inserts, and so forth.

veneer   Wood which has been peeled, sliced, or sawn to a uniform, and usually thin, thickness.

vertical integration   Controlling operations at different levels in the channel of distribution, i.e., controlling a sawmill which supplies a wholesaler.

wane   Bark or lack of wood on the edge or corner of a piece of lumber.

warehouse home center   A very large home center store carrying a wide product assortment mostly aimed at DIY customers.

INDEX